A History of Bodies, Brains, and Minds

AF729462

A History of Bodies, Brains, and Minds

The Evolution of Life and Consciousness

Francisco Aboitiz

The MIT Press
Cambridge, Massachusetts
London, England

© 2024 Massachusetts Institute of Technology

This work is subject to a Creative Commons CC-BY-NC-ND license.

This license applies only to the work in full and not to any components included with permission. Subject to such license, all rights are reserved. No part of this book may be used to train artificial intelligence systems without permission in writing from the MIT Press.

BY-NC-ND

The MIT Press would like to thank the anonymous peer reviewers who provided comments on drafts of this book. The generous work of academic experts is essential for establishing the authority and quality of our publications. We acknowledge with gratitude the contributions of these otherwise uncredited readers.

This book was set in Stone Serif and Stone Sans by Westchester Publishing Services. Printed and bound in the United States of America.

Library of Congress Cataloging-in-Publication Data

Names: Aboitiz, Francisco, author.
Title: A history of bodies, brains, and minds : the evolution of life and consciousness / Francisco Aboitiz.
Description: Cambridge, Massachusetts : The MIT Press, 2024. | Includes bibliographical references and index. | Summary: "A brief history of life and of the emergence of our species, focusing on the biological evolution of brains and conscious behavior and ending with some speculations about our future"—Provided by publisher.
Identifiers: LCCN 2023046299 (print) | LCCN 2023046300 (ebook) | ISBN 9780262049023 (paperback) | ISBN 9780262379359 (epub) | ISBN 9780262379342 (pdf)
Subjects: LCSH: Human evolution. | Brain—Evolution. | Human behavior. | Consciousness. | Human beings—Forecasting.
Classification: LCC GN281 .A229 2024 (print) | LCC GN281 (ebook) | DDC 155.7—dc23/eng/20240131
LC record available at https://lccn.loc.gov/2023046299
LC ebook record available at https://lccn.loc.gov/2023046300

10 9 8 7 6 5 4 3 2 1

To my family, and to all my students who have been a great source of inspiration.

Contents

Acknowledgments

First of all, I would like to thank all my teachers, advisors, students, and colleagues with whom I have engaged in sometimes vibrant discussions and conversations about the evolution of life, the brain, and consciousness. Directly or indirectly, they have all contributed to shaping this book. Special thanks are due to Florencia Álamos, Jorge Barros, René Bobe, Conrado Bosman, Ximena Carrasco, Pablo Fuentealba, Ricardo García, Rodrigo Henríquez, Leonie Kausel, Thurston Lacalli, Maëva Michon, Juan Montiel, Sergio Osorio, Bárbara Palma, Catalina Saini, Mary Ann Raghanti, Constanza Richards, Georg Striedter, Francisco Zamorano, and José Zamorano-Abramson. Felipe Serrano made all the book's illustrations including the cover, and Rodrigo Henríquez helped me struggle with the formatting requirements of the book. My son Anton Aboitiz contributed ideas to design the book's cover. Special thanks to Claudia Andrade, Pablo Fuentealba, and Paty Opazo, who covered my research and administrative duties and allowed me to focus on writing. The Pontificia Universidad Católica de Chile and the Faculty of Medicine also provided constant support for this project, permitting me to use my academic time in it. Likewise, I am most grateful to the editors of the MIT Press and the book's reviewers, who believed in the project. Finally, I am also grateful to my family and friends, especially my wife Carolina and my children, who have encouraged me to pursue this project, using much of family time.

Abbreviations

3Vi	third visual pathway
A	anterior
AC	anterior commissure
ACC	anterior cingulate cortex
AF	arcuate fasciculus
AFF	afferents
Am	amygdala
An	anus
ANR	anterior neural ridge
AO	apical organ
AT	anterior temporal lobe
Au	auditory areas
B	Broca's area
BG	basal ganglia
BNS	blastoporal nervous system
BP	blastopore
BYA	billion years ago
CC	corpus callosum
D	dorsal
DC	dorsal cortex
DL	dorsolateral prefrontal cortex
DMN	default mode network
DP	dorsal pallium
DVi	dorsal visual pathway
DVR	dorsal ventricular ridge
EEG	electroencephalogram

FEF	frontal eye fields
fMRI	functional magnetic resonance imaging
GMO	genetically modified organisms
GPRC	G-protein coupled receptor
GWS	global workspace theory
H	hippocampus
IIT	integrated information theory
IO	infundibular organ
IPS	intraparietal sulcus
IsO	isthmic organizer
IVF	in vitro fertilization
KYA	thousand years ago
L	lateral
LCd	laryngeal cortex dorsal
LCv	laryngeal cortex ventral
LH	left hemisphere
LP	lateral pallium
LTD	long-term depression
LTP	long-term potentiation
LUCA	last universal common ancestor
M	medial
MEG	magnetoencephalogram
Mo	mouth
MP	medial pallium
MRI	magnetic resonance imaging
MST	medial superior temporal area
MZ	marginal zone
MYA	million years ago
NA	nucleus ambiguus
NC	neocortex
NCC	neural correlates of consciousness
NCh	nerve cord
OB	olfactory bulb
OC	olfactory cortex
OF	orbitofrontal cortex
OL	olfactory cortex
OT	optic tectum

P	posterior
PAG	periaqueductal gray
PC	posterior cingulate cortex
PG	pineal gland
PP	posterior parietal cortex
REM	rapid eye movement sleep
RH	right hemisphere
RNS	reticular nervous system
Shh	sonic hedgehog
SLF	superior longitudinal fasciculus
SP	subpallium
STS	superior temporal sulcus
SVZ	subventricular zone
T	thalamus
ToM	theory of mind
TPJ	temporoparietal junction
V	ventral
Vis	visual cortex
VL	ventrolateral prefrontal cortex
VM	ventromedial prefrontal cortex
VP	ventral pallium
VVi	ventral visual pathway
VWFA	visual word form area
VZ	ventricular zone
W	Wernicke's area
ZLI	zona limitans intrathalamica

Introduction

The more we know about the universe, the more life on earth appears as an improbable event. We are living in Carl Sagan's "tiny blue dot" in a vast space filled by lifeless planets. But not only does life's origin and evolution appear to be unlikely—complex organisms, although quite abundant from our point of view, are represented by only a few small branches in the leafy tree of life—but the origin of our species, uniquely endowed with language, technological abilities, and ultimately consciousness is a highly mysterious event to most of us. Since we abandoned the view that we were created by God's design and are the center of the universe, we have been left with an existentialist discomfort regarding our origins and our place in nature. Who are we, a mindful species standing on a tiny planet, surrounded by other living beings but apparently all of us alone in this vast universe?

With this book, I intend to add a grain of sand to the discussions about our origins as part of the biological world, about the emergence of our minds and consciousness, and about our possible future as a species. Thanks to Charles Darwin and Alfred R. Wallace, all life on earth can be glued together through a single process, which is evolutionary theory. This will be the main thread that connects all themes discussed in this book, beginning with life's first attempts, passing through different stages in the evolution of animals and their brains, and ending with our species, provided with a not unique but highly remarkable brain that unfortunately may not guarantee our survival in the medium or long term.

This is a very unique and personal work; its contents will be largely based on my own work and ideas on the evolution of the nervous system and the human brain, but it will also include many substantial contributions from different authors that together may provide a unified perspective on biological and brain evolution. My aim is to stimulate discussions and hypotheses concerning our place in nature and our prospective survival as a species, which will hopefully motivate young scholars to ask themselves new questions, fueling emergent research lines and diversifying research. I hope it will resonate with undergraduate and graduate students in biology, neuroscience, and related areas, as they will probably be the most motivated to read it. I am thinking of my student years and how some books I read at that time made a profound impact on my development as a scientist. I hope this book will do the same for these students and young scientists. Yet, I think this book may also be interesting to professional researchers in the field and could also be read by science-educated people with a genuine interest in evolution and the brain.

Since writing about the history of life is a great challenge and could end up in a too long review, I have decided to make this through a relatively short narrative, since we humans are storytellers and understand the world through told histories. I will focus on a few key events in biological and brain evolution that represent turning points in the path toward sentient beings like us, starting from the origin of life. Accordingly, much data and evidence will have to be left out in order to make a relatively smooth and hopefully engaging story. Perhaps one of the most important aims of this book is to provide a sense of continuity from the emergence of life to the achievement of consciousness and the development of a technologically driven culture, with its good sides and bad sides. For this, at the end of each chapter, I make a brief summary of the main points raised in it. Finally, just a small caveat: since my training has been largely on anatomy, some parts of the book can be a little demanding to the non-expert. I have made my best to provide figures to help visualize the structures and relations described in it, within the space requirements provided by the editor.

The Book in Parts

The book is organized into fifteen chapters addressing what I consider to be the most relevant events and processes in the history that led to *Homo sapiens*, which is divided into three parts. In part I (chapters 1–2), I begin with some basic definitions. Chapter 1 contains an introductory review of evolutionary theory, followed by a discussion of what life is (chapter 2), where I discuss mechanisms of self-production, heredity, and homeostasis. In part II, I address the early stages of life and the origin of the nervous systems. In chapter 3, I narrate how the first cells may have originated, containing a basic set of genes and structures enabling them to maintain themselves and reproduce. In chapter 4, I describe how multicellular organisms came about and how the first neurons originated in the early animals. Afterward, I will talk about the properties of neurons and neural circuits, which serve as the fundamental building blocks for the organization of the nervous system (chapter 5). Part III (chapters 6–10) refers to the evolution of brain-bearing organisms, featuring at the end mammals and their cerebral cortex. Chapter 6 focuses on animals with bilateral symmetry and discusses invertebrate organisms, highlighting insects and cephalopods that display surprisingly complex behaviors. Chapter 7 reviews the origin and evolution of vertebrates through the elaboration of a new head including organized sense organs and paired cerebral hemispheres, the colonization of land, and the diversification of terrestrial vertebrates. In chapter 8, I discuss the expansion of the vertebrate brain, highlighting the brains of reptiles, birds, and mammals and the controversies regarding homologies between them. Chapter 9 discusses the diversification of mammals and their brains, emphasizing the role of olfaction and other senses, together with other unique characteristics that gave rise to the mammalian cerebral cortex. The latter is more extensively addressed in chapter 10, where I discuss the mosaic structure and connectivity of the neocortex, and especially its role in cognitive mechanisms that make up most of our minds. In part IV, I focus on our own species, its evolution, and our unique cognitive and behavioral abilities. Chapter 11 refers to the origin of *Homo sapiens* as a special primate that developed bipedality and an unusually big brain together with an exquisite hand dexterity associated with tool use and culture. In chapter 12, I discuss the origin of speech, highlighting an instinctive drive for communication in our species and the development of an auditory-vocal neural circuitry to acquire speech and music. I also address the role of manual and body signs in early hominin communication. Chapter 13 reviews two related aspects of human communication and social life, including the capacity to identify ourselves with others and the generation of a shared world through modern language. Chapter 14 addresses the emergence of consciousness, discussing some of the theories attempting to explain it (or dismiss it as an illusion), and again the role of language in amplifying our consciousness as a social phenomenon. Chapter 15 discusses the future of our species concerning biological evolution, cultural evolution, and technological development and its impact on brain health, well-being, and the survival of our species. At the end, I include a brief section about life as a cosmic phenomenon and the possibility of extraterrestrial evolution and intelligent life.

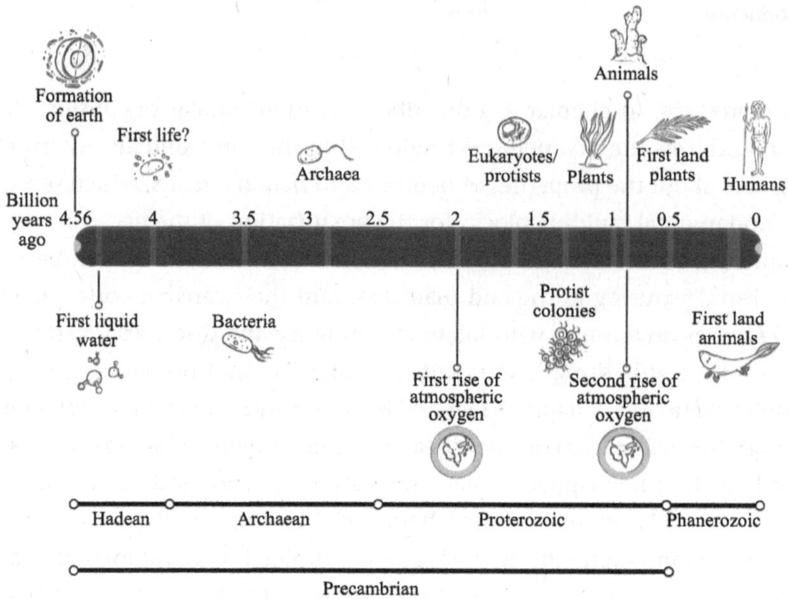

Figure 0.1
A timeline of life on earth.

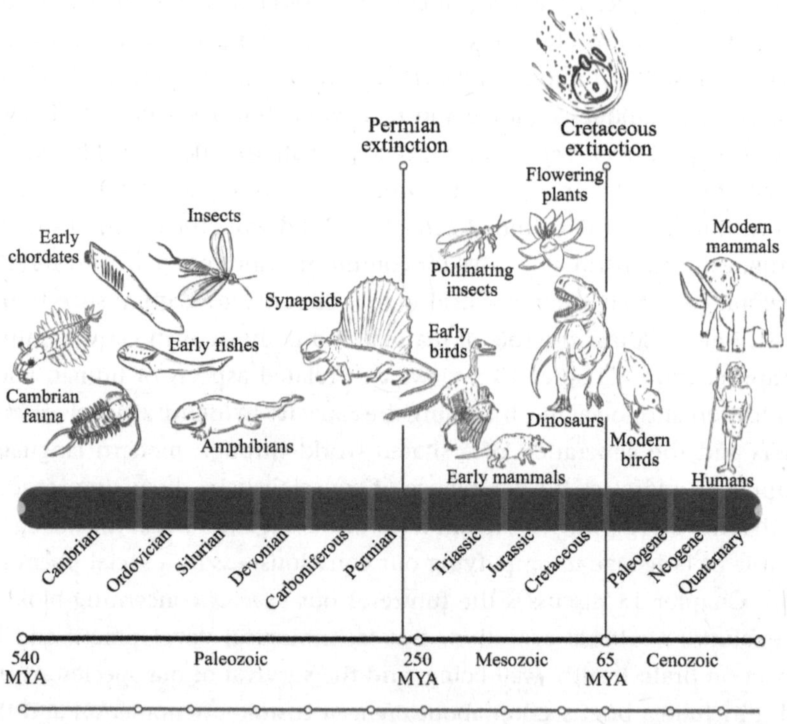

Figure 0.2
A timeline of complex animal life.

I Definitions

This section provides some definitions of the fundamental issues of biology: evolution and life. These will provide a framework for the subjects covered in the rest of the book. Firstly, the book starts with a strong assumption, which is that natural selection is the basic process underlying the origin of life and its evolution on earth. Natural selection implies the capacity for faithful copying (reproduction with inheritance) and the differential reproduction of distinct lineages, which may not necessarily be alive (like viruses). Genes, made of nucleic acids, are the basic units of inheritance, which can self-replicate, making copies of themselves, and provide the backbone for prebiotic and biotic evolution. Secondly, I provide a tentative definition of life, as self-replicating molecular ensembles embedded in self-producing cells. The combination of both molecular self-replication and cellular self-production gave rise to the bewildering diversity of life forms on this planet. In order to maintain themselves, living beings have developed intricate homeostatic mechanisms that enabled them to anticipate ambient alterations, thus providing the first elements to drive knowledge and cognition, and the subsequent evolution of a brain and consciousness. With these basic concepts in mind, in the rest of the book we can continue our journey through the history of life.

1 What Is Evolution?

Life thrives on this planet, and this is largely the result of evolution by natural selection, where all living beings have a common ancestry and have acquired different forms throughout history. Thus, most people understand evolution as the history of biological diversity, where species change over time, usually, but not always, adapting to the ambient conditions. Furthermore, biological evolution is typically considered to be a consequence of the existence of life, assuming that only life forms display evolution, and that evolution started after life's origin. However, before I start discussing about life itself, I will offer an account of the essentials of evolutionary theory, which is the principal axis along this book. I have made this choice, first because paradoxically there seems to be more agreement among scientists about what evolution is than about what life is. Secondly, according to several authors, evolution by natural selection may be a more ancient process than life itself, starting with self-replicating molecular ensembles where systems with better replicative capacities evolved into complex networks of chemical reactions that resulted in the first living cells. Yet, instead of talking about molecular evolution, here it will be easier to refer to the theory of evolution as it was first conceived by Charles Darwin and Alfred Russell Wallace, attempting to explain the diversification of species on earth. We will see that the same principles can account for prebiotic evolution, which will be especially addressed in chapters 2 and 3 when dealing with life's origins. This chapter describes the basic mechanisms of evolution and inheritance, the concepts of animal classification, and related topics that are the pillars supporting modern evolutionary theory.

A Theory of Change

Lamarck versus Darwin and Wallace

There have been two main theories attempting to explain the apparent miracle of biological evolution. The first proposal was the inheritance of acquired characteristics proposed by Jean-Baptiste Lamarck, who affirmed that the conditions of life-induced heritable changes ended up transforming the species into new forms. The second theory, which is accepted among most biologists today, is the theory of natural selection proposed by Darwin and Wallace.[1] According to this, individuals inherit their characteristics from their parents (which according to modern genetics are largely coded in the genes), but in some instances they spontaneously develop new, heritable characters, a phenomenon that Darwin called "descent with modification" and now is commonly referred to as mutations that alter the genetic code. A minority of these mutations specify characters that confer individuals increasing reproductive capacity relative to others in the species, which results in the spread of these genes and their characters in the population. In the long term, mutations related to some specific characters may accumulate, resulting in the evolution of new characteristics like limbs evolving from fins. Central to the notion of natural selection is that diversity is intrinsic to the species (members with different, heritable attributes are spontaneously produced) and that evolutionary change will take place only if some of the variants reproduce more than others.

Thus, Lamarckism and "Darwinism-Wallacism" differ on one essential point, which is how heritable variation originates. One the one hand, according to the former, species variation takes place as an individual adaptation that becomes inherited, providing a direct chain between lifestyle and heritable changes. On the other hand, in Darwinism-Wallacism, mutations or variants preexist or are produced randomly (they may be triggered by environmental events like UV irradiation, but these changes are not directed by the lifestyle of the organism). Note also that in Lamarckian evolution, directed variation is always adaptive—it corresponds to the organism's "needs"—while in Darwinian evolution, spontaneous variation is mostly nonadaptive (most mutations are neutral or harmful), and as said, only a few changes consistently confer reproductive benefit. Moreover, nonadaptive mutations may accumulate by chance, as in the statistical fluctuations of gene composition in a population, a phenomenon called genetic drift.

However, the increase in complexity of life on earth has most likely been the result of the action of natural selection on lineages from prebiotic ensembles to large organisms.

In the early twentieth century, Darwin and Wallace's theory of natural selection received enormous support from the rise of Mendelian genetics, since the rules of inheritance were totally obscure before that. But the final consolidation of natural selection as the key mechanism for evolution took place after the discovery of DNA as the material from which genes are made of, establishing the molecular basis for inheritance. All theories of biological organization are now required to include in one way or another the molecular interactions established by nucleic acids (DNA and RNA) and their products (proteins).

The Shade of Lamarck

Although Lamarck was wrong about the fundamental mechanism of inheritance, in my view, he raised some critical issues. First, he correctly pointed out that the behavior of animals can be fundamental to determining their evolutionary paths. James Baldwin presented this idea in a Darwinian context, claiming that the animal's conduct provided a framework into which selection acted. For instance, the behavior of stretching the neck in the giraffe's ancestors to feed on trees may have driven the selection of larger cervical vertebrae, as the longer-necked individuals were able to reach the higher leaves (recent findings suggest that fossil giraffe relatives used their long necks for wrestling as well, but the argument remains the same). Likewise, about 400 MYA some early fishes started using their fins to crawl on the sea bottom before they got out of the water and used them for terrestrial locomotion. These fish also evolved lungs from a rudimentary air bladder used to catch air to help flotation. Those animals who were better at crawling on the ground, or able to catch more air in their mouths, reproduced more, and their traits spread in the population. In subsequent generations, selection acted again on the descendants favoring the best crawlers and breathers and so on, promoting the evolution of limbs and lungs.[2]

This is neatly illustrated by an experiment in which researchers used some extant but "primitive" African fish called bichirs. Young bichirs were experimentally raised to live on land, taking advantage of their ability to breathe air through rudimentary lungs.[3] As each animal learned to crawl on the ground, it developed slight skeletal modifications in its fins that emulate the transformations that took place in the ancestors of terrestrial vertebrates. Now, suppose that this process went on through generations and the best crawlers benefited by having more reproductive capacity. Natural selection could act on the genes involved in these skeletal modifications, favoring those phenotypes that developed a better anatomy for crawling behavior. Thus, individual accommodations to circumstances provide the ground on which natural selection can act. Despite being wrong on the mechanism that generates variation and its inheritance, Lamarck foresaw a critical role of behavior in the generation of evolutionary novelties. In light of this, some authors consider organisms as agents of their own evolution by determining the path that natural selection will take. However, there is a twist to this argument as ultimately, adaptation results in the preservation of genes favoring these behaviors, discussed next.

Heredity

Genes

The discovery of DNA changed biology forever by establishing the principal mechanism of biological inheritance. DNA is a gigantic molecule organized in a double-helix structure built by four nucleotides (adenine, thymidine, cytosine, and guanine), whose sequence along the double helix determines the genetic code. DNA makes copies of itself during the reproduction of cells, and genes are passed on across generations indefinitely, specifying the main characteristics any species has. From the origin of genetics, a fundamental distinction has become widely accepted among biologists: the genotype, or the genetic endowment an organism has (its DNA), versus the phenotype, or the bodily, behavioral, and functional characters an individual possesses. The genotype determines the proteins with which the body is made, following a well-established sequence of events called gene expression. Very simply, gene expression consists of two steps: transcription, where DNA is copied into RNA (a DNA-like molecule), and translation, in which RNA is used as a template to build proteins (see figure 1.1). Notably, each combination of three pairs of nucleotides in the DNA or in the messenger RNA (called a triplet) codifies for a specific amino acid so that a sequence of triplets determines a particular string of amino acids to form a protein. Moreover, the genetic code is universal—that is, each triplet codes for the same amino acid in all cells from bacteria to humans—which strongly suggests that we all have a common descent.

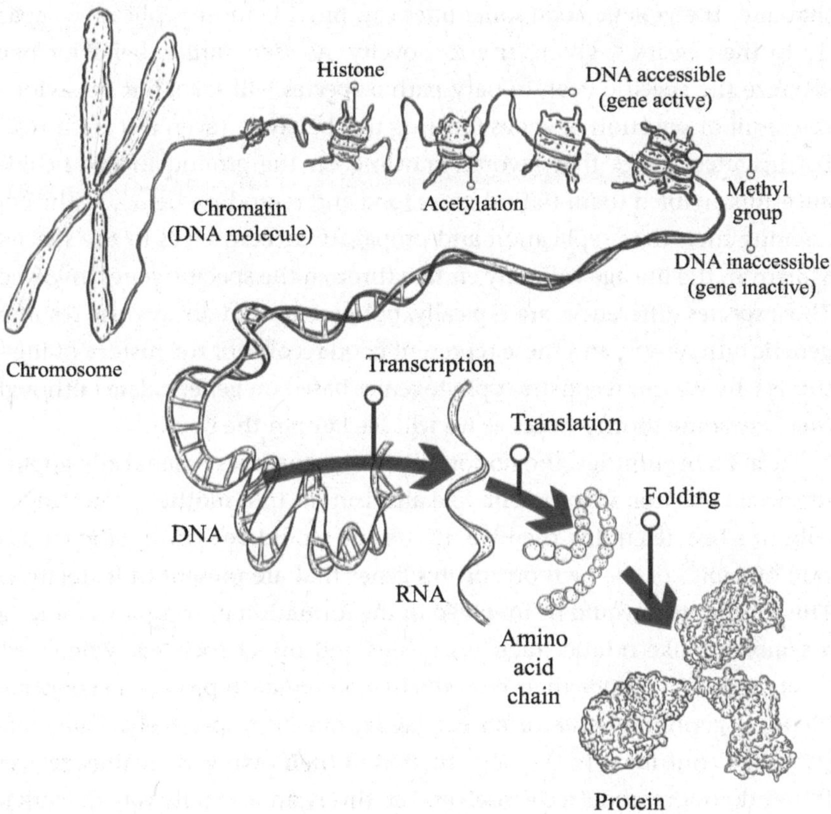

Figure 1.1
Steps involved in gene expression and epigenetic mechanisms. DNA is packed in chromosomes that are visible during cell division. The DNA string wraps around histones, but active genes unpack from these proteins. Other regulators of gene activity are acetyl and methyl groups as well as other molecules including small RNAs (not shown). In gene expression, the DNA sequence is copied into RNA (transcription), which codes for a chain of amino acids (translation) that becomes folded in three dimensions as a functional protein.

Selfish Replication

In his famous book *The Selfish Gene,* Richard Dawkins boldly proposes that genes make up the centerpiece of evolution, claiming that cells and multicellular organisms are simply vehicles that facilitate gene replication. In last instance, the struggle of life is not the survival of the species but rather each organism working to prolong the perpetuation of its own genes. In this way, genes can replicate indefinitely through generations, but mutations that alter the genetic code sometimes can provide more replicative capacity to their bearers, giving rise to novelty. As seen earlier, behavior may canalize the specific evolutionary path a species will take, but behavior is the result of selection of genes favoring this behavior (even if it is learned). For instance, genes that favored crawling on the ground in our fishlike ancestors enabled them to find more food and reproduce better, in the end favoring their own replication and propagating themselves in the species. Moreover, the lineage will only endure through the specific genes involved. Thus, species differences are typically, but perhaps not always, the result of genetic differences, and these represent good records of the history of life—this is why we can reconstruct phylogenies based on genetic data (although there are some thorny issues as we will see later in the book).[4]

From its beginnings, the notion of selfish genes was polemically applied to social behavior. Phenomena like altruism, as in a mother protecting its cubs or a bee defending the hive, in the end would be favoring the replication of copies of the own organism's genes that are present in its relatives. This mechanism would be involved in the formation of complex biological assemblages like multicellular organisms and insect societies, where cells or individuals sacrifice their reproductive potential to protect a genetically identical germline tissue or an egg-laying queen, respectively. Thus, relatives may contribute to the perpetuation of their own genetic lineage even if they do not reproduce themselves. Yet, this is not the only way to work in favor of others. There are many instances of cooperation between genetically dissimilar species, as in the case of symbiosis between gut bacteria and large organisms, where both species benefit from each other, or in cases of human cooperation in which non-kin establish trade exchanges. But in these cases, individuals are indeed favoring their own genes' survival through cooperation with non-kin, even if sometimes they have to carry foreign genes with them. There are also cases where strangers risk themselves to save unrelated individuals of different species, as when a hippopotamus attacks a crocodile predating on an antelope. Those instances can be seen as extensions of an altruistic behavior that has evolved primarily to protect kin. This view of evolution has been highly contentious especially among organismic biologists that reject what they think is a reductionistic approach. However, in my view, the selfish gene theory has strongly resisted the weight of evidence, and ultimately there is no better possibility at this point to account for diversity in biological evolution. Still, there may be a few exceptions to the primacy of genes in evolution, which I will discuss next.

Over the Genes

Despite being critical for inheritance, genes do not directly specify the phenotype; they only provide a blueprint for it, and there is a highly plastic and complex mechanism involved in the formation of the individual, including mechanical processes, increases in volume, tissue displacements, and so on.[5] During embryonic development, the ambient conditions may affect the activity of genes and other cellular components in such a way that the final phenotype is usually a mixture of genetic and environmental variables. Gene expression can be influenced by stress, temperature, and toxins, but also by the presence of neighboring tissues, by sensory stimulation, and even by the organism's behavior (remember the example of crawling fish). Nonetheless, this indirect and moldable influence on organismal development is sufficient for genes to drive natural selection.

More specifically, there are several mechanisms by which gene activity can be modulated. For instance, proteins produced by so-called regulatory genes may bind to DNA to enhance or repress the activity of some other genes. This is essential for embryonic development where distinct sets of genes are activated at successive stages of differentiation. Another important mechanism that modulates gene expression is epigenetics, where distinct molecules (not coded by regulatory genes) can bind to DNA, affecting gene activity. A first kind of epigenetic modulators are the histones, a set of large proteins to which the DNA molecule rolls around in dense packs (see figure 1.1). When packed with the histones, genes become silenced, but they activate again when the DNA thread is released from the pack. Histones are crucial for embryonic development, helping to silence large pieces of DNA during the differentiation of cells into distinct types, like neurons or gut cells. Furthermore, histones participate in cell division by arranging the DNA molecule in chromosomes that become separated in the two daughter cells. Secondly, small molecules called methyl or acetyl groups can bind to DNA, especially during stressful conditions, and also modulate gene expression. Finally, small RNA molecules are quite abundant in the cells and can also bind to and regulate the expression of genes or the activity of large RNA strings.

A combination of regulatory genes and epigenetic processes is crucial for normal development of multicellular organisms. Moreover, these regulatory events do not end with embryonic development, as they keep influencing gene activity in the adult, especially in the brain, promoting neuronal plasticity, learning, and memory, with some being associated with mental health. Thus, the combination of regulatory genes, epigenetic mechanisms, and environmental conditions results in a complex, dynamic lattice of interactions that is gradually transforming from the beginning to the end of our lives, framing our behavior as adults and ending up with senescence and finally death.

An important issue is that epigenetic modifications are usually not heritable in multicellular organisms. In the latter, epigenetic modifications are normally erased at the time of formation of reproductive cells (sperm and egg), and they are usually not passed to offspring. Thus, most of the network of genetic-epigenetic interactions that grows up during development is confined to our bodies and disappears at the moment of making our reproductive cells. In this way, however complex the mechanisms involved in making the individual phenotype can be, what is inherited through generations largely depends on the genetic makeup.

Lamarck Again?

However, there might be some exceptions to the previous statements. In the mid-twentieth century, Conrad Waddington proposed the concept of genetic assimilation, where phenotypic alterations due to environmental exposure could somehow become incorporated in the genes of complex organisms and be subject to natural selection. Perhaps some evidence on epigenetic inheritance complies with this notion. Recent reports indicate that some epigenetic effects can be passed across generations, especially in plants, where methylation changes can be traced along closely related lineages, but also in some animals like nematode worms, insects, and even rodents.[6] However, many of these findings have been obtained in the laboratory, and there is no data yet of epigenetic modifications enduring through longer times or of epigenetic mechanisms in the evolution of wild species.

Furthermore, inheritance of acquired characters may be common in unicellular beings and in some instances of cell division inside multicellular organisms, where epigenetic modifications can be transmitted to daughter cells at least through some rounds of cell division. For instance, mature epithelial and liver cells are able to undergo cell division during development, tissue regeneration, or tissue repair, keeping their epigenetic imprinting through reproduction.[7] Furthermore, cells directly transmit cytoplasmic and membrane components to their progeny, transmitting features of their differentiated state. While these phenomena are usually dismissed by evolutionary biologists, in my view, they may count as instances of non-genetic acquired inheritance and may have been a relevant mechanism in the origin of multicellularity. All in all, there is still little evidence of acquired epigenetic inheritance in animal evolution, although it may have contributed to the origin of multicellular organisms and to the differentiation of tissues, as we will discuss later (see chapter 4).

Phylogeny and Ontogeny

The Tree of Life

After discussing the mechanisms of evolution and heredity, I will now delve into the patterns that evolution generates. The main product of evolution is diversity, which implies that species diverge into an ever-branching tree of forms. One of the great problems in the history of biology has been how to group these different branches in a meaningful way. Just remember the ancient Greek anecdote where Plato proposed that man (humans) was a featherless biped, to which Diogenes brought a chicken with its feathers plucked off, saying "Here is a man." In the eighteenth century, Carolus Linnaeus classified living beings in inclusive taxonomical groups according to their similarities (dogs are canids, which are carnivores, mammals, vertebrates, and so on). Charles Darwin realized that this nested classification system implied that all life had a common ancestry and that the process of biological evolution had given rise to the great diversity of forms we see today. According to evolutionary theory, classification reflects a historical genealogy, or phylogeny of different groups and their relations to others. Each taxonomic group, no matter its size (for example, canids, vertebrates, or animals) is referred as to a "taxon" or a "clade," defined as a group of species that share a last common ancestor (they may have many common ancestors since the origin of bacteria, but the last common ancestor is a hypothetical stem species that gave rise to all species of the group and to no other species).

When studying evolution, a key aspect is the comparison and identification of similarities and differences between species. Species with similar characters will be grouped together and will separate from species with different characters. The similarity of some characters (for example, bearing wings) may imply that the respective species have a common ancestry and, in the case of birds, that their last common ancestor was a winged animal. In this case, the character "wings" is considered homologous in the group. However, wings may have appeared separately in two groups, like the wings of birds and bats. Since the last common ancestor of birds and bats did not possess wings, this similarity is said to be due to homoplasy or evolutionary convergence. Determining which shared features derive from common ancestry (homology) and which have evolved independently (convergence) is crucial for establishing animal phylogenies. While in the case of birds and bats it is very easy to tell the difference, as we will see in many other cases this is a quite difficult problem.

Some authors have proposed that homology can be found at different levels, from the microscopic to the macroscopic. For example, if an organ evolved separately in two sister lineages but its development relies on the same genetic networks in both cases (like the eyes of insects and vertebrates; see the next paragraph), it is said to be genetically homologous (the term "deep" homology is also used) but not morphologically homologous. That is, the last common ancestor may have possessed a set of genes that were later used for the development of similar organs in separate sister lineages. Likewise, an organ that was present in the last common ancestor and was maintained in two sister lineages might have evolved different genetic regulatory systems in each branch: that is, the genes controlling its development might have diverged in the two lineages. In this case, there would be morphological but not genetic homology. Note also that the levels of homology do not only apply to genes and morphology; we may also speak of homologous behaviors or functions if these traits were present in the last common ancestor. Thus, the establishment of homology at different levels, from genetics to morphogenesis to function and to behavior poses a daunting challenge for evolutionary biologists.

The evolution of the eye is a good example of genetic but not morphological homology. There is much diversity in eye shapes and structures, including the compound eyes of insects and the camera-like eyes of cephalopods and vertebrates. Underlying these diverse eye shapes, there is a common theme organizing eye development from flies to squids to humans. The genetic mechanisms of eye formation are highly similar in all these animals, depending on a similar set of regulatory genes for their development, especially a gene termed Pax6.[8] Mutations in Pax6 render impairments in eye formation in flies, humans, and mice. This gene has an extremely conserved function, as mammalian Pax6 genes, when inserted in fly eggs, were able to induce the differentiation of fly's eyes in different body parts of the developing larvae. To many authors, this implies that a Pax6-related genetic kit involved in the development of a light-sensitive cell type was present in the last common ancestor of all animals with eyes. These genes became recruited to make photoreceptive organs of different shapes and forms in different instances (arthropods, mollusks, and vertebrates), which separately evolved into eyes with different anatomies. This is an instance of genetic, or "deep," homology where the genes involved in eye formation are homologous across animal groups, but the eyes themselves may have appeared separately in each lineage and are therefore not homologous organs.

From General to Specific

An additional pervasive notion is that despite the increasing divergence of life forms, organisms always leave a record of their past history in their anatomy, in their embryonic development, and especially in their genes, which has been extremely useful for phylogenetic classification. In this line, it is important to recall the difference between the evolutionary history of an organism or its species (its phylogeny, which goes back to the origin of life) and its individual history (its ontogeny, which goes back to the moment it started its embryonic development, usually as a fertilized egg). In the nineteenth century, Karl von Baer and Ernst Haeckel emphasized the general concordance between the succession of embryonic stages during individual development and the species' ancestry. This observation led to Haeckel's formulation of the "biogenetic law," stating that the ontogeny of an individual recapitulates its phylogenetic history. For example, we as embryos start as a one-celled organism, then develop multicellularity and acquire a gut, a head, a tail, branchial slits, extremities, and so on. Many of these characters, like the tail and the branchial slits, are lost or become highly transformed during normal human development, but they reflect our evolutionary past as unicellular organisms or aquatic fish. While Haeckel considered that early embryos resembled the adult stages of ancestral forms, von Baer was more precise in asserting that early embryos resembled the early embryonic stages of ancestral lineages, rather than being comparable to their adult forms. That is, a human embryo resembles a fish embryo, not an adult fish.[9]

A related issue is that phylogeny is often depicted as a sequence of adult states, like the typical drawing of ape-to-human transformation. However, evolution is rather a sequence of individual ontogenies where one organism is formed as an early embryo, grows and differentiates, and then reproduces to make offspring, which grow to have more offspring, and so on (at least in multicellular plants and animals). Therefore, morphological changes in evolution take place through modifications of the embryological mechanisms that produce the adult form, rather than just changing the adult phenotype. Although the study of the evolution of embryonic mechanisms—colloquially termed evo-devo—has become a highly fashionable field today, this is not without controversy, especially when attempting to determine homology of organs across species where different criteria to establish this condition may be in conflict. Many scholars tend to focus on the similarities of adult characters to establish homology, as if evolution was a sequence of adult states, while others rely more on assessing similarities and differences in embryonic development. Whatever approach turns out to be correct in each specific case, in the end, a full explanation of evolution and diversity must account for how variations in developmental mechanisms gave rise to adult similarities or differences between species.

Saltation and Constraints

In the late twentieth century, some evolutionists, led by the late Stephen Jay Gould, advocated for a revision of the theory of natural selection,[10] firstly claiming that evolution may take place by saltatory or punctuated mechanisms, where new species would be produced in "jumps," with no intermediate stages. This seems to be contrary to the notion that natural selection is a slow process, successively accumulating small innovations until major changes are seen, like when a tiny sprout becomes a huge tree. Evidence for this has been observed when some species duplicate their genomes all at once (vertebrate origins are a notable example) or when animals add repeated elements to their bodies, like snakes or birds increasing the number of their body or neck vertebrae, respectively; they add them by integers. Another example is the loss of the tail by apes and humans, which apparently took place by the action of only one so-called jumping gene.[11] Yet, the possibility of punctuated evolution was in fact considered by early Darwinian evolutionists (see, for example, George Gaylord Simpson's notion of "quantum evolution"), who did not see this as a problem to the theory but rather as a particular instance of rapid evolutionary change.[12] Moreover, when speaking of the slow accumulation of small variations, Darwin was mainly referring to the acquisition of highly complex features like the eye, where intermediate conditions in complexity have been found in many instances. Among mollusks, the evolution of image-forming eyes has been tracked in great detail from a simple pigmented spot in the skin to a cup-shaped structure, then to a cup provided with a rudimentary pupil, and finally evolving a lens and a cornea to maximize light penetrance (gradual evolution can also be observed in the evolution of compound eyes of arthropods).[13] Notably, each of these stages provides an improvement in the image-forming capacity of eyes, which complies with the gradual evolution of complex organs by natural selection.

Another criticism to natural selection was that embryology is a highly directional and constrained process that makes only some transformations possible, resulting in the appearance of traits that may have not been directly selected for. In other words, morphological evolution would be dictated by so-called developmental constraints that guide embryological transformations rather than on the selective pressures imposed on the species. Conceptually, it is quite clear that embryological variants provide the rainbow of possible anatomical innovations for natural selection to act on. The point in question is the range of possible variations development may provide, which in some cases may be quite wide but in others may be more constrained, with only few variants available, that may affect other characters beside the primarily selected ones. This is an empirical rather than theoretical question. Moreover, even if in certain instances there is a highly constrained set of morphological variants, these will likely be inherited by genetic mutations and only some of them will be eventually successful at reproduction, which complies with the mechanism of natural selection. Summarizing, the evidence of punctuated equilibria and the argument for developmental constraints may add important insights to the theory, but they do not represent a major challenge to it.

Finally in this line, a great deal of evidence has shown that there are non-adaptive forms of evolution, as in the previously mentioned case of genetic drift. This is based on the fact that many genetic changes are simply neutral for adaptation: that is, they do not confer any positive or negative reproductive advantage. These neutral mutations may become common in the population just by chance, especially when populations are small. Interestingly, neutral genetic changes tend to accumulate over evolutionary time, leaving a record of the distinct lineages and their relations to others. This phenomenon has been widely used to track phylogenies in different species, although not without discussion, as I will discuss throughout the book.

Perspective

Biological evolution is commonly understood as the process by which living beings emerged and diversified in the history of our planet. The evidence strongly favors Darwin and Wallace's theory of evolution by natural selection against Lamarck's theory. The former implies that there is an intrinsic variation of heritable traits within a species and organisms possessing distinct traits are able to leave more descendants than others. Inheritance is a key aspect of evolution, which is mostly carried by the genes (DNA) that are transmitted across generations. There are other instances of heredity like epigenetic modifications or cytoplasmic heredity, but these are largely found in cell division, and there is still little evidence for it in complex animals.

Since life on earth shares a common evolutionary ancestry, species can be classified in a nested pattern where larger categories contain smaller categories, forming an ever-branching tree of life. A critical issue in classifying organisms is determining homology, or similarity of characters due to common descent. Nonetheless, homology may be found at different levels, from genes to anatomy to behavior, and homology at one level does not imply homology at other levels (as shown in the example of eye evolution). Another crucial aspect of evolution is that despite their profound diversification, complex organisms, especially animals, retain signs of their past history (their phylogeny) during their individual development (their ontogeny). Consequently, there is usually a concordance between the events that took or take place in both the evolutionary history and the embryology of a species, which can be of great relevance when attempting to reconstruct the evolutionary history.

2 What Is Life?

After reviewing the basics of evolutionary theory, we now come to define what life is. The notion of life (besides our own) is probably a very ancient one partly because we feed on living beings and can be food for others, which may make the difference between the subject's life and death. Before we are one year old, we learn to distinguish living from nonliving stuff. However, despite our basic intuitions, making an explicit definition of life has been a tremendous challenge.[1] In high school, we learn that living beings are made up of cells, composed of complex organic molecules and DNA that together generate processes like metabolism, growth, reproduction, and the capacity to respond to external stimuli. The *Oxford Dictionary* entry for life is "the ability to breathe, grow, produce young, etc. that people, animals and plants have before they die and that objects do not have." A commonly cited definition is that provided by NASA in its program to search for extraterrestrial life, as a "self-sustained chemical system capable of undergoing Darwinian evolution." While these are in fact lists of typical characteristics of living beings, none of these proposals define life as a process. In this chapter, I will discuss some early and more recent attempts to define what life is, and I will propose a definition of it that curiously is as old as biology itself but has been overlooked by many scientists, perhaps in their attempt to characterize it in terms of physical-chemical mechanisms.

From Soul to Molecules

The Ghost in the Machine

Notwithstanding the early attempts to rationalize the anatomy and characteristics of living beings, life itself has been historically defined as a spiritual phenomenon. The notion of an immaterial soul, or a spirit tied to the essence of life, has endured for most of our history and remains consciously or unconsciously in many people today. According to Aristotle, the soul had at least three components: plants bearing a vegetative or nutritive soul, animals having in addition a sensitive or appetitive soul, and only humans being endowed with a rational soul besides the other two. This view pervaded classical thought until the seventeenth century, when a more rationalistic perspective was raised by René Descartes.[2] Descartes proposed dualism, asserting that the body and soul (or the mind) were separate entities, where the pineal gland was the site in which the mind exerted control over the body. It is commonly said that he claimed that as opposed to humans, animals lacked a soul. However, he seems to have recognized that they had a different kind of soul than humans, a mechanistic soul being capable of movement but lacking reason or morality, and especially the ability to communicate their thoughts to others. To Descartes, animal behavior could be explained from the physical properties of their organs. Furthermore, he proposed an analogy of living mechanisms to rudimentary automata that were known at the time, consisting of interlocked pieces that executed some simple movements.[3] Descartes's mechanistic perspective is usually considered a turning point in the scientific descriptions of life, implying to some that life could be produced or explained in physical terms. While simple automata have been created by humans since the ancient times, rationalistic thinking associated to them became highly influential after Descartes, as exemplified by the "digestive duck" designed by Jacques de Vaucanson in 1739, which appeared to process food and then defecated a green substance. While this and similar machines were clearly hoaxes, they pointed to the notion of life as resulting from physical mechanisms in the line of Descartes's rationalism and gave early ground to the emergence of robotics in the twentieth century.[3]

In parallel to these mechanical considerations, in the seventeenth and eighteenth centuries, many authors started considering that beyond mechanics, living stuff was chemically different from lifeless matter. An important insight in this line was provided in the eighteenth century by Georges Buffon, who asserted that life was made of "organic molecules" (now known as carbon-based molecules) that assembled into combinations that enabled them to make copies of themselves. On the other hand, several scholars defended the vitalistic theory, claiming that life was characterized by a vital impulse in a way similar to Aristotle's notion of the soul and that there was a discontinuity between living and nonliving stuff, with no possible intermediate stages.[4] According to vitalists, the basic properties of life, irritability and regeneration, needed no further explanation, just like gravity was a property of planets.

A substantive scientific advance derived from the invention of the microscope, which led to the postulation of the cell theory by the 1830s, asserting that cells are the fundamental building blocks of all living beings and that all cells derive from other cells. Thus, the cellular theory became a bold statement about the universal structure of life and pointed to the minimal unit that could be considered alive (viruses are not alive). However, despite its universality, the cell theory did not point to any underlying process generating the properties of life. In fact, the microscopic resolution of the time did not allow to distinguish organelles or components inside the cell, and the material composing it was referred to as "protoplasm," a sort of living substance that was the basis of life itself, which contributed to maintain the vitalistic theories.

Perhaps the strongest blow to vitalism was provided by the nascent biochemistry in the late nineteenth and early twentieth centuries. Chemists became quite successful in producing compounds like urea or acetic acid, which were considered to be exclusively biological products. Furthermore, the isolation of an enzyme from yeast that triggered fermentation from sugar to ethanol and many other discoveries strongly prompted the emergence of biochemistry as a science, together with a chemically based understanding of life. A great figure in this line was Hans Krebs, who unveiled the energy-producing metabolic cycles inside the cell in the early twentieth century. The Krebs cycle, of which I will speak later, contributed to destroy the vitalist principles, replacing them with chains of chemical reactions that gave rise to the different properties of cells (see figure 2.1).[5]

The Heat of Life

Another line of mechanistic thought was based on physical-chemical concepts associated to the functioning of heat-producing engines. In the 1940s, the outstanding physicist Erwin Schrödinger bravely asked himself "What is life?," arguing that living beings apparently contradict the second law of thermodynamics (which states that energy degrades into useless forms), being able to decrease entropy to generate order or complexity at the expense of releasing heat into the environment.[6] Organisms are supposed to accumulate "negative entropy" (also referred to by

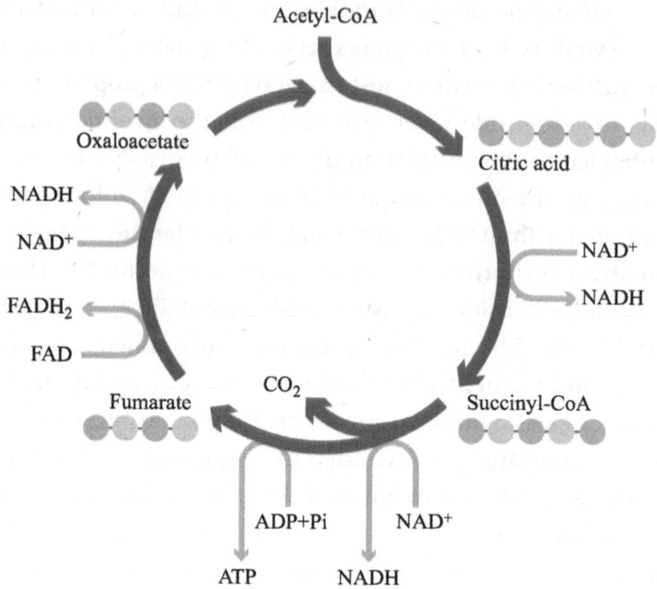

Figure 2.1
The Krebs cycle, or the tricarboxylic acid cycle. It consists of a series of chemical reactions of simple organic molecules, including oxaloacetate and citric acid, and provides energy in the form of ATP for cellular functioning.

some as information), which provides them with an internal organization. Schrödinger's thermodynamic approach still sparks much interest among physicists, and since his book was published, a series of intriguing theoretical possibilities of physically based biological evolution have been proposed that are not always in line with his original views. For example, according to Kate Jeffery and Carlo Rovelli, highly ordered organizations can evolve spontaneously as when water and oil tend to separate in two phases.[7] In this example, the separate state ("ordered") is statistically more probable than the mixed state ("disordered"). This process results from the properties of water and oil molecules and their interactions, without need of external work, and yields an increase in entropy. Likewise, stars, galaxies, and life itself would be highly ordered spontaneous configurations that increase entropy simply by being more probable states than a homogeneous distribution would be, given the physical conditions where they exist (gravity for stars and galaxies; macromolecular attractions and reactions in the case of life). Although I confess this is a quite intriguing perspective that deserves investigation, in this book I will hold to the classical view of living systems as systems that store energy to be released for their own maintenance. The thermodynamic approach has been crucial to the development of theoretical models of self-organization systems that may resemble the living state, and for studies of the origin of life as we will see in the next chapter.

The Living State

Body Builders

Notably, the term "biology" was introduced in 1802 by Jean-Baptiste Lamarck, which has granted him the name of "father" of biology despite his infamous theory of the inheritance of acquired characteristics. In my opinion, one of Lamarck's greatest, but largely ignored, achievements is that, together with coining the term "biology," he proposed a mechanistic definition of life as a self-producing process. He specified the four faculties common to all living beings: (i) feeding; (ii) "building up their bodies, that is to say, forming for themselves the substances of which they are made . . . and which are mainly supplied in the form of food"; (iii) development and growing; and (iv) reproducing themselves.[8] Lamarck displays these faculties as a sequence where feeding is first, then is the construction of the body, which produces as a consequence development and growth, and finally is reproduction. Thus, obtaining nutrients may be a requirement for life, but it does not make life itself, while development, growth, and reproduction are consequences of the living condition. What is left as a central mechanism for life is self-production.

After Lamarck, many other authors subsequently mentioned the capacity of self-production as a critical property of life, but these ideas never went further than isolated comments.[9] In the 1970s, the neuroscientists Humberto Maturana and Francisco Varela went deeper on this idea, formally proposing a theory of the living organization that defines living beings as self-producing machines. They introduced the term "autopoiesis" (from the Greek, self-production) to describe the process by which the components produced by the organism participate in a network of production of new components and so on, defining also its borders as a unit in space (see figure 2.2). Self-production indicates that there are no external forces or

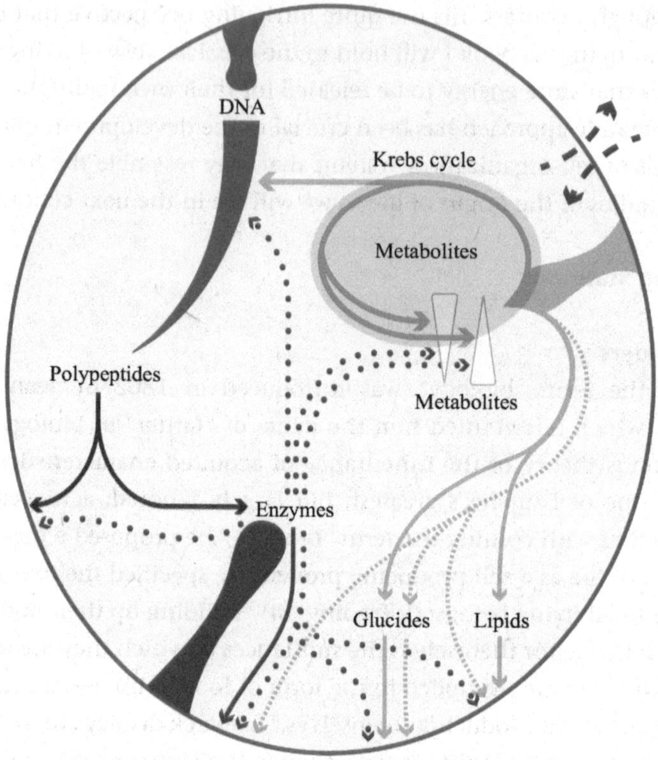

Figure 2.2
The self-production cycle, or autopoietic cycle, according to Maturana and Varela. I have highlighted the Krebs cycle as a critical component of this process.

design giving rise to this organization, but rather that the living being construes itself through the properties of its molecular components and the interactions between them.[10]

Self-production or autopoiesis is perhaps the best definition of the living state we can have at this point. Nonetheless, in my opinion, there are two critical elements of life that are not fully accounted for by this definition: reproduction and the capacity to evolve may require a specific extra mechanism to take place that is not given by the self-production process itself. Like I said in the previous chapter, this property is faithful inheritance, which is most largely accounted for by the self-replicating DNA molecule, which in addition orchestrates self-production in living cells. In the next section, I will address some issues related to autopoiesis in the cell, its origins, its maintenance, and the basic capacity to respond to stimuli (the latter two were considered critical properties of life according to vitalists). At the end of the chapter, I will come back to the problem of heredity in self-producing systems.

Self-Production in the Cell

Self-production is an energy-demanding process. Catalyzing energy-liberating reactions and storing chemical energy to be used for building new molecules are critical functions to maintain the living organization. Thus, living beings rapidly developed mechanisms incorporating energy-rich nutrients or absorbing external energy (like light), storing this energy in chemical bonds, and releasing it to keep their functions ongoing. Metabolism is the network of chemical reactions that directly or indirectly participate in the storage and liberation of energy required for building new components (nucleic acids, proteins, or membranes) of the system. An essential component of metabolism is the previously mentioned Krebs cycle, a circular set of chemical reactions involving carbon-rich molecules, one of whose first intermediate products is citric acid, the same found in lemons and oranges. The Krebs cycle takes place in cells during respiration (with modifications in bacteria and anaerobic cells) and produces the energy-keeping molecule called ATP (adenosine triphosphate) by breaking down organic molecules and liberating CO_2. This is a major mechanism to store energy for body building, but this cycle can also go in reverse or be incomplete, using only some of its reactions as in anaerobic bacteria.[5] Going backward at the expense of ATP, the Krebs cycle can start with hydrogen and CO_2 and use its intermediate products to build proteins, fatty acids, and even nucleotides, all the basic building blocks of life. Therefore, this little engine contains all the elements to build components and to obtain energy for these processes, making it a pivotal aspect of self-production in most cells. In the next chapter, we will see that this process, or parts of it, may have been essential for life's origin in the early earth. Nonetheless, there are other mechanisms that cells can use to extract energy, and to build complex molecules. In the absence of oxygen, reactions like glycolysis (the fast breakdown of glucose used by our muscles during intense exercise) or fermentation (in which ethanol is produced as a byproduct, as in wine and beer production) are alternative ways to liberate energy. Finally, a fundamental mechanism to build complex organic molecules is

Calvin-Benson's cycle, used by photosynthetic plants to fixate CO_2 and to generate glucose. This chemical pathway was possibly late to appear in life's history, but most life on earth now depends on the energy-rich compounds provided by algae and plants in photosynthesis.

Thus, the energy stored by the metabolism is then used to construct proteins, fatty acids for membranes, and nucleic acids (genes). These components are assembled in a complex but highly robust large-scale network in which the genetic material of the cell determines the structure of proteins that participate in the self-producing network, and in turn, proteins, nucleic acids, and other molecules regulate the activity of genes, specifying which genes and proteins will be active at a given place or time (see chapter 1). In this way, genes coordinate the process of self-production but are also regulated by the products they contribute to generating.

Before Self-Production

Maturana and Varela claimed that the autopoietic organization is an all-or-nothing condition: either a system is self-producing and encapsulates itself or it is not and dissolves. However, this may not have emerged from nothing, but rather there was probably a continuum of processes before its achievement. A step toward self-production may be found in so-called autocatalytic chemical reactions, where the reaction product contributes to generate subsequent reactions of the same kind, in an ongoing cycle (see figure 2.3).[11] The mathematician Alan Turing was a pioneer of this field, proposing in the 1950s models of oscillatory reaction-diffusion chemical reactions that yielded self-organized patterns (these models have been recently used to describe pigmentation patterns like leopards' spots, butterflies' wings, and fingerprints). Further studies in autocatalytic networks were carried by the Nobel laureate Ilya Prigogine and collaborators, who made theoretical models for autocatalytic processes that fitted simple but self-maintained chemical reactions that could be maintained in a state of circular flux with little input.[12] The Krebs cycle is an example of an autocatalytic process when running forward, but some of its backward reactions can also be described as autocatalytic.

Thus, autocatalysis has been widely proposed as a key step in life origins, where primitive proteins and nucleic acids organized in networks where nuclei acids produced proteins and these in turn participated in the synthesis of nucleic acids and small proteins, and so on. Simpler instances could

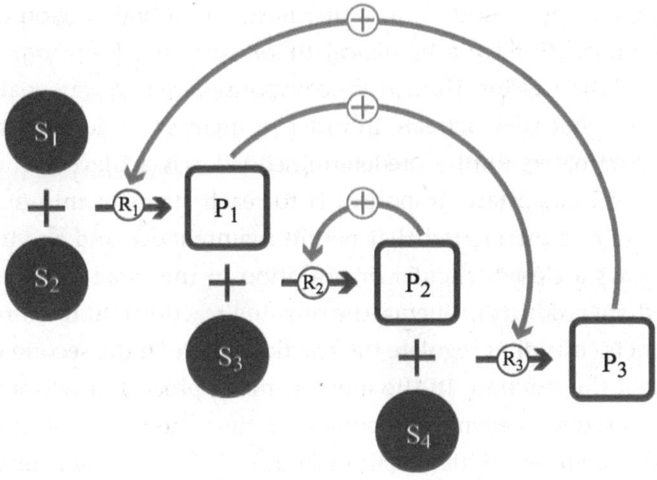

Figure 2.3
An autocatalytic network. The products of distinct reactions catalyze the reactions that gave rise to them. Ss are reactants, Ps are products, and Rs are chemical reactions between S and P. Different Ps may influence other reactions backward or forward.

be ensembles of proteins and RNA called ribozymes,[13] which we will discuss in the next chapter. Adding more reactions to this incipient network would have formed integrated systems that regulated themselves and maintained their integrity through time, provided they took place in an encapsulated space.

Maintenance and Responsiveness

Keeping in Balance

One problem is that these networks of chemical reactions are usually unstable and last for a very short time, being vulnerable to minimal ambient fluctuations. Only those systems that were able to keep within range critical parameters required for these reactions to occur were able to support themselves through time. This process, termed homeostasis (or maintenance of the internal state), is based on a variety of mechanisms that detect when some parameter goes out of levels and generates a response to restore these levels. Central to homeostasis is the concept of feedback, where the product of some reaction inhibits the same process that gives rise to it, like an engine that shuts off when having generated enough product or when generating too much heat, as in a thermostat. In a brain region called the hypothalamus, we have a biological thermostat that keeps our temperature around 36° Celsius. Homeostatic responses are in a way goal-directed in the sense that they activate in order to increase or decrease the value of some parameters until a predetermined value is achieved. The general tendency of homeostatic responses is to reach states of minimal energy expenditure and interaction that permit maintenance and survival. Feedback displays a closed, circular organization in the same sense as autocatalysis, where products influence the ongoing reactions. In the former case, the products control or regulate the reaction, while in the second case they may propel the reaction. In the more complex process of self-production, where the products become components of the system and contribute to its construction, autocatalytic reactions and feedback mechanisms drive the flux of reactions so that the system can maintain itself or grow, and at the same time work within certain boundaries. (However, developing a neat boundary in living systems requires an additional component, the cellular membrane that I will address in the next chapter.)

Irritability and the Beginnings of Knowledge

To maintain homeostasis, very early on, living beings evolved sophisticated molecular sensors and effectors generating metabolic responses and movement. Bacteria, the simplest living beings, are able to sense environmental alterations and drive their behavior in search of food or to avoid toxic substances. Likewise, plants obviously respond to light and water but can also detect mechanical stimuli, respond to infection by pests, and communicate with other plants nearby. Their responses are produced by growth, by releasing chemical signals, and in a few instances with movement. None of these beings has a nervous system, much less a brain, yet they are able to react in different ways to the environment and to alterations of internal conditions.

A critical aspect of homeostasis maintenance is to respond to changes early enough that the perturbations can be more easily compensated, saving energy and avoiding risk. Imagine yourself driving your car. You don't want to wait until the car is getting to the borders of your lane before making a correction maneuver. Just a small deviation of your intended direction makes you to rapidly correct the course of the vehicle to stay in range. An experienced driver does this largely unconsciously, but someone learning to drive or under the influence of alcohol will have a much slower reaction and a higher risk of accidents. Living beings do exactly the same: they anticipate the incoming changes by detecting subtle alterations before they get too large. Computationally, this process has been termed predictive coding and refers to mechanisms that anticipate future events to facilitate and minimize energy loss in the achievement of some goal, which in the short term may be maintenance of specific parameters like blood glucose levels but in the long term refers to staying alive (and perpetuating one's genes).[14] All living beings, from the simplest bacteria to the more complex organisms like us, have developed mechanisms to sense and react to small environmental alterations before it is too late.

Importantly, together with being able to anticipate future events, organisms may accumulate knowledge about repetitive situations to dampen their reactions and save energy, or amplify them in the presence of threatening stimuli. Although all living beings may be able to predict incoming threats and adapt to events on the basis of past experiences, among multicellular animals, a new organ has evolved for this function: the nervous system, including specialized sensory and motor or effector components. The nervous system is built atop the self-preserving body homeostatic mechanisms, sensing changes as early as possible to predict coming events and generating fast behavioral responses that are ultimately oriented to maintain their internal conditions. In addition, nervous systems may keep a more sophisticated memory of past situations than purely biochemical mechanisms so that increasingly effective responses can develop in the presence of repeating situations. The generation of a sensory-nerve-muscle network has two main effects in the multicellular organism. First, it accelerates its dynamics as the velocity of cellular communication becomes much faster. Second, for the same reason, distant parts of the organism become rapidly connected as a global nerve network starts controlling the whole state of the animal. Animals respond quickly, but because of this, they spend more energy and need to find their food quite often. In some groups, this has resulted in an expanding spiral of complexity leading to the development of complex sense organs, brains, and motile apparatuses.

Constructing the World

More complex nervous systems may generate sets of responses to the outside world based on past experiences and the anticipation of future events, which enables them to maximize their resources. These configurations become increasingly elaborate as complexity unfolds through evolution, and many speak of complex organisms generating models or representations of their surroundings, which is a step toward experiencing the environment as an external reality. Note however, that the notion of representation can be misleading, as it does not entail any specific neuronal mechanism but rather refers to a "map" of the world, which someone is able to observe (but there is no inner spectator in the brain). As I will explain in subsequent chapters, animals create their own world rather than just perceiving it, by organizing their sensory and response mechanisms into coherent wholes (from now on, I will use this interpretation of "representation"). From our anthropocentric perspective, we tend to think that the reality we experience is the real world and that animals model their reality similarly to us, but this is in fact an internal construction that allows us to move through it. The facts that the worlds of different species coincide in important ways (for example, a rat's way through a labyrinth must comply with the labyrinth made by the researchers testing it) and that these models allow us to survive without getting hit by cars and other events imply that there is an outside world to which the experiences we have must fit, but we cannot say much about its intrinsic nature or its ultimate reality. In fact, all our descriptions of the world, including the animals we study and their observed behaviors, are part of the reality we are constructing on the basis of our perceptions, behaviors, memories, and feelings. Moreover, it makes no sense to think about the fabric of reality because the only way to access it is through our senses and behavior, which basically respond to the maintenance of our selves and are the result of our evolutionary history. This view is hardly new, dating at least from Immanuel Kant in the eighteenth century, but has gained increasing acceptation in neuroscience, especially in relation to issues like consciousness. To insist, sometimes neuroscientists tend to think about the brain as an information processing system that captures an external reality and is devoted to solving cognitive or logical problems, minimizing the fact that the brain is an extension of the organism's homeostatic mechanisms, contributing to its survival and the maintenance of its lineage, and all our perceptions of the world are ultimately directed to these functions.

Self-Producing Lineages

The Reproduction of Self-Production

In this chapter, I have argued that self-producing or autopoietic systems may be the best representatives of life as a process, featuring maintenance, growth, and a basic responsiveness to the environment. But Maturana and Varela went beyond these features and claimed that all properties of life derive from the autopoietic organization, including reproduction and evolution. In this view, reproduction is seen as simply the fission of a self-producing network into two halves, as long as both daughters contain the necessary elements to maintain themselves through time. Likewise, evolution consists of the diversification of the self-producing networks throughout history according to a process called "natural drift." However, like I have said, this proposal downplays the crucial participation of the self-replicating genetic material in the organization of self-production and heredity, and the necessary role of natural selection in evolution. Only with a specific mechanism of inheritance, self-production systems can sustain themselves across generations, producing long-lasting lineages and heritable variants (mutations), yielding diversity and differential reproduction among the descendants. Next I will illustrate this argument using two examples: fire and synthetic cells.

Richard Dawkins proposed the example of fire as a self-maintaining system that is able to reproduce. Fire consists of the production of incandescent particles that generate new incandescent particles at the expense of carbon combustion. Although much simpler than living cells, burning flames display a basic autocatalytic (or if you will, a self-producing) organization, where ongoing reactivity is triggered not by the chemical products of the combustion reaction (CO_2 and H_2O) but by the heat released in the reaction. A flame has limits in space as particles that cool down are released out of it; flames can also grow in size and reproduce by generating new flames nearby. However, the reproduction of flames lacks a critical feature: the fire generated by, say, a yellow flame (as in a burning match) may come to generate a blue flame in the oven. The color of the flame depends on the material that is being burned and is not transmissible from fire to fire, unless it is cremating the same material. In other words, fire can reproduce but does not display heredity beside the basic fact that new flames also produce incandescent particles. Therefore, fire cannot evolve.

A more complex condition relates to the division of protocells, membrane-bound vesicles that are devoid of complex biological processes that can divide generating "daughter" vesicles. Still, the lack of inheritance mechanisms severely limits their possibilities to evolve. In this line, the recent manufacturing of synthetic cells in laboratories may provide an interesting example. These human-made cells were produced by mixing proteinaceous and lipidic components of bacteria with a chemically synthesized genome including some 400–500 genes considered by the researchers as indispensable for life (I will come to this again in chapters 3 and 15). The cells became able to maintain themselves, achieving a state of self-production and being capable of growing and dividing, even producing cellular aggregates. However, division was not even in these cells; daughter cells were smaller and had strange morphologies compared to the parental ones. Only after adding seven additional genes (some but not all of them known to be involved in cell division), cells became able to produce daughter cells practically identical to their parents.[15,16] With a minimally robust gene pool, these synthetic organisms have been able to undergo natural selection and adaptive evolution. Thus, although the genetic material is needed for faithful reproduction, not any genetic combination will guarantee this process, but instead it requires a specialized set of genes that probably evolved through natural selection, where lineages of cells better able to produce offspring similar to them were able to survive for a longer time. All living cells on this planet probably derive from a cell able to faithfully reproduce its offspring.

Perspective

This chapter proposes a definition of what life is and how it is organized, considering both the intrinsic workings of a living system and its capacity to evolve. The minimal unit of life as we know it is the self-producing or autopoietic cell, provided with a molecular self-replicating hereditary device (DNA or RNA). Cellular self-production and molecular self-replication may represent the key processes accounting for life as a dynamic process and its evolution on earth, respectively. A critical aspect of life is the maintenance of the internal state or homeostasis. In an obvious sense, autopoietic systems are homeostatic as long as they maintain themselves through the self-production cycle. However, in the primitive earth conditions, not all of these early systems were equally robust to maintain self-production for long, and probably most of them simply became extinct. Only those able to regulate distinct physical-chemical parameters critical for the self-production process survived. Through reproduction, heritable changes, and natural selection, some of these systems evolved strong homeostatic networks, eventually becoming the modern cells that exist now.

Besides homeostasis, developing a memory of past events may contribute to generating early responses. Thus, the evolution of regulatory networks implies the generation of a time-arrow as past experience is accumulated in order to make predictions about the world. All these processes contribute to the formation of a rudimentary self that endures solely by virtue of the interactions between its parts. While all living beings are endowed with molecular networks that enable them to maintain themselves and reproduce, most animals have developed a nervous system that greatly increases the memory and prediction capacity of relevant events and is able to generate fast behavioral responses to them. Thus, the nervous system amplifies and accelerates the regulatory and homeostatic capacities of animals, setting them in a new timeframe and providing them with a network for systemic regulation, and in addition permits increasing perceptual and behavioral capacities.

Having provided some conceptual scaffolding about what is evolution and what is life, in the rest of the book I will start narrating the history of life. The next chapter refers to the origin of life on earth and the early evolution of unicellular beings, in order to discuss in the subsequent chapter the emergence of multicellularity and the first animals.

II Beginnings: From Cells to Neurons

This section refers to the origin of life and the early evolution of cellular organisms. Life on earth resulted from the conflation of many physical-chemical circumstances, giving rise to self-replicating and self-producing molecular ensembles that, by natural selection, gave rise to the first living cells and the astounding diversity of life forms on this planet. The early cells proliferated and conquered the world, some of them being able to perform photosynthesis, which catapulted evolution into increasing levels of complexity. In this process, the symbiotic relation between different types of cells enabled the emergence of complex cells and gave rise to multicellular organisms including plants, fungi, and animals. Plants and fungi have evolved sophisticated cellular communication systems and are able to "cross-talk" with other individuals, but they lack a nervous system. Animals were the first multicellular predators, specializing in capturing other organisms to feed on them. Soon after their emergence, they developed two major innovations: an internal cavity, the gut to digest food, and a nervous system that coordinated gut and body movements to capture prey. This enabled them to live free lives, self-propelling to reach new environments and colonizing the world. The section ends with a chapter discussing neurons, the building block of nervous systems. Neurons originated by recruiting several cellular processes into a highly differentiated cell type that makes up an interconnected network driving behavior and body homeostasis. As it develops in an embryo, this network unfolds into a highly organized ensemble that is able to modify its structure and activity according to circumstances, providing a substrate for the evolution of cognition.

3 The Birth of Cells

As far as we know, the only planet that harbors life is earth. Terrestrial life depends on the presence of carbon, hydrogen, nitrogen, and oxygen among other elements, and takes place in liquid water, where carbon-based organic molecules react continuously to participate in the self-production process. Water is common in the universe but mostly in the form of solid ice, on planets so cold that early life there is impossible. For a planet to have liquid water on its surface, there must be an appropriate combination of temperature and pressure, which depends on many factors like the planet's mass, the size of its star, and the distance and speed of the orbit the planet makes around it. The region around a star where liquid water can exist on a planet's surface is called the Goldilocks or habitable zone. In our solar system, the only planet in the habitable zone at this point is earth. Yet, the evidence indicates that Mars had liquid water and rivers about 4–3 BYA, and might have sustained microscopic life at some point in its history. Furthermore, there are places outside the habitable zone where liquid water may exist below the surface, heated by the celestial body's core, as in Europa and Enceladus, moons orbiting Jupiter and Saturn, respectively.

According to one hypothesis, water on earth was present inside rocks since its early beginnings, but it may have largely vaporized during the moon-forming impact some 4.5 BYA and been brought back to earth later again in ice-containing meteorites. The early earth was covered by a vast sea of magma that eventually cooled, forming the earth's rocky crust. Concomitantly, vaporized atmospheric water and emerging water from the earth's inside condensed into oceans about 4.2 BYA. The earliest evidence of life comes from 3.5 BYA in fossilized bacteria called stromatolites, although some findings suggest earlier dates, up to 4 BYA or more. From then on, cellular life evolved slowly during the Precambrian age until the arrival of multicellular organisms at the final stages of this period, around 1 MYA.

Genesis

A Warm Little Pond

One thing is having a planet with liquid water, and quite another thing is to develop life forms on it. Curiously, the notion of the spontaneous emergence of life from inanimate matter was relatively common in pre-Darwinian times (the theory of spontaneous generation). This concept was debunked by Louis Pasteur, who showed that life always derives from other life, providing indirect support for a common ancestry of all life. However, there must have been a point in history where life emerged from scratch. Darwin first proposed that life could have begun in "some warm little pond" with ammonia, phosphoric salts, light, electricity, and other factors that underwent complex chemical reactions. Following Darwin, Alexander Oparin and J. B. S. Haldane argued that life could have emerged from inorganic molecules in an early earth environment subjected to high temperatures and radiation. Particularly famous was the experiment performed by Stanley Miller and Harold Urey in 1953, who used a mixture containing ammonia, hydrogen, carbon dioxide (CO_2), and nitrogen diluted in water, which was bombarded by strong electrical discharges, heat, and ultraviolet light, conditions considered to be similar to those of the early earth. This treatment produced a series of organic molecules that would have served as the basic components for life's emergence.[1]

A more recent proposal sustains that life arose in submarine volcanoes or hot water vents, heated by the underlying magma and creating a highly alkaline environment. These conditions would have released a molecule called carbonyl sulfide that could trigger the formation of small organic proteins. This view has been supported by the discovery of extremophile microorganisms living in these harsh conditions. Another view considers that instead of the deep oceans, the cradle of life would have been terrestrial volcanic environments with shallow water pools. This approach considers conditions similar to those seen now on Mars to propose that the molecular feedstock for organic molecules was another molecule, hydrogen cyanide. Of course, there is an extraterrestrial hypothesis for the origin of organic molecules, falling to earth in comets and meteorites together with water. This is supported by recent findings of all the nucleotides required for nucleic acids being present in meteorites that formed 4.5–5 BYA. This finding also suggests that DNA and RNA could have been or could be present on other planets! A highly speculative hypothesis, called the "Goldilocks Universe," proposes that in the very early universe, some 10–17 MY after the Big Bang, the whole universe had the right pressure and temperature conditions for liquid water to exist. Complex molecules could have been formed in these conditions, seeding the basic elements (like nucleic acids) for the origin of life, that lay dormant to be activated in different galaxies of the expanding cosmos.[2]

Notably, since organic molecules have characteristic three-dimensional shapes, many of them come in two alternative configurations that are mirror images of each other, like the left and the right hands. This phenomenon, called chirality, implies that when assembling into large molecules or when interacting with other chemicals, left and right versions may not be compatible with each other. In fact, biological molecules only have one of the two versions, like all amino acids are of the "left-handed" group and all sugars are of the "right-handed" group. Why this turned out to be the case could have been just chance, or there could be slight functional or energetic differences between both forms. Anyhow, if we ever encounter carbon-based extraterrestrial life, we cannot be sure that its molecules will have the same chirality, or whether they will be molecular "mirror images" of ours.[3]

A critical issue in this context is the origin of nucleic acids and DNA. A widespread hypothesis states that nucleic acids first evolved as strains of RNA, which may have made copies of themselves, perhaps aided by different chemical compounds, including small proteins, amino acids, or other nitrogenated molecules that would form self-replicating complexes. In this view, ribozymes, which are RNA molecules with catalytic activity, have been proposed as key for the early emergence of life. Thus, self-replicating complexes including small proteins and RNA strings, perhaps reminiscent of the ribosomes that translate the mRNA code into protein chains in living cells, may have represented the early prebiotic self-replicating molecular machineries. Considering this evidence, Manfred Eigen proposed that short nucleic acid sequences formed cooperative populations of self-replicating chemical "quasispecies" that underwent natural selection, favoring the more stable and better replicating ensembles.[4]

Yet, self-replication needs energy. Perhaps the most basic metabolic process is the Krebs cycle, or some of its constituent reactions. Although the Krebs cycle usually requires oxygen to run, some of its basic reactions can take place in the absence of oxygen, aided by minerals that are present in hydrothermal vents. According to some biochemists like Nick Lane, the Krebs cycle is the basic engine for self-production, and its reactions might have been able to form the basic nucleotides, amino acids, and fatty acids, providing a scaffolding for incipient self-producing networks to evolve.[5]

A long-standing controversy among early life researchers refers to what came first: metabolism or self-replication capacity. Like other disputes of this kind in evolutionary theory, this will probably last for very long. Perhaps both processes, molecular self-replication and energy storage/liberation cycles conflated to make the earliest self-producing machineries that worked as the precursors of early life. The first cells that were able to undergo biological evolution were probably the result of an alliance between autocatalytic metabolic networks and self-replicating molecules that somehow regulated the activity of these networks and provided a rudimentary mechanism for inheritance.

Keeping It All Together

A big complication of the early self-production systems is that their reactants would quickly wash away if not contained in a closed environment. A major innovation in cell origins was to keep all components inside a membrane-bound compartment. Besides avoiding diffusion, this protected the self-replicating network from parasitic molecules that might take advantage of it for their own reproduction, as modern viruses do when they infect cells. Thus, the acquisition of membranes that encapsulate all these reactions marks the origin of cellular life. How this took place in the history of earth is again a source of much speculation with no definite answers. A current hypothesis proposes that multiple layers of lipids formed in the bottom of hydrothermal pools in land volcanoes and RNA self-replication took place between these layers. Then, the lipid layers would release droplets with RNA molecules inside, forming self-replicating protocells.[6]

But even inside the cell, different components needed to assemble in aggregates to maximize their probability of interactions and to avoid interference from other reactions taking place in the cell. While more complex cells evolved a cytoskeletal network and a set of internal membranes that partly solved this problem, how proteins and other molecules self-assemble in subcellular structures remains a major mystery for biologists. One answer may come from thermodynamic processes in which different molecular components spontaneously condense in superstructures or separate like oil drops in water (see chapter 2),[7] providing a plausible mechanism for an early compartmental organization inside the cells.

Early Cells

LUCA and Their Kin

According to some authors, the first living beings made up a population of cells containing nucleic acids, proteins, and other compounds that divided but also could fuse among themselves, making up a network of lineages that exchanged materials, rather than a branching tree as we are used to seeing in genealogic diagrams or in depictions of the "tree of life." Thus, evolution was basically communal where proto-genes and other components could be largely shuffled from cell to cell. At some point, a threshold would have been reached where cells acquired a more integrated organization, assembling themselves as discrete membrane-bound systems where the vertical transmission of genes (from parent to daughter cells) became the norm.[8]

From this population emerged a group of cells that gave rise to all presently living beings, from bacteria to humans. This group is called the Last Universal Common Ancestor (LUCA).[9] LUCA is believed to have been an anaerobic organism or rather a diverse population of organisms, perhaps living in warm hydrothermal vents some 3.7–4 BYA or before. It probably required no more than the hydrogen provided by the hydrothermal vents and CO_2 to fuel its metabolism, like present day methanogen microorganisms do. LUCA possibly possessed some five hundred genes and the proteins they codified, indicating that it was a rather complex kind of cell.

Prokaryotes

LUCA gave rise to all existing cells, which diversified into the six major categories or kingdoms of life. The most basal are two kingdoms of prokaryotes: bacteria and archaea, consisting of free-living or communal cells lacking a cellular nucleus.

Archaea are bacteria-like cells, some of which live in extreme environments like hydrothermal vents. Many living prokaryotes have whip-like flagella that enable them to swim around and a molecular skeleton that participates in cell division and other functions. The earliest prokaryotes in the fossil record consist of Hadean stromatolites (there are still some living stromatolites in different parts of the world).[10] Like LUCA, the early prokaryotes were most likely heterotrophic, that is, they obtained energy by degrading complex nutrients. In the early earth there was too little oxygen to break the organic molecules, and energy liberation was largely accomplished by anaerobic processes like fermentation or using sulfur, nitrogen, or carbon dioxide instead of oxygen. Some evidence suggests that the first aerobic life originated about 3.1 BYA, perhaps in microscopic oxygen-rich pockets inside rocks, and may have remained like that for many million years.[11] A fundamental innovation took place some 2.3 BYA with the great oxidation event, produced by the expansion of cyanobacteria that evolved the capacity for photosynthesis. Aerobic cells proliferated by feasting on glucose-rich photosynthetic cells, starting the first food chains. Photosynthesis provided a major energy input for life and promoted the evolution of most life forms on the planet.[12]

Enduring Symbiosis

The remaining four kingdoms of life (protists, plants, fungi, and animals) are grouped together as the eukaryotes and derive from a unique mixture of bacterial and archaean cells (see figure 3.1). Protists are unicellular organisms that sometimes aggregate in colonies, while plants, fungi, and animals are multicellular organisms, each having evolved from separate protist ancestors. Against common intuition, fungi are closer to animals than to plants, both being related to a branch of protists termed Opisthokonta, so-called because they swim propelled by a backward flagellum. Among Opisthokonta, Holozoans represent a group composed by animals and a few branches of closely related protists (like choanoflagellates, which will be discussed in the next chapter) but not fungi.

The earliest fossil eukaryotes date from about 2 BYA, emerging after three probably interrelated critical events. One was the production of a complex intracellular membrane system that wrapped the DNA inside a nucleus but also produced a membranous intracellular structure called the endoplasmic reticulum, where the synthesis of proteins and other vital processes take place. Likewise, their membranes underwent a major increase in electrical excitability amplifying the number of specialized membrane proteins

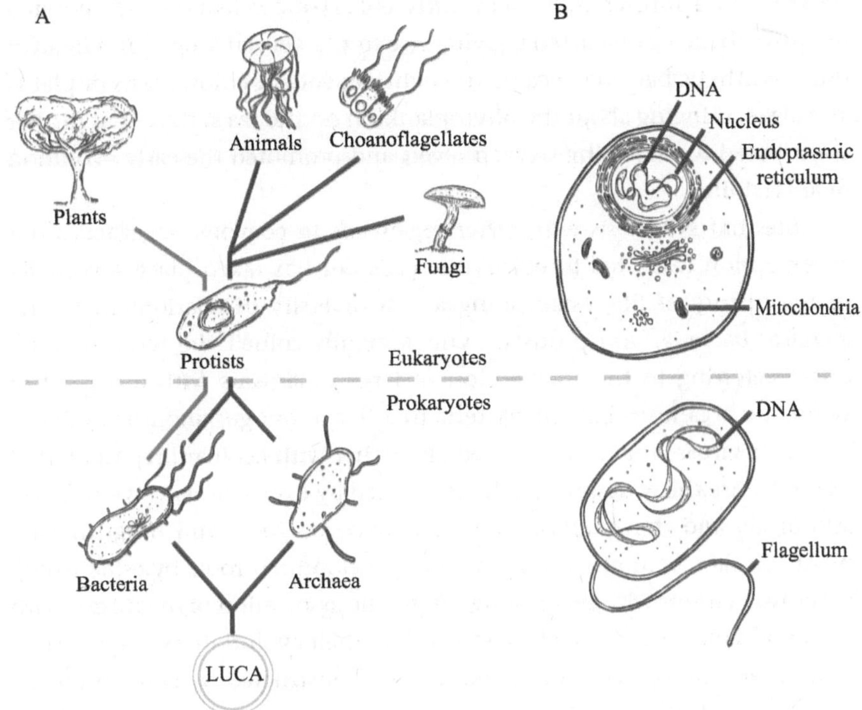

Figure 3.1
The six kingdoms of life. *A:* LUCA is the last common ancestor of all life, giving rise to bacteria, archaea (both prokaryotes), and protists, plants, fungi, and animals (all eukaryotes). Choanoflagellates are a branch of protists close to animals (together they are called holozoans). Eukaryotes were formed by a fusion between archaea and aerobic bacteria. Plants were formed by a fusion between photosynthetic bacteria and protists. *B:* Eukaryotic cells display internal membranes and a nucleus that encapsulates the DNA, while prokaryotes do not.

that allow the exchange of ions between the cytoplasm and the exterior. Secondly, eukaryotes developed a highly complex molecular cytoskeleton that enables them to display more elaborate behaviors than prokaryotes. The third, and perhaps the most important event, took place when a group of archaean cells called Asgard archaea, engulfed smaller anaerobic bacteria that displayed a highly energetic metabolism, which eventually became the aerobic mitochondria present in all eukaryotic cells. Complexity dramatically increased with the origins of eukaryotes, as the number of cellular genes of all kinds practically doubled just after the symbiotic relation started.[13] In a further step, some early eukaryotic cells also incorporated photosynthetic cyanobacteria, giving rise to the vegetal kingdom. Engulfed photosynthetic bacteria became the light-dependent chloroplasts of plants and algae, bringing about the phytoplankton on the sea surface that further contributed to increasing oxygen levels and promoted the early evolution of animal life.

Note that symbiosis with other organisms to coevolve together is not an exception restricted to eukaryotic cells but has taken place repeatedly in the history of life, establishing a vast diversity of relations including archaea, bacteria, and protists. Lynn Margulis coined the term "holobionts," referring to the coevolution of large organisms with a myriad of symbiotic organisms like the bacteria that live in our gut, mouth, and skin, as well as viruses that have evolved altogether with cellular life. Additional examples are the widespread lichens, formed by associations between fungi and algae, and the symbiotic relations between roots and fungi in land plants. In fact, land plants were able to develop their roots by establishing close associations with preexisting fungi. The association of organisms with their microbiota can be tracked way back to their evolutionary history, such that an accurate phylogenetic tree can be constructed by comparing the microbiota of different species. In a way, organisms and their symbionts become an evolutionary composite that responds to evolutionary pressures and undergoes adaptation as a whole system.[14]

Cells as Gene Propagators

With the onset of cellular life, genetic replication became a highly stable process, and genes could spread to different environments. Cellular reproduction was a fundamental mechanism by which genes and other components became faithfully transmitted to the descendants and spread in the early earth. As I mentioned previously, in addition to vertical gene transfer that takes place from parent to progeny during reproduction, genes have additional ways of propagating themselves through horizontal gene transfer, where cells exchange genes either directly or indirectly through viral infection. Horizontal gene transfer is common in prokaryotes, in protists, and even in multicellular organisms including vertebrates, even transferring genes between species so separate as plants and insects.

While gene sharing is good for gene propagation, it also implies a big risk for cells and for the genes inside them. Since their origins, living cells have been exposed to self-replicating parasitic molecules like viruses that take advantage of their machinery to spread themselves. This way, cells elaborated mechanisms to reject these intruder molecules by recognizing them as foreign to their body. Bacteria have developed the so-called CRISP-R system that copies pieces of the RNA sequence of viruses infecting them and inserts this copy in their own DNA, as a sort of memory of the infection event. In a subsequent infection, the inserted DNA is copied into RNA that binds to an enzyme that cuts RNA; when the bacterial-produced RNA assembles with the viral RNA, the enzyme cuts the latter into pieces, inactivating it. This is perhaps one of the first mechanisms involved in fighting infection and, more so, a mechanism to distinguish self from nonself. In chapter 15, we will come again to the CRISP-R system as it has been taken advantage of by scientists to cut the DNA into pieces and delete or insert new genes in plants, animals, and even in our species.

Furthermore, taking advantage of gene transfer processes, there are the so-called selfish or mobile genes that can jump from one region of the chromosome to another, and perhaps more relevant from one cell to another, favoring their own replication similarly to how biological and computer viruses spread. According to recent studies, viruses may have originated several times by detaching from the host DNA, possibly as mobile genes. Conversely, some mobile genes may derive from viruses that infected the cell's DNA in the past. Furthermore, selfish genes may increase their fitness by attaching to other genes that provide some advantage to the cells bearing them, like, for example, antibiotic resistance in bacteria. Notably, some of these genes carry the CRISP-R system to destroy other mobile genes that may compete for their attachment sites. As it usually has no adaptive function for the organism, selfish DNA mutates rapidly and accumulates as noncoding DNA especially in complex species, whose largest DNA fraction is the so-called "junk DNA" deriving from degenerate mobile DNA that has lost its coding ability (still, some of these DNA segments can sometimes be recovered by cells to make up new genes). Junk DNA tends to grow in complex cells without limits as long as it does not interfere with the activity of the functioning genes and the cellular functions (sometimes it may even contribute to gene regulation). Thus, DNA and RNA can proliferate and evolve beyond the genes' and the cells' boundaries, being nonetheless dependent on the self-producing machinery of cells for their replication.

The Spark of Life

Receptors, Channels, and Pumps

While genes benefited from cells for their propagation, the latter had to keep their boundaries tight in order to survive and be good replicators, which was provided by the cell membrane. This also becomes the interface with the environment, acting as a sensorimotor organ that establishes physical contact with objects, prey, and other cells. Key to all these processes are all the membrane-bound proteins that serve as receptors for substances, as attachments to the substrate or to other cells, and especially those that regulate the flux of electrically charged ions across the membrane. In all cells, the cytoplasm accumulates negative ions relative to an excess of positive charges on the outside, a voltage difference that is concentrated in the cell membrane. For instance, positively charged ions like calcium and sodium have higher concentrations on the outside than on the inside, and, if allowed, they will tend to move in to eliminate the gradient. An exception is potassium, a positive ion required in large quantities for many cellular processes, which is more concentrated inside than outside the cell. In the cellular membrane of all cells, there are complex proteins called "pumps" that transfer ions against the electrochemical gradient at the expense of energy (for example, moving sodium outside and potassium inside), contributing to the maintenance of the membrane electrical potential.

Additionally, there are specialized membrane proteins called ion channels that open a miniature hole in the membrane and permit the passive flux of ions, particularly calcium and sodium to the inside, and potassium to the outside, tending to diminish the ionic differences between both sides of the cellular membrane when they open. Ion channels are usually closed, but with certain stimuli they can open for a short time to allow ionic flux across the membrane, which may trigger distinct cellular mechanisms. Membrane ion channels can be of several kinds: ligand-gated receptors that open or close after binding to signaling molecules; mechanically gated channels, which open after physical stretching of the membrane, as when a cell swells due to excessive water intake; or voltage-gated channels, which usually open when the electric voltage across the membrane decreases below a threshold. Apparently, the first voltage-gated channels carried potassium ions to the cell's outside, as is observed in some modern bacteria. Voltage-gated channels are fundamental for the electrical properties of neurons but are also critical in other cellular processes like the rapid electric changes taking place when a sperm fertilizes an egg, blocking the entrance of additional sperm.

In this context, calcium is a potent activator of several key processes, including metabolic regulation, cell motility, secretion, phagocytosis, electrical activity, gene activity, and cell division, so its internal concentration must be tightly regulated. Thus, calcium is a master regulator of many functions contributing to the cell's homeostasis and may be among the first mediators of behavior and responses to environmental stimuli. Modern cells have developed distinct mechanisms to store calcium in vesicles, releasing it only in the necessary conditions and quantities. Calcium was very scarce on the early earth, and consequently the first cells contained very little of it. However, due to geochemical erosion, calcium began to wash out from the earth's crust, accumulating in the sea. This led to an increase in this ion's concentration in the water, which forced cells to develop mechanisms to maintain their low concentration of calcium.[15]

Behavior without Neurons

Another critical element for cellular survival (and gene proliferation) is the capacity to sense the ambient and interact with it in order to obtain nutrients and escape from danger. Prokaryotes have membrane molecules that act as receptors of external stimuli, of which the most well-known is the light-sensitive bacteriorhodopsin, which is distantly related to the eukaryotic (and vertebrate) visual pigments. Some unicellular protists, algae, and even certain fungi have evolved subcellular eye-like structures to maximize light absorption, usually for photosynthesis. Eukaryotes increased the diversity and complexity of their receptor molecules, for instance, by amplifying a special family of transmembrane receptors, the so-called G protein-coupled receptors (GPCR) that include the opsins and the bacteriorhodopsin. In higher animals, GPCRs are involved in different kinds of perceptual mechanisms, including vision and chemo-sensation as well as in hormone reception and neurotransmission, and participate in learning as we will discuss in the next chapters. Thus, while membrane sensory molecules were probably present in the earliest cells, they were significantly expanded with the origin of eukaryotes, and furthermore in multicellular evolution, especially among animals.[16]

While the responses produced by a cell may be highly varied, involving secretion of substances or facilitating chemical reactions, here I will specifically refer to behavior as biologically powered displacement of the organism or its surroundings, like phagocytosis in amoebas or generating water currents for filter feeding as in ciliate protists. Behavior is most useful for catching food, approaching mates, and evading damage. Additional forms of behavior are related to growth, as can be observed in plants moving to the light or digging their roots. However, these phenomena usually take place in a longer timeframe than what interests us. Like any definition in biology, there are apparent exceptions to this proposal as for example the ability to change colors in some complex species, particularly chameleons and cephalopods, which might arguably count as behavior. I will consider these as physiological responses of behavioral relevance, rather than behavior per se. Cells have evolved complex molecular mechanisms to move around, ingesting food and expelling waste across their membranes, as in endocytosis (or pinocytosis, depending on the size of the transported substance) where the cell engulfs some particles or even liquid that becomes stored in a vesicle inside the cytoplasm. Secondly there is exocytosis, where internal vesicles fuse with the cellular membrane and release their contents to the outside. These processes are tightly regulated by the cytoskeleton that interacts with membrane proteins in order to produce the necessary shape changes during the cell's behavior.

Regarding movement, free-living unicellular organisms including bacteria and protists display a variety of behaviors like swimming toward and ingesting food or moving away from noxious stimuli. Bacteria are able to swim in one direction and change direction in response to different chemical signals. They switch direction by a behavior called "tumbling," produced by modulating the movements of their flagella. LUCA probably displayed a motile flagellum, enabling it to swim around like bacteria do. Among protists, we may have all observed videos of amoebas slowly extending their pseudopods (tentacle-like protrusions) to chase smaller free-living cells, or white blood cells moving after bacteria in the blood serum. Likewise, *Paramecium* is a complex free-living cell that swims around powered by thousands of cilia all over its body and displays avoiding and approaching reactions to a variety of physical and chemical stimuli (see figure 3.2). These responses are associated to electrical membrane charges produced by voltage-gated calcium channels. For this reason, it has deserved the nickname of "swimming neuron." Another example of complex behavior is provided by protists of the genus *Euplotes*, who use some fourteen "legs"

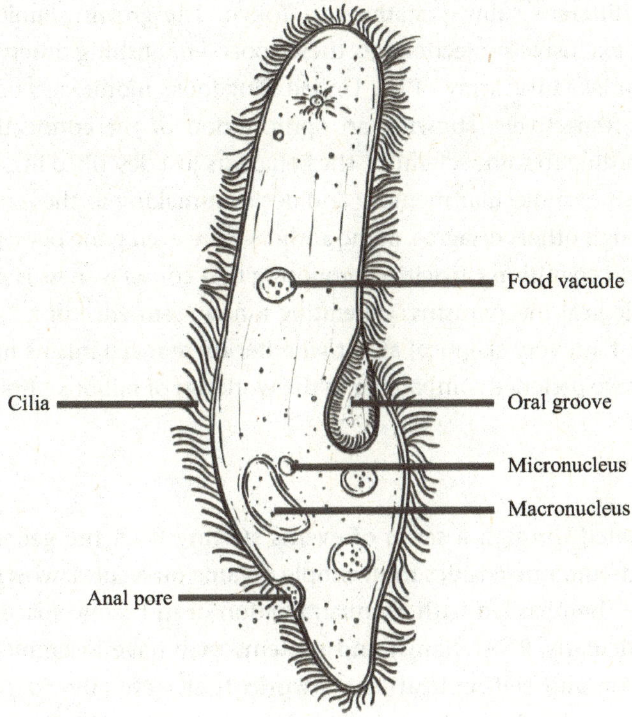

Figure 3.2
Paramecium.

supported by microtubules to swim and walk on surfaces, moving these extensions with great complexity and very flexibly according to the substrate they move on.

Finally, the syncytial slime mold *Physarum* (syncytia are gigantic multinucleated cells, like the fibers of our skeletal muscles) is a protist that grows like a fungus with "tentacles" up to several centimeters long that are used to catch bacteria and other protists as food. As these tentacles grow in a substrate, they encode a memory about the location of nutrients in the filamentous skeleton, whose tubes grow in thickness when in vicinity of an already visited nutrient source, and orient their growth toward the nutrient location. In addition, they can manage to detour complex pathways in search of nutrients, as if running through a maze. Furthermore, in a series of clever experiments, slime molds were grown over culture wells with food spots in different places representing a map of different UK cities, or of the different subway stations in Tokyo. The growing mold quickly developed extensive projections to these spots, establishing a network that closely mimicked the array of the United Kingdom's motorways or the subway map, respectively, showing an optimization of the connective pathways. According to some scientists, the behaviors just described imply simple mechanisms of molecular memory and decision-making in these enormous cells, although others disagree. Some authors have even gone beyond simple learning and cognition capacities, proposing that consciousness is grounded in cell biological mechanisms, generating a nano-sentience or a "senome." Personally, I am very skeptical about whether these mechanisms may make a subjective experience comparable to the workings of billions of neurons.[17]

Perspective

Life originated through a series of events starting from the generation of amino acids and nucleosides from simple organic molecules, whether in an early planet bombarded with cosmic radiation or in cosmic space. Nucleic acids (particularly RNA chains) and proteins may have assembled within autocatalytic and self-replicating networks that were able to propagate within the early earth environment, possibly in extreme volcanic conditions. A critical event in life's origins was the formation of a lipidic membrane that encapsulated all these chemical reactions in a restricted space, making up the first cells. The last common ancestor of all cells, LUCA, was an already complex organism or population of organisms provided with some hundreds of genes coding proteins involved in self-production. Particularly, membrane-bound proteins served many functions including cell adhesion to the substrate and the transfer of ions across the membrane, providing the basis for cell aggregation capacity and electrical excitability, respectively. These early achievements provided the foundations for the eventual evolution of the nervous system.

Early cells became subdivided into two large domains, bacteria and archaea, both bearing a prokaryote organization (without a cell nucleus). Cyanobacteria evolved photosynthesis, which provided a rich source of energy and oxygen for life to evolve. Eukaryotes (cells provided with a cell nucleus) emerged from a symbiotic fusion between Asgard archaea and bacteria that supplied energy, making up the first eukaryotic cells. Later, some eukaryotic cells engulfed photosynthetic bacteria, giving rise to plants.

Algae and plants dramatically increased oxygen concentration in the atmosphere, providing a further increase of energy for the emergence of more complex life.

Behavior and motility were probably present in LUCA and the earliest cells, who may have been able to sense and react away or toward different kinds of stimuli by virtue of molecular motors like flagella or contractile proteins inside them. Thus, a nervous system is not a requirement for behavior. The next chapter refers to the origin of multicellularity and the emergence of animals, emphasizing the differentiation of the first nervous systems.

4 Sticking Together

Since life's beginnings, cells quickly found ways to establish aggregates that increased their survival capacities, making sometimes very large colonies that cooperate metabolically and exchange genes and nutrients. However, a further and much more radical increase in biological complexity was achieved by organismal multicellularity in which different cell types began to specialize in diverse functions generating a new level of organization. Curiously, the evolution of multicellular organisms has been a very rare event in the history of life, taking place only among eukaryotic cells and only three separate times there: plants, fungi, and animals. Multicellularity probably appeared during the late Precambrian period, leading to a great diversity of macroscopic forms in the million years to come. The evolution of multicellularity created the biosphere as we know it, making life a macroscopic phenomenon and generating large-scale fluxes of nutrients and energy that gradually modeled the earth's landscape.

Organismal multicellularity implies a progressive commitment of cells to living together. Bacterial or protist aggregates, and even simple multicellular organisms like sponges and some medusae, may reversibly dissociate to liberate the individual cells and reassemble again in other circumstances. On the other hand, separating the individual cells from a complex multicellular organism usually results in the cells' deaths, indicating an irreversible dedication to collective life (still, early embryonic cells of complex organisms can also reassemble in some circumstances). As we will see, the transition to this stage was probably a slow and gradual process. Associated to this increasing commitment to living together, cells evolved communication mechanisms that enabled them to interact between themselves and to collectively react to the environment. However, animals developed a particular tissue, the nervous system, that enabled them to live a free life and feed on other multicellular beings.

Strength in Numbers

Cooperative Cells

While the simplest forms of cellular colonies are found in bacteria, these already show a great degree of cooperation and communication between the individual components. Colonial bacteria may release signaling molecules (hormones or neurotransmitters also found in animal brains) to their surroundings, producing different kinds of responses in their neighbors. When forming such aggregates, bacteria operate as a network, coordinating the activities of thousands (or millions) of cells, which allows them to adapt to different environmental circumstances. Like bacteria, protist cell assemblages work as integrated networks that can synchronize their cellular activities to respond to different kinds of stimuli. A common example of such colonies is *Volvox*, a protist alga that forms spherical, hollow aggregates containing small gametes inside, male and female, for sexual reproduction. Furthermore, social amoebas are able to form multicellular "slugs" that move toward light sources and build spore-containing structures. But as mentioned, although protists have evolved colonial aggregates more than twenty times in evolution, in only three instances has this achieved organismal levels in which embryonic development and cell differentiation is the norm.[1]

Making an Individual

Unicellular beings can aggregate in two ways, either by coalescing different cells or by having a parental or stem cell that produces daughter cells that keep bound to each other. The first case takes place in bacteria and in some protists where cells disassemble and later assemble in a new cluster. Thus, grouped cells can be genetically different among themselves, which enhances intercellular conflict and is highly susceptible of defeat or parasitism (some cells may quit the colony, simply be hitchhikers in the group, or tend to predate or eliminate other cells). For this reason, although being very common, this kind of aggregate never went too far in evolution. On the other hand, when daughter cells are genetically identical (they are all clones of the parental cell), they might cooperate better because what is good for one is good for the survival of all the cluster's genes. This is the rule in all complex multicellular organisms (plants, fungi, and animals), which since very early have developed mechanisms to block the ingression of parasitic foreign cells, distinguishing self from nonself and presaging the evolution of an immune system.[2] However, in clonal multicellularity, cells are likely to undergo mutation as they reproduce, originating variants, some of which may tend to disrupt the stability of the aggregate. This is what happens in cancer, for example, when cells acquire mutations that block the regulatory mechanisms limiting their proliferative capacity, eventually destroying the organism.

In order to maintain the genetic blueprint across generations, many organisms "sequester" in very early development a specialized population of stem cells, the germline, that undergo a slow rate of mutation and capitalize the reproductive capacity as gametes or reproductive cells to preserve the lineage. The rest are the somatic cells (like skin, gut, muscle, and neurons in the case of animals) that devote their lives to the higher-level organism. (To illustrate this point, recall the case of some social insects that have a reproductive caste surrounded by sterile workers and soldiers dedicated to the maintenance of the queen and its descendants.)

To the famous biologist August Weismann, the sequestration of germ cells for individual reproduction implied a radical shift in the evolution of multicellularity. From this moment, reproduction would have become not just a matter of cell division and cell aggregation in colonies but of the buildup of a multicellular organism that behaves and accommodates to the environment to preserve its germline, transmit it to its descendants, and so on. Nonetheless, things are not so clear-cut, as germline sequestration seems to be only one mechanism by which multicellular organisms reproduce. Some simple multicellular organisms like sponges can reaggregate after their cells have been dissociated, indicating that all cells can contribute to another individual. Furthermore, in plants, fungi, and some animals, reproductive cells may differentiate late in development, emerging from undifferentiated somatic cells (stem cells) that remain in different tissues and serve as a reservoir for regeneration and development. Furthermore, although reproduction through a germline is a common character of multicellular organisms, many of them can also reproduce simply by fission in two halves, or by budding sprouts that separate from the parental body and become separate individuals. Thus, the exclusive reproductive dependence on Weismann's germline appears to be present only in higher animals, possibly those endowed with large brains.[3]

Do Plants Feel?

An additional issue is that when organisms become too large, a critical problem is maintaining communication between its different parts in order to coordinate responses to the environment. As I have said, many of the plants' behaviors are slow responses produced by directed growth to approach light (shoots) or water (roots), or to evade noxious conditions. Some authors have also suggested that the roots provide an interface to communicate and exchange a large amount of nutrients and material with fungi, other plants, and soil microorganisms, generating highly complex cooperative ensembles that respond to environmental fluctuations. These ensembles may cover a whole forest, where trees may recognize their offspring and transmit nutrients and information to them.[4] Furthermore, plants' bodies have been proposed to be organized through their vascular systems and by specialized cellular mechanisms including neuron-like electrical and chemical signals that provide them with sensing capacities, even displaying long-term goals. Finally, plants display some limited forms of fast behavior such as responding to touch like *Mimosa* leaves, which rapidly retract at touch as a predator defense, or the carnivorous Venus flytrap, whose leaves are able to catch insects as they stimulate mechanosensory hairs on the surface. These responses are mediated by neuronal-like electrical membrane properties and by synaptic-like communication between specialized cells, reminiscent of rudimentary neural networks. More recent studies have reported that plants are able to make and perceive sounds, but the significance of these phenomena is not clear.[5]

Fungi, which are not merely mushrooms (these are just the reproductive organs) but the disseminated underground network of fine filaments called hyphae, sometimes symbiotic with plant roots, display complex route-finding and foraging behaviors as they extend their processes. The network of hyphae is densely interconnected by molecular transport systems along their length, but they also display fast electrical transmission at rates close to those of animal sensory axons.[6] As in unicellular and colonial behaviors, some authors have argued that these mechanisms may render a degree of intelligence or consciousness in plants and fungi, although many more are highly skeptical about this and consider it a misleading anthropomorphic view of nature. A rule of thumb when dealing with behavioral mechanisms is to try the simplest explanation first, and if this does not suffice, higher-level alternatives must be considered.

Predators

Almost Animals

The highest step in organismal complexity is undoubtedly provided by animals, or metazoans, which derive from a branch of protists (holozoans) that separated from the ancestors of fungi at least 750 MYA. Notably, holozoans (and particularly animals) diverged from their sister group, the fungi, by acquiring a series of new genes that were important for the achievement of complex multicellularity. The closest relatives to animals are a group of protists called choanoflagellates, endowed with a long flagellum and a collar of small cilia around it that is used to trap bacteria brought to them by currents generated by the flagellum (shown in figure 4.1). Noticeably, choanoflagellates closely resemble the choanocytes of marine sponges, which are ciliated cells located in their internal epithelium that generate water currents to circulate oxygen and nutrients. Like the previously mentioned *Volvox*, choanoflagellates sometimes aggregate in spherical colonies for sexual reproduction, usually with the flagella directed to the outside in order to swim by coordinating flagellar movements. The aggregates of choanoflagellate cells depend on the secretion of several proteins forming an extracellular matrix, that is, an extracellular "skeleton" to which cells adhere to keep together. Furthermore, they can shift in shape from a spherical organization into a two-dimensional sheet by modulating the composition of this extracellular matrix.[7]

Interestingly for our purposes, choanoflagellates display other characters that may provide some insights on the protist–animal transition. Firstly, the multicellular sphere can react to light, opening and turning inside-out, acquiring a bell shape where the flagella are located inside the bell. This organization is not too good for swimming but is excellent to trap bacteria (this is reminiscent of the choanocytes inside the sponge's cavities). The capacity to generate a curvature in both directions (cilia out and cilia in) has been considered to be a precursor of one of the most basic embryonic mechanisms of animals, the cavitation of the early mass of cells to form the primitive gut (see the next section). In addition, choanoflagellate cells can

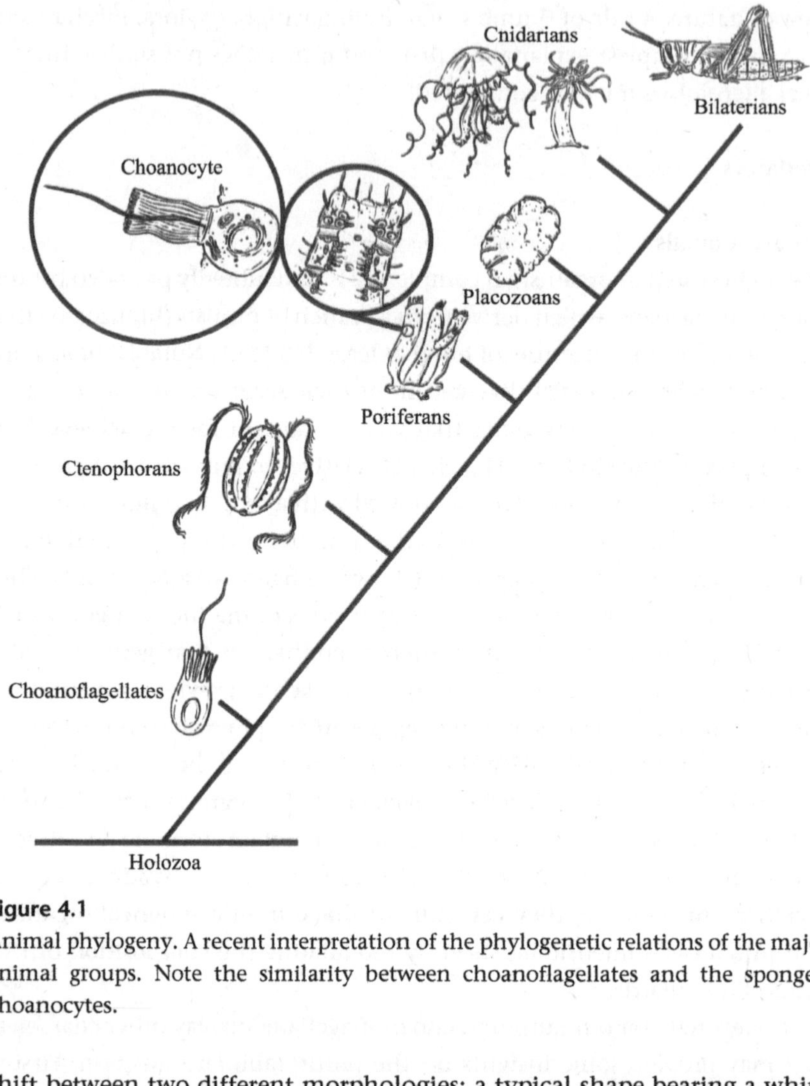

Figure 4.1
Animal phylogeny. A recent interpretation of the phylogenetic relations of the major animal groups. Note the similarity between choanoflagellates and the sponge's choanocytes.

shift between two different morphologies: a typical shape bearing a whip-like flagellum, similarly to some animal epithelial cells, or a spherical amoeboid morphology, like animal cells inside the tissues or the embryo. This provides a starting point for the differentiation of body cell types in animals. Finally, choanoflagellates also display a rich secretory mechanism provided with abundant vesicles that release neurotransmitters and other chemicals to the medium, providing signals to their neighbor cells when living in aggregates. Notably, the organization of their secretory system resembles that of the neurosecretory vesicles in animal neurons, possibly representing an ancestral mechanism for the origin of synaptic communication.

What, If Anything, Is an Animal?

Perhaps the most general characteristic of animals is that they acquire their nutrients from their surroundings, either by suspension or filter feeding or by ingestion. Protists do the same thing, but animals perform these functions at a multicellular level. By contrast, plants and fungi directly absorb nutrients from the substrate or from organisms they grow on, respectively. Thus, eating, in a general sense, is a key feature that defines animals living today. To eat, animals usually invest energy in moving toward and capturing food, or in attracting food to themselves in order to internalize nutrients. In my view, the nervous system evolved in this context, primarily organizing the animal's behavior to catch food.

Living animals range from the simplest marine sponges, jellyfish, and other microscopic organisms to more complex species provided with a head and bilateral symmetry, including worms, crabs, snails, fish, and of course us. The earliest recorded animals are marine fossils dating from the Ediacaran geological period (the last period of the Precambrian eon) about 650 MYA or before. Ediacaran fossils represent a remarkable diversity of life forms, some of them similar to modern jellyfish (*Cyclomedusa* and *Auroralumina*) and sponges (*Fractofusus*), biomineralized forms (*Cloudina*), some crawling and burrowing worm-like creatures (*Yilingia* and *Ikaria*, possibly the first animals with bilateral symmetry), and others suggesting molluscan affinities (*Kimberella*) (see figure 4.2). However, a large fraction of the Ediacaran forms were sessile organisms (attached to the sea bottom) that are not classifiable in any living animal group and probably represent the first experiments in animal life. These organisms did not have distinctive organs, mouth, or head, and some had leaf-like morphologies that maximize body surface and are optimal for capturing particles like microscopic algae that float in the sea. One such species was *Charnia*, a feather-shaped sessile form that is presumed to be an early metazoan. These organisms were initially termed "Vendobionta," but this classification is no longer valid as they may not form a taxonomic unit. That is, they may have had separate origins from different protist ancestors. Despite not representing a natural

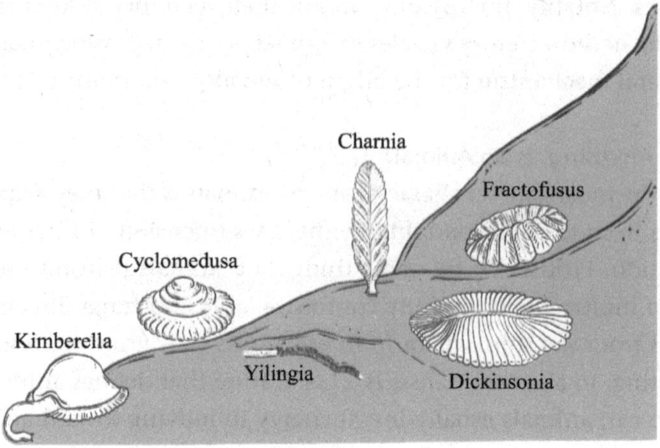

Figure 4.2
Ediacaran fossils mentioned in the text.

phylogenetic group, some vendobiontan organisms may represent an early, possibly suspension-feeding stage before the origin of modern animals.[8]

Modern marine sponges, or poriferans, are perhaps ecological equivalents to the sessile Ediacaran biota. They are also quite simple animals, with an amorphous body capable of very restricted movements. Apparently in contrast to vendobiontans, sponges bear an internal cavity bounded by two cellular layers, somehow comparable to an external skin and a "gut" lining the inside cavity, both separated by an aqueous substance called mesoglea. The internal epithelium contains the flagellated choanocytes (that resemble the choanozoan protists discussed previously), which generate a water current that enters through small pores and exits through a large opening called the osculum. Choanocytes secrete neurotransmitters like GABA and nitric oxide to the extracellular milieu, triggering contraction of a network of cells that expel water from the inside cavity. This kind of signaling resembles the action of endocrine systems in higher animals that liberate hormones to act on distant tissues. In a closer step to a nervous system, the so-called neuroid cells of the sponge's internal lining display secretory vesicles and extensions that make contacts with the digestive choanocytes. The neuroid cells are related to digestive mechanisms rather than sensing external stimuli, supporting the notion that the nervous system originated in association with feeding behavior.[9]

Guts and Neurons

Pouched Animals

We have seen that sponges display a two-layer organization that generates water currents inside them, but some early animals went a step further, using the internal space to digest large food particles. This innovation took place through a developmental innovation called gastrulation, in which a sac-like cavity is formed in the early embryo and gives rise to the intestine of modern animals (see figure 4.3A). This was perhaps the most fundamental

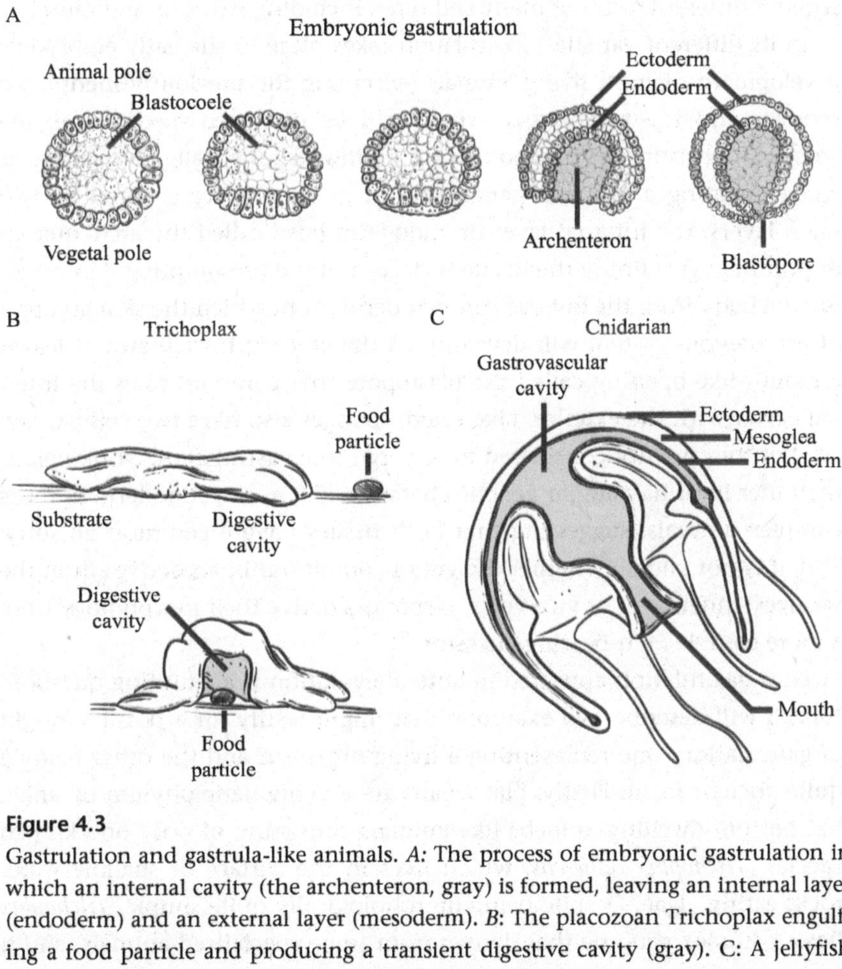

Figure 4.3
Gastrulation and gastrula-like animals. *A:* The process of embryonic gastrulation in which an internal cavity (the archenteron, gray) is formed, leaving an internal layer (endoderm) and an external layer (mesoderm). *B:* The placozoan Trichoplax engulfing a food particle and producing a transient digestive cavity (gray). C: A jellyfish showing its internal, gastrovascular cavity (gray) layered by the endoderm.

innovation of animal evolution, providing not only a scaffolding for body architecture but also enabling animals to move and find prey.

Gastrulation was made possible by a unique feature of animal cells: as opposed to colonial bacteria, plants, and fungi, in which individual cells have sacrificed their motile capacities to become part of the group, animal cells retained the mobility and morphological plasticity of free-living cells, possibly aided by the multiple gene gains acquired in holozoan and early metazoan evolution. Cell plasticity enabled the generation of amoeboid movements, cellular displacements, and changes of shape that provided the basis for embryonic development starting from gut formation and the subsequent differentiation of many cell types including neurons and muscles.

In its different varieties, gastrulation takes place in the early embryonic development of most living animals (excepting the previously mentioned sponges and the placozoans, which will be discussed next). Here, the embryo, consisting at this point of a hollow cellular ball, invaginates as when deflating a football, generating an internal space covered by two main layers: the internal layer or endoderm (also called the archenteron, or primitive gut) lining the inside surface, and the presumptive skin covering the body from the outside (the ectoderm, from which the skin and part of the nervous system will develop). As the embryo invaginates, it leaves a mouth-like opening called the blastopore that communicates the internal cavity with the exterior. Like I said, sponges also have two cellular layers, but they are not considered to undergo true gastrulation. Nonetheless, the inner layer has similar genetic characteristics as the endoderm of more complex animals, suggesting that both tissues have a common ancestry. Still, it is not known whether the gut of complex animals derives from the sponges' inner layer or vice-versa, if sponges derive their morphology from a more complex gut-bearing ancestor.[10]

How gastrulation appeared in animal evolution is a daunting question. Next, I will describe two examples that might testify for a possible origin of gastrulation, one representing a living organism and the other being a quite ancient fossil. Firstly, placozoans are an enigmatic phylum of small, flat bottom-dwelling amoeba-like animals consisting of only one known species, *Trichoplax adherens*, which lives in the surface of shallow water rocks eating algae. Despite being morphologically quite simple, *Trichoplax* has a complex genome that derives from some specialized animals, and it may not represent the ancestral animal condition but rather might reflect an evolutionary simplification of some more complex ancestors. *Trichoplax* has an amorphous flat body covered by a surface of ciliated cells and does not develop a head, tissues, or organs, nor does it have an intestinal cavity (but it has cells expressing neuropeptides, similarly to neurons). Still, the cells in the ventral or inferior side may resemble gut cells, and those in the dorsal side are more skin-like. It does not display gastrulation during development, but its feeding behavior, called "temporary gastrulation," might be reminiscent of (but not ancestral to) the feeding behavior of some early metazoans (figure 4.3B). When encountering a food particle, placozoans engulf it to form an internal vesicle to digest it, resembling the cellular mechanism of amoeboid phagocytosis and the formation of digestive vacuoles. Secondly, on the fossil side, *Dickinsonia* is a 560 MYO large (could grow to more than one meter), flat ovoidal organism segmented transversally and provided with a midline defining a longitudinal axis, resembling modern animals with bilateral symmetry. Similar to present-day placozoans, *Dickinsonia* displays presumed intestinal diverticula, and its traces left in the fossil sand suggest that it ingested food engulfing it in vesicles, perhaps representing the earliest evolutionary rudiments of gut formation.[11]

In any case, the acquisition of a mouth and an internal cavity to digest large food particles liberated animals from strict filter or suspension feeding behavior and allowed them to capture larger prey, diversifying in multiple morphologies and behaviors, which ended up replacing the "vendobionta" fauna. As said before, the nervous system, tightly coupled with sensory and motor or secretory systems, originated in close relation to the early gut to support this basic function: capturing prey. In fact, the nervous tissue starts differentiating in the embryo during or shortly after gastrulation. Some authors consider that escaping from danger, particularly predation (the fear response in humans) represents the most ancient function of the nervous system. However, what would the first nervous system–bearing animals be afraid of, if they were the only multicellular predators? Of course, avoiding danger, reproduction, and protection from physical disturbances are also very basic behavioral functions, but plants and fungi cope with these issues without neurons. I prefer a scenario where once animals started moving to catch their food, they benefited from fast responses to different stimuli, including danger, and from the capacity to search and find mates.

Jelly Bodies

Among true gut-bearing animals, the simplest condition is observed in jellyfish and sea polyps (called cnidarians) and by a group of jellyfish-looking luminescent marine animals termed comb jellies (or ctenophores) (see figures 4.1 and 4.3C). Cnidaria have a complex life cycle with alternating generations, one being free-swimming sexual medusae, the next generation being asexual polyps that bud off young medusae, and so on. On the other hand, ctenophora, or comb jellies, are superficially similar to jellyfish but are usually provided with eight rows of ciliated cells ("combs") in their skin that they use for swimming. Both animal groups are typically provided with long tentacles with urticant cells called cnidocytes (cnidarians) or sticky cells called colloblasts (ctenophores) that are used to catch prey and as a defense against predators.

Cnidarians and ctenophores display a gastrula-like, bell-shaped body consisting of two layers, skin and gut, separated by an aqueous mesoglea. The internal cavity of polyps and jellyfish has one opening derived from the embryonic blastopore, which works as mouth and anus, ingesting and expelling water and nutrients to and from the internal cavity according to body contractions and expansions. Ctenophores are a little more complex, with a ventral mouth and a complex digestive canal, and have developed dorsal anal pores from where water and waste come out. Another crucial characteristic shared by jellyfish and comb jellies is that they usually have a so-called radial symmetry, with a circular body as seen from above, that can be subdivided into a series of similar "cake piece" segments. More interesting to us and in clear contrast with sponges, both groups also display a well-defined sensory-neuro-muscular network, located both in the skin and gut, that controls body physiology as well as feeding and locomotion.[12]

Interestingly, both cnidarians and ctenophores appear to derive from sessile polyp-like animals, but it is not clear that they are directly related. According to some recent genetic analyses, ctenophores were the first branch to separate in animal evolution, while the less organized placozoans and sponges would have evolved after them. Furthermore, cnidarians or jellyfish would have arisen even later, having a direct common ancestor with the more complex bilaterian animals (with bilateral symmetry like worms and us). Thus, the gut and nervous system of jellyfish and bilaterians have probably a common origin, but ctenophores could have evolved a gut and nervous system very early and perhaps separately from the other groups.

In other words, there are two possibilities, the first being that gastrulation and the nervous system originated twice, one time in the origin of ctenophores and the other in the common ancestor of cnidarians and bilaterians. The alternative hypothesis is that gastrulation and the nervous system were formed once, very early in the history of animals, and were retained in ctenophores, cnidarians, and bilaterians but became lost in other early animal groups like sponges and placozoans. This perspective implies that sponges and placozoans "went back" in evolution, losing these features. We will probably need much more evidence and different levels of analysis to discern between both possibilities. The rise of genetic and molecular evidence has promised to solve these disagreements, but as more data become available, more complex statistical analyses are needed, and the results depend on the algorithms used and the assumptions the researchers make to perform their studies. As long as there is no consistent agreement on these issues, controversies will remain.[13]

Sensing and Moving

I discussed previously the possibility that the earliest nervous systems were involved in controlling rhythmic movements of the gut cavity. Furthermore, a recent study reports a neuropeptide-driven, satiety signal in cnidarians that is comparable to that of bilaterians, again supporting the notion of an ancestral gut-regulatory function of the nervous system. However, of more interest to us is that besides controlling the gut, the differentiation of a sensorimotor apparatus that coordinates the body parts enabling directed locomotion and capturing food was a fundamental innovation. These behaviors are executed by unicellular animals using relatively simple mechanisms, but large animals needed an additional system to transmit signals across their bodies, producing responses in distant parts of it. The differentiation of the nervous system provided a network coordinating different body parts, allowing animals to actively capture food and live a free life.

Perhaps a first stage in sensorimotor coordination was the division of labor between sensory cells and effector systems including contractile muscle-like cells. Recent studies indicate that both in jellyfish and comb jellies (cnidarians and ctenophores), neurons have a common embryonic origin with sensory and secretory cell types, suggesting that in both cases neuro-secretory cells became chemically coupled to rudimentary muscle cells to generate fast responses to environmental stimuli. As the sensory cells became able to extend long processes to contact distant effectors, the organism could grow to larger sizes while maintaining its behavioral capacity. Furthermore, if two or more neuro-secretory cells became coupled in a chain, the organism could grow even larger while at the same time generating a primitive network for transmission that could activate muscles in different body regions.

Thus, the nervous system of ctenophores and cnidarians is a diffuse network located between the skin and gut all over the body and is connected with sensory cells and contractile, ciliated, or secretory cells that generate rhythmic movements for feeding or locomotion. The ctenophore neural network has been recently characterized in great detail, showing that it secretes neuropeptides (very small proteins), which to the authors is consistent with a neuropeptide origin of synapses. However, more recent work indicates that the ctenophore nervous system is largely syncytial, that is, neuronal processes are continuous from cell to cell, without synaptic connections between them except to the effector systems. According to some researchers, at least the cnidarian neural network is not totally homogeneous, containing two main components located in opposite positions of the body: the blastoporal and the apical nervous systems (see figure 4.4). The blastoporal nervous system makes up a ring around the mouth/anus, forming a network controlling food intake and locomotion. On the other hand, the apical component relates to the apical organ, a cluster of ciliated sensory and endocrine cells located at the opposite side of the blastopore that regulates body physiology. According to one view, the central and peripheral nervous systems of more complex animals derive from these two networks present in cnidarians: the apical organ might have been involved in the differentiation of the head, part of the brain and sensory organs of bilaterians, while the blastoporal nervous system would have given rise to the central nervous system or nerve cord, and to the peripheral and enteric nervous system of bilaterians (see chapter 6). Again, not all researchers agree with this view, especially regarding the proposal for homology between the jellyfish apical organ and the bilaterian sense organs and brain. Most likely, this controversy will also remain for some time.[14]

Perspective

The consolidation of cells in assemblies that increase their survival chances may be as old as life itself. In this way, bacterial biofilms and protist colonies have evolved mechanisms to increase their homeostasis and respond

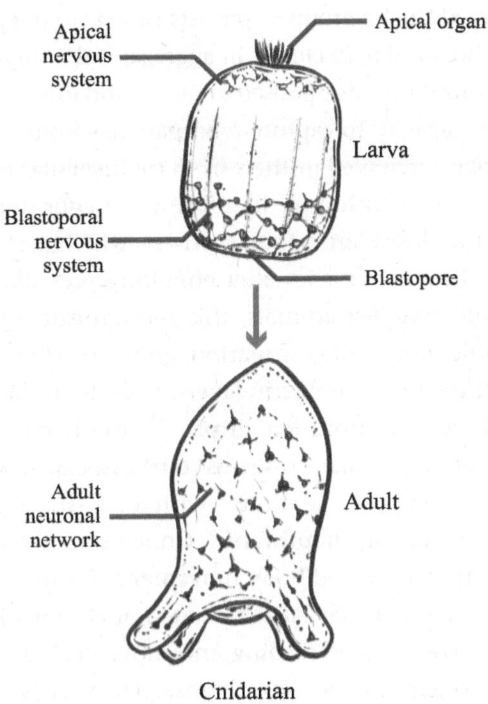

Figure 4.4
The reticulate nervous system of jellyfish. This consists of a diffuse network that subdivides into apical-sensory and blastoporal-motor components in the larva and the adult. Data from Arendt 2021.

to the environment as a collective. Yet, achieving multicellularity is a complex process where, first, the cell aggregates emerged as genetically identical clones of a stem cell, instead of being formed by different cells that happened to merge together. This implied a basic mechanism of self-nonself discrimination where foreign cells became rejected from the group, which may have also paved the way for the evolution of the immune system of animals. Furthermore, evolving multicellularity implies a radical change in reproductive strategy, where individual cells sacrifice their evolutionary potential, specializing to fulfill functions benefiting the group, while other cells (whether they are pluripotential stem cells or the specialized germline) maintain the reproductive capacity of the whole organism. However, this process did not occur at once, and there are many intermediate stages among plants, fungi, and animals.

Animals emerged from a group of protists similar to the present-day choanoflagellates, who are able to cluster in aggregates that may resemble some early animal organizations. As opposed to plants and fungi, animals are characterized by their capacity to capture food particles from the environment. The earliest animals originated in the seas as multicellular aggregates, growing to large sizes as suspension-feeding forms. A major step was the origin of a primitive gut to digest larger food particles, which perhaps started with the formation of internal vesicles after engulfing prey like modern placozoans do. In more complex animals, this mechanism may have evolved into the embryonic process of gastrulation, giving rise to a true gut and the differentiation of the two main germ layers: endoderm (primitive gut) and ectoderm (primitive skin, from which part of the nervous system would emerge). The onset of gastrulation seems to be associated with the acquisition of a radially symmetrical anatomy, with a bag-shaped internal cavity and an opening working as mouth/anus. Concomitantly, a primitive neuronal network that coordinated body movements for feeding and locomotion evolved together with specialized ciliated or contractile cells. Yet, this network not only coordinated feeding and movement but also contained a sensory apical organ coupled to neurosecretory cells regulating body homeostasis and general behavioral states. This organization is observed in the larva and adult state of jellyfish and comb jellies, respectively. In the following chapter, we will discuss the properties of neurons as the basic building blocks of the nervous system, to return in chapter 6 to the evolution of animals and the process by which the nervous system condensed into a central cord in some bilaterian groups.

5 The Butterflies of Mind

Santiago Ramón y Cajal, one of the founders of neuroscience, referred to neurons as the "butterflies of the soul." Neurons are quite unique cells, whose ability to engage in highly intricate networks is based on three principal characteristics: firstly, their morphology stands out with a cell body with protruding dendrites that receive input from other neurons, and a sometimes very long axon supported by specialized cytoskeletal proteins called microtubules, that ends in a terminal synaptic arbor (see figure 5.1). Secondly, their cellular membrane is highly excitable to transmit electrical signals. Finally, they establish specialized connections or synapses between them.[1] In this chapter, I will briefly review the key properties of neurons, their evolutionary and developmental origins, and their ability to form neural circuits and the grounds for perception, action, and memory, which will provide the basis for many discussions in the following chapters of the book.

A Unique Cell Type

Neuronal Ramifications

Dendrites make up the receptive component of neurons, and they typically branch into different shapes that have been used to classify neurons into different types. These protrusions usually receive synaptic inputs from the axons of other neurons, triggering an electrical signal that reaches the cell body and is relayed to the axon, which connects to other cells. There has been a long tradition derived from computer science to treat neurons like binary devices that may or may not transmit the signal to the next neuron. However, this misses the fine but crucial dynamic aspects of neuronal processing at a cellular level. In fact, much of neuronal processing takes place at the dendritic level, where each dendrite may be capable of performing

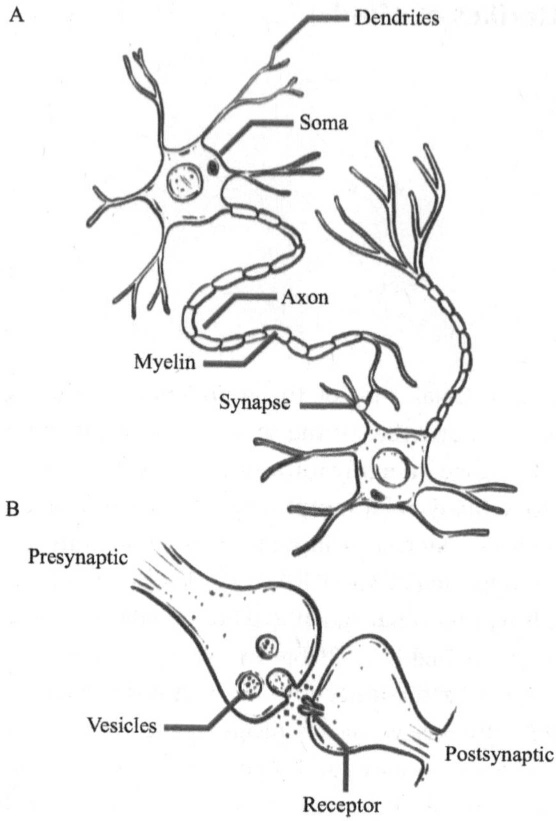

Figure 5.1
Anatomy of a neuron. *A:* The fundamental components of neurons. *B*: An inset displaying a synapse.

complex operations independently of the others. Likewise, much of the electrical activity that is experimentally recorded in the human brain and becomes associated to cognitive phenomena (like the electroencephalogram, EEG) derives from small, graded currents generated at the dendritic level. However, to convey the signal to other neurons, dendritic activity must eventually sum up enough electrical power at the axon's initial segment to produce the all-or-none action potential that is transmitted along the axon.[2]

Long-Distance Transmission

As said, the axon connects different neurons, and connects neurons with effector organs like muscles or endocrine glands. Due to their sometimes impressive length, axons can communicate very distant regions in animals' bodies or brains. For this purpose, neurons in jellyfish and bilaterians evolved the auto-propagated action potential, perhaps one of the most important acquisitions in the evolution of nervous systems. As said, the action potential is an all-or-none intense but transient current that rapidly propagates along the axon without decrement, being as robust when it begins as in the distant regions that it reaches. This is due to an exquisite molecular mechanism by which the small membrane currents at one point in the axon induce the opening of voltage-gated channels in adjacent regions, in a domino sequence that rapidly runs from the axon's initial segment to its synaptic terminals. Notably, and unlike a domino effect where the pieces stay down after they fall, the channels shortly close after they open, and the membrane quickly reestablishes its in-out voltage difference to be ready to fire again. This is especially important because while, as said, the action potential is an all-or-none phenomenon, neurons can increase the intensity of the signal by producing rapid bursts of impulses that are transmitted in sequence along the axon.

To increase excitability and especially axonal conduction, the origin of neurons was tightly coupled to an important diversification of genes coding for voltage-gated channels. While comb jellies (ctenophores) relied more on the expansion of potassium channels, jellyfish and bilaterians augmented their potassium and calcium channels as well, providing an even more rapid propagation of the electrical stimulus. A final upgrade was achieved in bilaterians where sodium channels have gradually replaced calcium channels in the axonic membranes, further increasing signal transmission velocity.[3]

Connections

Another critical feature of neurons is their ability to form synapses to transmit the electrical excitation to other cells. Synapses are composed of a presynaptic structure (usually an axon) that transmits a signal to a contiguous postsynaptic structure (neuron, muscle cell, or endocrine cell). Synapse rudiments have been observed in choanoflagellate protists and sponges, and the basic cellular and molecular elements involved in synaptic transmission may date even to LUCA. Structurally, there are two main kinds of synapses: electrical and chemical. Electrical synapses consist of a direct contact by "gap junctions" (see chapter 3) that enable the passage of ions and small molecules across the membranes of both cells. However, the majority of synapses are chemical, where the axon terminal releases neurotransmitters that are captured by specialized receptors in the postsynaptic cell membrane.

Neurotransmitters are a variety of substances that are used for chemical signaling in the cells of most animals and in unicellular organisms as well. This further supports the view that intercellular chemical signaling is probably a feature as old as life itself, and that neurons have used preexisting elements to evolve their communication mechanisms.[4] Chemically, neurotransmitters can be amino acids, amines, neuropeptides, and other compounds. Aminoacidic neurotransmitters are usually classified as excitatory (glutamate and aspartate) and inhibitory (GABA and glycine). Monoamines (nitrogenated compounds like dopamine, norepinephrine, serotonin, and others) and other neurotransmitters like acetylcholine usually work as neuromodulators in the central nervous system, participating in alertness, motivation, learning, and other functions in many animals. Some of these substances like norepinephrine and acetylcholine are important neurotransmitters in the peripheral nervous system as well. Notably, while acetylcholine has been found in prokaryotes, monoamines have been found only in bilaterian animals. Finally, neuropeptides are involved in endocrine functions, or are co-liberated with other neurotransmitters, especially in inhibitory synapses. In many species, neuropeptides participate in regulation of food and water intake as well as in social, parental, and sexual behavior, and in other functions critical for homeostasis and survival.

The other component of neurotransmission are the receptors for the different neurotransmitters that have coevolved with them and are as ancient as cell signaling is. There are several kinds of neurotransmitter receptors, both slow-responding metabotropic ones (triggering metabolic processes) and fast ionotropic receptors (opening ion channels) that electrically activate the membrane. GABA receptors are widespread among animals, and receptors similar to these have been described even in bacteria. On the other hand, glutamate receptors are found in metazoans (they belong to the GPCR receptor family mentioned in chapter 3) and may have arisen from the fusion of bacterial potassium channels and membrane-binding proteins. The so-called NMDA receptors are a particular class of glutamate receptors that participate in learning and neuronal plasticity in vertebrates but are also present in other bilaterians and in jellyfish, who are also able to learn by associations.[5] Finally, the presynaptic and postsynaptic surfaces are armored by a complex scaffolding of membrane and cytoskeletal proteins, which derive from ancestral components involved in exocytosis and cell adhesion in unicellular organisms (see chapter 3). Neurons seem to have recruited separate postsynaptic components several times, depending on the specific neurotransmitters and receptors involved.[1,4]

From Sensation to Behavior and Back

Once they were able to connect with each other, the first neurons in ancestral organisms quickly assembled themselves in chains that connected sensory and motor elements, establishing the first sensorimotor reflexes that allowed them to detect prey. Adding neurons to these pathways, this elementary circuit evolved into basic networks driving behavior. However, the nervous system is not organized in just a one-directional chain from sensation to behavior. Equally important, the execution of behavior generates new sensory stimuli that modulate the motor programs to be executed next. In other words, effectors send backward signals to the nervous system informing it about changes that occur during the execution of behavior, in a cyclical process termed reafferent organization. For example, when you extend your arm to grasp an object, you see the movements of your hand, but also the internal proprioceptors of your muscles and joints are sending signals to the brain, all of which help tune the arm's movement so you can successfully grasp the object. Some authors argue that such reafferent circuit organization provided a basic sense of self in early animal evolution.[6]

Closed Circuits

Furthermore, a cyclical organization is evident at all levels inside the nervous system, from microcircuits to large-scale neuronal networks, where circuits are usually arranged in closed loops that keep an internal dynamics (see figure 5.2A). Thus, neural activity is sustained in reverberant ensembles at all scales, whose activity can be modulated by external (sensory) inputs but keep an internal organization that enables them to respond to these stimuli. Moreover, the proper functioning of these neural networks tightly depends on the balance between excitation and inhibition. Too much activity could be as bad as too little. Since very early, neural networks managed to keep this equilibrium in a dynamic way, assembling circuits comprising interconnected excitatory and inhibitory neurons that work in an alternating excitation-inhibition oscillatory balance. Conrado Bosman and I (and several other reports) have highlighted an oscillatory pattern of excitation and inhibition as the hallmark of neural circuits at different levels in all nervous systems from jellyfish to humans, where local circuits consisting of a few neurons tend to oscillate at fast frequencies and large, distributed neural ensembles encompassing the whole network usually oscillate at slower frequencies. An example of a primitive oscillatory neural network can be provided by *Hydra*, a cnidarian polyp with one of the simplest nervous systems among animals (see figure 5.2B). It has recently been found that this animal displays spontaneous low-frequency oscillations in its nerve net, quite similar in pace to those observed in complex brains at rest. Moreover, the *Hydra* nerve net consists of different components that work at different frequencies to generate behavior: one is active at rest and is also involved in body elongation, and the other two participate in body contraction in response to stimuli (circular and longitudinal contraction, respectively). Furthermore, *Hydra* individuals can be experimentally disaggregated to individual cells and then allowed to reaggregate forming an individual again. In this process, neurons start forming small circuits that coordinate their activities and then couple with other local circuits to finally merge in a body-wide synchronic ensemble. Such spontaneous low-frequency

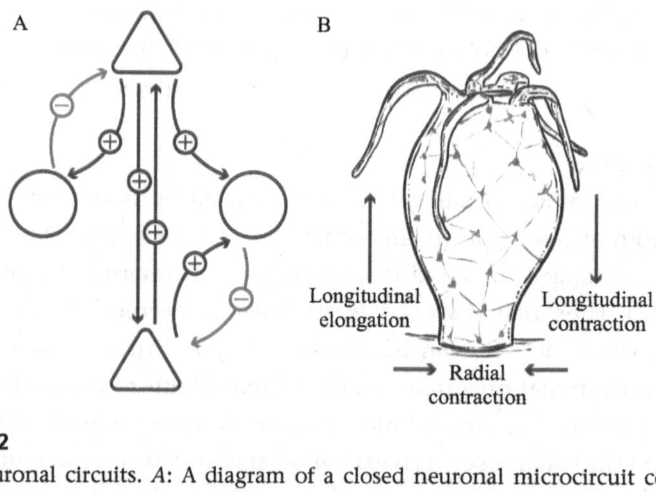

Figure 5.2
Basic neuronal circuits. *A*: A diagram of a closed neuronal microcircuit containing excitatory neurons (triangles) and inhibitory neurons (circles), which maintains an oscillatory balance of activity. *B*: The Cnidarian *Hydra*, showing its distributed nervous system that oscillates at different frequencies generating longitudinal elongation, longitudinal contraction, or radial contraction depending on the activity patterns.

activity is present in all living organisms from unicellular to multicellular, serving as an organizer or template to pattern nervous system organization. We will see throughout the book that oscillatory activity pervades biological processes and especially neuronal activity at quite high levels.[7]

The Other Cells

Finally, the nervous system not only contains neurons but also glia, cells that provide developmental, mechanical, and physiological support to neurons (astrocytes), facilitate electrical conduction by producing myelin (oligodendrocytes in the central nervous system; Schwann cells in the peripheral nervous system), and some of them provide immune support (microglia). Some glia, particularly astrocytes, even have roles in mechanisms of neural plasticity and processing. Glial cells may have originated very early in animal evolution, working as a lipid store for neurons, but became lost in many lineages. A particularly relevant glial cell type is the so-called radial glia, which serves as a stem cell or a progenitor cell for neuronal and glial production in central nervous system development (see the next section). Possibly, radial glia were present in the origin of bilaterian animals, perhaps being also the ancestral glial cell that gave rise to other kinds of glia. A later acquisition in nervous system evolution is myelin, which wraps the axon, increasing its conduction velocity. Myelin is found in jawed vertebrates and in some invertebrates. In the former, a retroviral jumping gene that is essential for myelination in modern members of this group may be ancestral to them and may have been involved in vertebrate myelin origins.[8]

The Assembly of a Brain

A Neuron Is Born

While we have discussed the basic properties of neurons and neural circuits, it is also important to have a notion about the mechanisms that give rise to them during individual development and how these assemble into a mature nervous system. Neurons are produced by quite similar mechanisms in jellyfish and in bilaterians, which strongly supports their common origin. In these groups, the generation of neurons, termed neurogenesis, is controlled by a family of regulatory "pro-neural genes" and a small set of other regulatory genes that together drive neuronal differentiation. These genes are also present in ctenophore (comb jellies) development, but it is not known if they participate in neuronal differentiation. In sponges, proneural genes are found in sensory cells, but again it is not known if this reflects a primordial animal condition or is a derived simplification from a nervous system–bearing ancestor.[9]

During embryonic development, neurons emerge in a sequential process where undifferentiated cells gradually acquire neuronal characters by expressing a variety of neuron-specific genes. In jellyfish and in some bilaterians, embryonic neural progenitor cells distribute and disperse across the body surface (and the gut), producing neurons and giving rise to a reticular nervous system under the skin. Alternatively, neurons may originate from a restricted region in the embryo to become a central nervous system or nerve cord, as happens in more complex bilaterian animals. In many bilaterians, the cellular progenitors that give rise to neurons are the radial glia mentioned in the previous section. These cells first divide through so-called symmetrical divisions, where one stem cell gives rise to two equal stem cells that keep dividing for a few rounds (in vertebrates this process continues through many rounds, yielding large numbers of progenitors). At some point, progenitors start dividing asymmetrically, that is, one of the daughter cells remains as a dividing stem cell and the other starts differentiating as a neuron, a process that sometimes repeats over several rounds of asymmetric cell division. The newborn neuron detaches from the epithelial surface to which progenitors are usually stuck and many times undergoes a migration to its final, adult position where it establishes its connections with other neurons, sensory cells, or effector cells (muscular or endocrine). A more detailed description of neurogenesis in the vertebrate brain is provided in chapter 8.

As far as we know, differentiated neurons are not able to divide anymore, and it had been long thought that we are born with a fixed contingent of neurons that would only decrease as we age. However, in many animals, a population of neural stem cells may remain throughout life, continuously producing neurons for repair. Adult brain neurogenesis has been also observed in songbirds and mammals (however, in humans, there is still conflicting evidence about its presence).[10] In the adult mammalian brain, there are two small regions, one in the lateral cerebral ventricles and the other in a region of the hippocampus called the dentate gyrus, that can produce neurons that become added to olfactory and memory circuits, respectively. The functional significance of these new neurons is not totally clear yet, but evidence suggests that they participate in mechanisms of neural plasticity and other functions.

Another common assumption is that neurons are exclusively derived from the external or ectodermal embryonic layer. However, as discussed in the previous chapter a series of studies from jellyfish to mammals have reported that a significant population of neurons derives from the internal endodermal tissue, suggesting a double origin of the neuronal phenotype, one deriving from the visceral epithelium making up an enteric neural network that coordinates gut movements, and the other deriving from the ectodermal epithelium, coordinating behavior and hormonal regulation. Nonetheless, at all times, these two components make up a tightly interconnected lattice that works coordinately to maintain body homeostasis.[11] Finally, a characteristic of neural development is that newborn neurons usually migrate, in most animals very short distances, but in vertebrates they may do so for quite a long distance to reach their final position where they coalesce with similar neurons to form a ganglion, a nucleus, or a layered structure. This process will be further discussed in chapters 7 and 8.

Pathfinding

Like I said, as neurons develop in the embryo, they must extend their axons and dendrites to establish connections with other cells, which is orchestrated by a complex molecular machinery where growing axons follow chemical cues to find their final targets. Like neuronal migration, this process is regulated by sophisticated molecular signaling systems that guide the early neurons and their axons to their site of destination.[12] Axons grow similarly as plants and roots do, branching in different directions according to specific clues that guide their path. They use other cells (like the radial glia) or other axons as a substrate to attach as they grow, following chemical attractants and being repelled by other compounds. Along their path, they encounter signals like highway posts that enable them to switch directions and find their final targets. Just like a highway system, there are major routes through which many kinds of axons travel, and then smaller and smaller routes that are specific for increasingly smaller groups of axons destined to innervate particular places, resembling when you drive from outside town to your own neighborhood, using different routes and signals that let you know when to turn (see figure 5.3A).

The mechanisms described generate what Gerald Edelman has called a "primary repertoire" of connectivity that becomes present at birth and provides the basis for coherent behavior.[13] The primary repertoire largely determines the development of stereotyped and species-specific behaviors that are so-to-speak wired in the genes. These include simple reflexes like the knee joint reflex, locomotory behaviors, ingestion and defecatory conducts, and basic reproductive behavior. Many of these reflexes, or fixed action patterns as defined by ethologists, need to mature through some practice and require a degree of plasticity to develop (at least in higher vertebrates), but their core organization is strongly set by the species' genetic endowment.

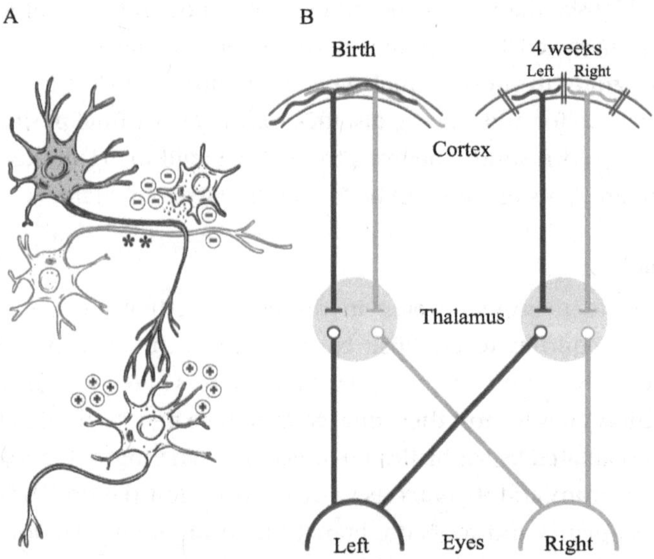

Figure 5.3
Axonal growth and plasticity. *A*: A growing axon (dark) follows different cues that direct its growth. For instance, it may use an older, grown axon as a substrate (**), then be repelled by chemical signals that force it to change direction (—), and finally become attracted to a cell that turns out to be its final target (+++). *B*: Sensory plasticity in brain development. The visual pathway from the eyes to the cerebral cortex is depicted here, in which the left side represents the condition at birth and the right side indicates the condition after one postnatal month. At birth, projections from both eyes innervate wide and overlapping regions, but they subsequently retract, segregating from each other as seen to the right.

Plasticity

Sculpting the Brain

Especially in the large brains of vertebrates, connections undergo an extensive process of reorganization after they are formed. Firstly, there is an excess of neurons and connections produced between many different brain regions and also with body organs. In late development, this excess results in the death of many neurons (particularly in the peripheral nervous system), or in the retraction of many axons (in the central nervous system). These massive regressive periods are produced by the competition between axons for limited amounts of "trophic" factors secreted by their targets (as in the peripheral nervous system)[14] or by the selection of circuits according to functional requirements (cells firing more similarly tend to connect while those firing asynchronously tend to diverge their projections). For instance, in the early-developing cerebral cortex, visual and auditory sensory inputs overlap extensively, but later on they segregate into visual-projecting regions (the visual cortex) and auditory-projecting regions (the auditory cortex). The initial exuberance of connectivity in brain development can be seen as a way to compensate for early injury. For instance, when the visual projections are eliminated in early development, the auditory projections remain in the visual area, transforming it into an auditory-like cortex, which may partly be why early blinded subjects develop enhanced hearing capacities.

Use It or Lose It

As I mentioned, another mechanism for refining the connectivity is provided by neural activity itself, which builds over the shoulders of the genetically established primary repertoire. Cells that fire together tend to stay connected. This is a basic principle of neural plasticity (but not the only one), which was initially postulated by Donald Hebb in the 1940s. Hebb claimed that learning mechanisms were based on the modification of the synapses through a quite simple process: if a presynaptic neuron is effective in changing the activity of the postsynaptic one, the synapse will be reinforced. If it is unsuccessful, the connection will retract. Subsequently, David Hubel and Torsten Wiesel earned the Nobel Prize in 1981 partly for unveiling this mechanism in the development of the visual system of the cat.[15] Cats receive projections from both eyes in their visual cortex, which at birth are largely overlapping. When the kittens open their eyes and become exposed to light, these projections segregate in alternating stripes along the cortex's surface, as in a zebra pattern where the "black" stripes receive axons from one eye and the "white" stripes receive axons from the other eye (see figure 5.3B). When newborn kittens were deprived from vision in only one of the eyes, the stripes receiving the functional eye's projections remained expanded while those receiving the deprived projection shrunk dramatically. This implied that the neurons corresponding to the functional eye were effective in stimulating the cortical cells and became strengthened, while those from the deprived eye retracted as they were unable to stimulate the cortical neurons. Noticeably, eye deprivation had no effect if it was performed in adult cats, being only effective in very young kittens and only during a short, critical period lasting a few weeks after birth. After this period, the plasticity of the nervous system becomes more limited. There are mechanisms of plasticity in adult brains, but these processes are much slower and more restricted than in the newborn. Critical periods are fundamental for the maturation of instinctive behaviors, but apparently also for complex behaviors like language acquisition (which is a combination of instinctual drive and learning). Critical periods for brain development are of special importance for humanitarian issues, especially in low-income countries where infants and young children may suffer lack of sensory, cognitive, and motor stimulation, generating a vicious circle where poverty impairs brain development in a way that makes it increasingly more difficult to overcome this condition.

The mechanisms of neuronal plasticity were found to rely on specific molecular signals, depending on the NMDA receptors described earlier, which according to the patterns of pre-synaptic activity, may strengthen (long-term potentiation, LTP) or depress (long-term depression, LTD) specific synapses for a long time.[16] To many scholars, LTP and LTD represent main molecular mechanisms (but again not the only ones) for long-term neural plasticity and learning. They participate in the remodeling of the brain according to sensory stimulation in infants, in the transformation of short-term memories into consolidated long-term memories in the adult, and in many other processes occurring in complex nervous systems. In addition, dysfunctions of NMDA receptors have also been linked to a variety of neuropsychiatric disorders, from epilepsy to schizophrenia to Alzheimer's disease.

Self-Made Brains

Through these simple mechanisms of synapse modification according to neural activity, the brain can assemble a network that makes each of us unique while at the same time conserving structural and functional properties that enable us to survive in the world. Edelman proposed the term "neural Darwinism" to describe the process by which the brain self-assembles on the basis of the selection of those circuits that end up being more robustly wired and providing a fitter cognitive and behavioral outcome.[13] A key notion in this process is the previously mentioned reentrant signal indicating that the output of the circuit conveys a retrograde signal that feeds back on the former, thus providing a clue about the circuit's fitness (in fact, the word "circuit" refers to circularity). As mentioned earlier, this takes place at all levels, from the small microcircuit to the large-scale circuits and to the execution of behavior that produces a set of signals that are fed back into the sensory systems.

Going further on this line, the neuroscientists Maturana and Varela (mentioned in chapter 2) argued some years before Edelman that the brain works as a closed system mainly driven by its internal processes.[17] In other words, every change in neural activity leads to another change in neural activity and so on, in an ever-cycling loop. Thus, external stimuli may modulate or trigger distinct processes going on in the network (for instance, amplifying or inhibiting specific oscillatory dynamics) rather than the brain being a passive receiver of external information. Nonetheless, despite its closedness, the system's activity must comply with the structure of the outside world; otherwise it ultimately disintegrates (recall the example of the rat and the labyrinth in chapter 2).

The Role of Stimuli (and Behavior)

Learning

Once mature, animals need to move around their environment, which is not always safe, and there is scarcity of food. Although they are born well equipped to survive, in an unstable environment, they need to make accurate predictions by detecting the underlying regularities of their surroundings. That is, they must learn how to behave in different circumstances. One of the first things animals must learn is whether they should avoid, approach, or simply ignore the stimuli they receive. That is, they need to ascribe some value (positive, negative, or neutral) to the novel events they perceive in order to orient themselves toward or away from them. Among the simplest forms of learning are the processes of habituation and sensitization, which were elegantly described in the marine snail *Aplysia* by Eric Kandel and collaborators.[18] Habituation is a form of simple learning where presentation of an innocuous stimulus, like softly touching the snail's respiratory siphon (the opening to the gill cavity), normally produces a reflex defense reaction, in this case consisting of gill retraction. However, if the stimulus is repeatedly presented, the snail habituates to it and ceases retracting its gills in response. It gets used to it. The opposite mechanism is sensitization, where repetition of a noxious stimulus like a brief electric discharge triggers an increasingly stronger reaction.

The Associative Brain

Another form of learning is by association of two stimuli that coincide in a temporal window, as in Pavlov's dogs who were trained to link the sound of a bell (the conditioning stimulus) with the subsequent delivery of food (the unconditioned stimulus). This is called conditioned or associative learning, and despite its simplicity, it is considered to be one of the most fundamental mechanisms for learning, from jellyfish to higher animals. Kandel showed the cellular mechanisms of associative learning by repeatedly pairing the soft touch stimulus with the electric current in *Aplysia*, which ended up triggering a complex defensive response just with the presence of a soft touch. Like sensitization, associative learning also depends on the strengthening of synapses, including NMDA receptors activation. Associative learning is the basis of reward-mediated and fear conditioning, where rewarding or punishing stimuli are repeatedly associated with stimuli or events taking place close in time to the reward/punishment. More generally, associative learning enables animals to identify cues that predict future events that are of relevance to them, like feeding or danger like a smell or a sound before the appearance of a prey or a predator.[19]

In many animals, a fundamental element in associative learning is the neurotransmitter dopamine, which modulates the response to both reward and punishment quite similarly across animal groups from bugs to humans. In vertebrates and insects, dopamine participates in forming predictive or causal associations between relevant stimuli and subsequent outcomes (say, the bell or a light that predicts the delivery of food), enabling the animal to use previous experience to anticipate events that are relevant to it. Additionally, dopamine, together with other neurotransmitters, has been found to play a role in more elaborate cognitive and behavioral mechanisms of higher animals like motor learning, motivation, attention, memory, social behavior, and even interoception, aspects that probably rely on this basic predictive and associative role.[20]

Associative learning may also take place between a stimulus and a behavior as in operant or instrumental conditioning. In this case, the reward does not come automatically after the conditioning stimulus (the flashing light), but the animal must find its way to it, like learning to press a lever that delivers food after a light turns on. This kind of behavior has been typically observed in vertebrates, but there are also reports in insects, cephalopods and even in cnidarians. The association between stimulus and response creates a habit, which repeats when being exposed once and again to the stimulus. A further elaboration of this conduct comes when, say, the experimenter hides the lever to be pressed and the animal has to develop a new strategy to obtain the reward. In other words, goal attainment becomes a flexible behavior rather than the stereotyped response produced by conditioned habits. Some authors assert that in these conditions the animal must have somehow internalized the intended goal, pointing to an incipient mental event.

Finally in this context, a more elaborate form of learning is spatial learning, as in the experiments of a laboratory rat solving a maze. Closely related to spatial learning is the capacity to remember events and make use of them for further behavior, like remembering the different choices the animal made when learning to go through the maze (Did I turn left or right at this point, and what happened next?). I will address these functions again in chapter 10.

Perspective

This chapter highlights the basic properties and functions of neurons and the nervous system, which are fundamental for themes that will be discussed in subsequent chapters. The molecular machinery to build neurons and synapses was coopted from proteins, cytoskeletal elements, and secretory mechanisms that were present in unicellular organisms. A key feature of neurons and some sensory and muscular cells is the capacity to generate self-propagated action potentials, which permit the rapid and strong transmission of electrical signals for long distances. Through synapses, neurons can organize in extensive networks controlling body states and driving behavior, learning, cognition, and ultimately our sentient minds. These networks rely on an ongoing balance between excitation and inhibition that generates a dynamic oscillatory pattern that is characteristic of neural systems.

While there is a tightly orchestrated molecular guidance mechanism that drives the development of the basic connectivity of the nervous system, this seems to be more important in small animals with relatively few neurons. In large brains like those of vertebrates, these cues provide a rough scaffolding that is later refined partly by competition for limited trophic support and subsequently by mechanisms dependent on neuronal activity and sensorimotor coordination that depend on critical periods to adapt to the ambient conditions. There are two basic and interrelated processes driving most forms of learning: activity-dependent synapse stabilization and associative learning. Provided this background on neuronal function, I will go back in the next chapter to the history of animals to discuss how brains evolved together with a complex set of body innovations in bilaterian organisms.

III The Rise of Bodies and Brains

The third section addresses the evolution of complex animals with bilateral symmetry, provided with a distinctive head and a tubular gut with two openings (mouth and anus). For the purposes of this book, the most interesting components of bilaterians are their heads, containing the sensory organs, the mouth, and the brain. Bilaterians have diversified in distinct lineages, yielding some highly intelligent animals like arthropods (especially some insects) and cephalopods (octopuses and squids). Vertebrates, allegedly the most complex animals endowed with the highest intelligence, emerged from a side branch of burrowing worms that developed a swimming, tadpole-like larva that gave rise to the first fishes. Fishes became extremely successful during the paleozoic, filling all the seas, but also some of them ventured onto land, following the earliest terrestrial plants and invertebrates. The full colonization of land implied a series of biological transformations, including the expansion of the brain. While the brain of vertebrates develops according to a highly conserved genetic schedule, there is an enormous diversity in brain forms and sizes across vertebrate groups, whose interpretation has led to intense controversies among evolutionary neuroscientists. In this context, mammals arose through a complex sequence of bodily and neural innovations, featuring the sense of smell and the emergence of a large brain containing the cerebral cortex, a character unique to vertebrates. Animals with large brains like mammals have evolved sophisticated cognitive capacities, including a strong memory and the ability to construct internal, coherent models of the outside world. Yet, a cerebral cortex is not necessary for the evolution of cognitive abilities. Birds have also evolved large brains, and their cognitive abilities rival those of mammals, without having a cerebral cortex proper.

III The Rise of Bodies and Brains

[illegible]

6 Heads, Mouths, and Anuses

Back to the history of life, some 550 MYA, the Ediacaran period was followed by the Cambrian period, beginning the Phanerozoic eon that is characterized by the increasing abundance of animal fossils. Although bilaterian animals made their debut in the Ediacaran period, the Cambrian was marked by a drastic increase in biological diversity, where worm-like animals with bilateral symmetry populated the seas. By that time, a variety of creatures could be found especially in river deltas, which have been suggested to represent a cradle for biological diversification in many instances of life's history. Like the Ediacaran fossils, many of these new animals were related to animal groups existing today (crustaceans, snails, squids, vertebrates, and others), but some strange forms like *Opabinia*, *Anomalocaris*, and *Hallucigenia* were at first sight very unlike any living form, as if they had come from another planet (see figure 6.1). Nonetheless, increasing fossil and anatomical findings have shown that many of these strange animals are indeed classifiable within extant taxa, and as said, the major Bilaterian groups possibly made their first appearance during the Ediacaran period. Indeed, instead of an abrupt Cambrian explosion or animal "Big Bang," the evidence suggests that in the Ediacaran-Cambrian transition there was a gradual replacement of the fauna, with increasing numbers of bilaterian animal forms while the "vendobiontan" animals described in chapter 4 tended to disappear. The origin of Bilateria marked an amplification of morphological diversity that filled the seas with all kinds of different creatures until today. Furthermore, this event paved the way to the formation of complex brains and behavior, and the emergence of intelligent life.[1]

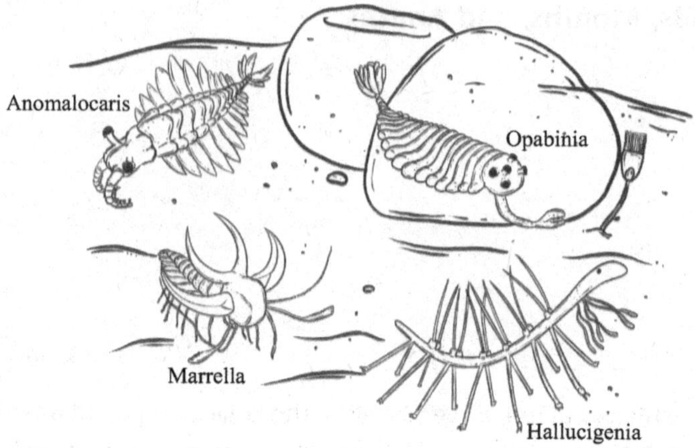

Figure 6.1
Some early Cambrian animals.

Body Elongation

Through-Guts

Most animals that we know are bilaterians, with symmetrical left and right sides that are joined in the midline and a distinctive head and a body behind it. In addition, most bilaterians display a through-gut with a mouth and an anus, as opposed to the pouched or gastrula-like gut of jellyfish. Furthermore, beside the ectoderm and the endoderm, bilaterians have developed a third embryonic layer, the mesoderm, that gives rise to muscles, a circulatory system, and internal fluids like blood and lymph. Finally, instead of a diffuse nerve network covering the body as in jellyfish, bilaterians are typically conceived as displaying a central nervous system or nerve cord, running along the middle of the body from head to back, that includes a brain connected with sensory systems in the head region (but there are several exceptions to this). The emerging evidence of strong genetic similarity in the development of these body traits, from flies to humans, prompted some authors to suggest a common origin of them across Bilateria. They proposed the term "Urbilateria" to describe the first bilaterian animal, which was conceived as a worm provided with a head, a through-gut, a segmented body, and a central nervous system including a small brain and a nerve cord (see figure 6.2).[2]

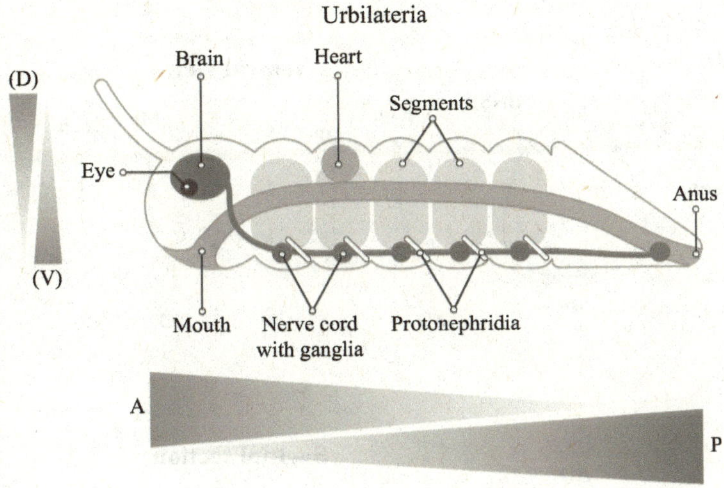

Figure 6.2
Urbilateria. The hypothetical animal is depicted with a tubular gut with mouth and anus, a distinct head, a brain, a nerve cord, and a segmented body. Besides the left and right sides, bilaterians are organized along two main axes: antero-posterior (or rostro-caudal) and dorso-ventral. The former runs from "A" to "P," and the latter from "(D)" to "(V)." (D) and (V) are in parentheses because in vertebrates this axis is inverted, dorsal becoming ventral and ventral becoming dorsal (see next chapter, figure 7.6). Both in the antero-posterior and the dorso-ventral axes, complementary gradients of morphogens (shown in opposing triangles in each axis) pattern body organization.

However, the transition from a bell-shaped organization as in jellyfish to the worm-like tubular organization of bilaterians is one of the most enigmatic processes in the history of animal life, and controversy abounds on this topic (see figure 6.3). There are two main theories for this event; the first is the Gastraea theory, which is compatible with the notion of Urbilateria. In this scenario, a gastrula-like larva would have suffered body elongation and the ancestral blastopore opening would have stretched longitudinally, fusing the lips on the sides and leaving two openings at the extremes: mouth and anus. The Gastraea theory also specifies the generation of a head at the apical region, and importantly the generation of a central nerve cord at the sides of the elongated blastopore, deriving from the blastoporal nerve network of jellyfish (see chapter 4).[3]

An alternative hypothesis, the "planula" theory, proposes that the ancestral bilaterian had a sac-like gut in its early beginnings and later a

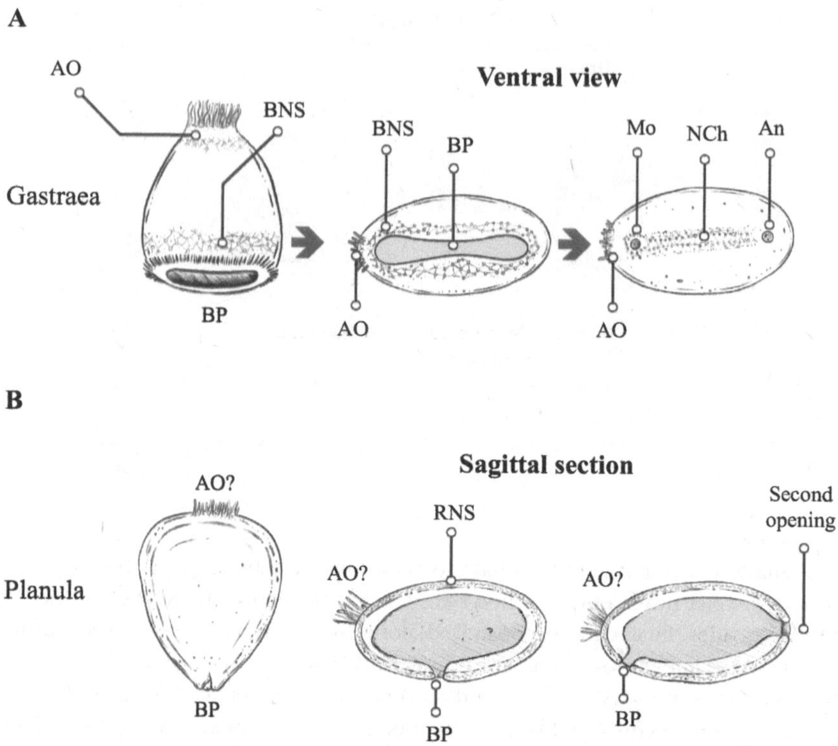

Figure 6.3
Theories of the origin of bilaterians. *A*: The gastraea theory, in which the blastopore (BP) elongates and fuses in the midline, leaving two openings: mouth (Mo) and anus (An). The blastoporal nervous system (BNS) becomes the nerve cord (NCh), and the apical organ and nervous system (AO) make the sensory organs and part of the brain. The drawings to the right show the larva from below (ventral view), exposing the elongated blastopore. *B*: The planula theory in which the blastopore remains as a single opening and a second opening is produced to make the one-way gut. The gut is depicted as a gray shadow in a sagittal section through the midline that separates left and right sides. According to this theory, the early planula would have had a reticulate nervous system (RNS) spread in the body surface, similar to that of cnidarians.

second opening was produced to make the tubular gut. This perspective received support from the recent classification of some bottom-dwelling marine worms called Xenacoelomorpha, as the most basal bilaterians (see figure 6.4). Xenacoelomorpha have a distinctive head and rudimentary sensory organs, but they display a pouched gut with only one opening located in a "belly button" position that gives it an appearance similar to the

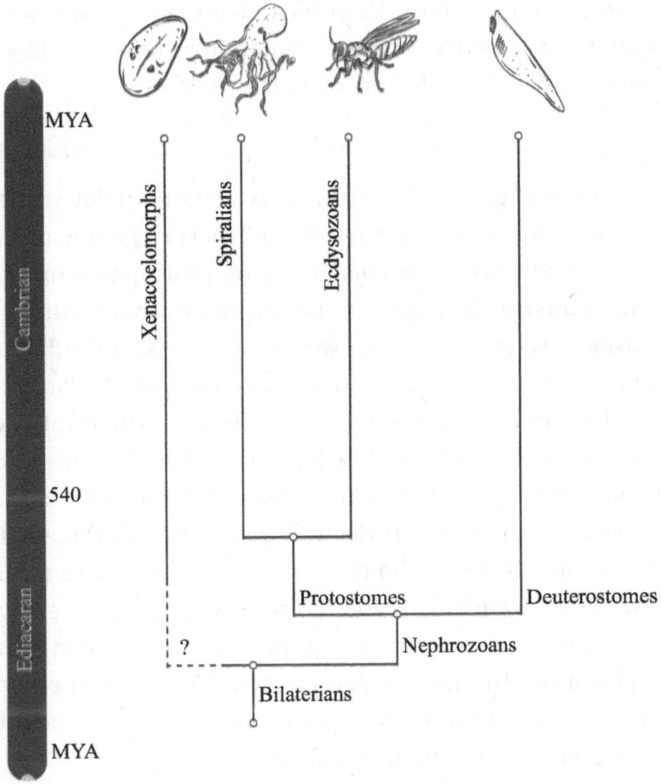

Figure 6.4
Phylogenetic relations of bilaterians. The position of xenacoelomorphs (marked by a question mark) is depicted as ancestral, but there is discussion on this point.

hypothetical planula stage. According to this view, the rest of the bilaterians, excepting xenacoelomorphs, are still considered to be monophyletic, forming a clade termed Nephrozoa. The modern planula hypothesis suggests that the ancestor of Bilateria had a jellyfish-like diffuse nervous system, and that the central nerve cord could have arisen several times in nephrozoan evolution. Nonetheless, the basal position of Xenacoelomorpha is still a contentious issue and the final word on this topic has not been said yet. Some studies have classified these animals as a branch of early deuterostomes, a clade related to vertebrates (see below).[4] This is reminiscent of the problem of finding the phylogenetic position of other key animals like the comb jellies (Ctenophora) discussed in chapter 4. In any case, whether or not xenacoelomorphs are the most basal bilaterian branch, there seems to be agreement that the presence of a head, containing a brain and aggregates of sensory cells, is an ancestral feature of all the clade.

Entry or Exit?

When attempting to classify the different branches of Bilateria (let us forget about xenacoelomorphs for now on), things keep getting complicated. Bilaterians are subdivided in two major groups, protostomes (meaning first mouth) and deuterostomes (second mouth). Protostome animals include insects, crustaceans, round and flat worms, mollusks, and other less well-known animals. On the other hand, deuterostomes include starfish (echinoderms) and the vertebrates among other species, and will be fully addressed in the next chapter. The classical dichotomy asserts that in protostomes, the embryonic blastopore usually becomes the mouth (the anus appears as a second opening), while in deuterostomes the blastopore is usually related to the anus, and a second mouth is formed to make a through-gut. Although this phylogenetic division has been amply supported by genetic data, there are many exceptions to the rule, in which some basal protostomes display deuterostome-like development.[5] Thus, the mechanisms by which a through-gut was initially formed remain an enigma, and this may have occurred more than once in evolution.

Body Patterns

Axes and Segments (1)

Despite these outstanding early embryological differences, in later developmental stages, genetics has evidenced a highly conserved anatomical organization in all bilaterians, where the main dorso-ventral (back-to-belly) and antero-posterior (mouth-to-anus) body axes are established by a complex network of regulatory genes that remains surprisingly similar all through the group. These main axes arise during gastrulation and are driven by specific "morphogenetic centers" located at the extremes of these axes (anterior, posterior, dorsal, and ventral, respectively; see figure 6.2). These centers secrete substances called morphogens that diffuse along the respective directions, being more concentrated near their source of production and progressively dilute as they diffuse away from it. This produces opposite concentration gradients of the respective molecules along their respective axes (say, from front to back and from back to front in the longitudinal axis). Consequently, each body region can be viewed as possessing a unique combination of different molecules secreted by these morphogenetic centers. Subsequently, a specific group of regulatory genes is activated in different body regions, driving cell differentiation and providing an identity to the different parts of the body from mouth to anus and from back to belly. I will speak about the dorso-ventral axis in the next chapter, while here I will focus on differentiation along the antero-posterior axis.

Shortly after the establishment of the longitudinal axis, many bilaterian embryos become organized as a series of repeated units along their length, like a caterpillar (figure 6.2). This feature is called segmentation and is present in different groups including earthworms, arthropods, and chordates (vertebrates). This process is largely driven by the so-called homeotic genes, which determine the cellular and anatomical characteristics of the different segments from mouth to anus. The discovery of homeotic genes took place when observing mutant flies that developed four wings instead of two (*Bithorax* mutation) or grew limbs in the heads where antennae should be (*Antennapedia* mutation). It was soon found that all these mutations corresponded to a set of genes containing in their sequence a short segment termed the homeobox, which codes for a DNA-binding aminoacidic sequence that works as a strong regulator of the activity of other genes, driving cell differentiation during development. Furthermore, each of these genes was active in specific segments of the body determining their respective identities, and the genes were also linearly organized along the chromosome so that their ordering was concordant with their sequence of expression along the body's longitudinal axis. Subsequent studies determined that Homeobox genes have similar roles in the development of animals as diverse as flies and mice (they are important for development in cnidarians too), indicating that they are master regulators of body patterning. Furthermore, homeobox-like genes are also present in plants and protists, suggesting that they originated in the context of cellular plasticity and differentiation.[6]

Notably, bilaterian homeobox genes become active at specific embryonic stages across the different species, which makes it possible to compare the different animal groups with respect to their body development. These homeobox-expressing stages are collectively known as the phylotypic stage, which coincides with the moment where the embryo first displays the major characteristics that allow it to be recognized as a member of a major clade (arthropods, mollusks, vertebrates, and so on). Thus, within each major animal group, the phylotypic stage not only represents the moment of homeobox gene expression but is also when the embryo develops the main morphological features of its class (in vertebrates, this corresponds to the stage where branchial pouches appear in the embryo).[7]

Nerve Cords

Together with all these body innovations, a major acquisition of bilaterians is the origin of a brain and a central nervous system. As depicted in the model of Urbilateria, in many bilaterians, the nerve cord is organized as a series of ganglia—related to the body segments—interconnected by nerve fibers along the animal's length. Animals with nerve cords or central nervous systems are considered to display more complex behaviors than those with diffuse nervous systems, but there may be exceptions. In the nerve cord, neurons usually arrange forming so-called central pattern generators, which are oscillatory circuits driving basic behaviors like locomotion and food ingestion. In addition, connections between the left and the right halves of the cord (called commissures) coordinate movements in both sides of the animal. Interestingly, the molecular mechanisms involved in brain and nerve cord development, and in commissure formation, are highly conserved across Bilateria, being largely dependent on a small group of genes. This again suggests that there was an ancestral genetic kit that was used for building a central nervous system in different animals, but it is unclear whether these morphological traits were acquired only once or on several occasions.[8]

Anteriorly, the bilaterian nerve cord joins the brain, usually a large ganglion receiving sensory projections from the head and in close relation to the mouth, particularly in protostomes where it develops around the esophagus. Let me recall that according to one hypothesis the sense organs and the brain of bilaterians would derive from the ancestral apical organ and nervous system, respectively, which would have fused with the nerve cord deriving from the blastoporal nervous system. Needless to say, although there is genetic evidence supporting this proposal, it is not shared by all scholars in the field.

The nerve cord and brain are not the only neural components of Bilateria, as there are also extensive autonomic peripheral and enteric nervous systems that probably derive from the diffuse neural network in the jellyfish-like ancestor. In many invertebrates, the peripheral nervous system emerges from a "lateral neural border" surrounding the embryonic nerve cord.[9] In the next chapter, we will see that some of the major innovations that gave rise to the vertebrates rely on the growth and differentiation of the neural crest, possibly homologous to the invertebrate's lateral neural border.

Philosophical Bugs

Exploring the World

Endowed with these body and neuronal innovations, bilaterians spread through the seas and colonized land diverging into many lineages, some of which achieved high levels of complexity and evolved sophisticated behaviors. In this process, the protostomes subdivided into two major clades, the "ecdysozoans" on one side and the "spiralians" on the other. Ecdysozoans include insects, crustaceans, and nematode worms, while spiralians include flatworms, earthworms, and mollusks. I will now speak of the protostome radiation, while the history of deuterostomes will be the subject of the next chapter.

Among protostomes, animals with complex nervous systems have evolved among crustaceans, spiders, and especially insects on one side, and in cephalopods on the other. Importantly, all these animals have evolved locomotory and feeding appendages: arthropods have paired appendages, and cephalopods have tentacles that give them liberty to move and catch food. The development of limbs, mouthparts, wings (in insects), and olfactory antennae crucially depends on homeobox genes, and particularly the genetics of limb development is strikingly similar in arthropods, cephalopods, and vertebrates, which once again implies a common genetic (but perhaps not anatomical) ancestry.[10]

Together with complex behaviors, specialized visual and chemosensory organs have evolved in these animals. Complex invertebrate animals have paired eyes (sometimes several pairs of eyes like spiders) and mobile olfactory or taste organs (in the mouth and antennae) that enable them to explore their immediate surroundings. Concerning vision, there are two main kinds of photoreceptor cells: rhabdomeric that are highly sensitive and used for rapid vision, and ciliary, related to bright-light vision. Cnidarians use primarily ciliary receptors, and many invertebrates have both kinds of photoreceptors, but arthropods and cephalopods tend to rely on rhabdomeric photoreceptors (vertebrates have ciliary receptors, although deuterostome ancestors used both types). The morphological evolution of eyes, whether convex (compound eyes) or concave (camera-like eyes), results from a common functional requirement: establishing a topographic map of space on its surface. Compound eyes do so by tending to a spherical shape where light beams from different places tend to hit the eye perpendicular to its surface, and camera-like eyes develop a small opening (the pupil, comparable to the diaphragm of an optic photographic camera) that concentrates the entrance of light like an hourglass and distributes the light beams to different positions in the retina, making up an inverted map of space.[11]

Topographic vision must have been a critical achievement associated to more complex brains and behavior, as this is the first instance where an "image" of the space surrounding the animal is formed, permitting it to locate visual stimuli in the environment. Furthermore, with increasing visual acuity, it becomes possible to identify objects and the relations between them. This may provide the initial scaffolding to generate a unified configuration of the environment that is independent of the animal's position (an allocentric representation), coding for the location of different objects and for the animal's place in relation to them. Furthermore, since eyes are typically paired, each eye has a different perspective of the visual scene, but these perspectives usually overlap in the front of the animal (what is called central or binocular vision, as opposed to peripheral vision where only one eye dominates). Merging the "images" from both eyes at the center of the visual field implies a neural mechanism to comply with these different perspectives, a process that gives rise to depth perception (try to estimate distances with one eye closed, for example). Furthermore, head and eye movements transformed vision from a rather passive sense to an instrument to explore the immediate surroundings.

In addition, eyes benefited strongly from the capacity to distinguish distinct light wavelengths, including ultraviolet light (many invertebrates, including cephalopods, also distinguish polarized light), by diversifying the opsin photoreceptors. Still, increasing types of photoreceptors is not the same as having color vision. For instance, the mantis shrimp has twelve different photoreceptor molecules responding to different wavelengths, but each group of photoreceptors works as a separate channel triggering specific behaviors, like attacking, finding food, or mating. Proper color vision, as in bees, is acquired when the activities of different photoreceptors is balanced against each other, which generates a unified chromatic space. The benefits of color vision may relate to seeing the world in conditions of changing ambient illumination, but may have been also exploited as social signaling mechanisms or as a food-searching strategy like insects pollinating flowers.[12]

Besides vision, chemo-sensation is an important sense for orientation. Olfaction provided by the antennae is a critical sense for arthropods, especially insects, as it works as a distant sensor that enables the animal to find the path to food or mates. Octopuses have specialized olfactory organs below the eyes, and their suckers are provided with exquisitely sensitive touch and taste terminals, which enable them to detect the chemical composition of their environment. Furthermore, their arms are provided with light-sensitive receptors that they use to explore the illuminating conditions outside the cave they usually live in. Insects and other arthropods also display sound sensitivity, and there is evidence of hearing capacities in cephalopods too.[13]

Tiny Brains

The brains of insects are quite small, no bigger than a pinhead, and the largest may have not many more than two hundred thousand neurons compared to the estimated eighty-six billion neurons of the human brain. This certainly makes one wonder how these animals can develop relatively complex behaviors, and perhaps more importantly, what are the minimal neuronal elements to have cognition or even some rudimentary sentience. The brains of insects (and many arthropods) are organized into a group of ganglia, of which the "mushroom bodies" have been most related to cognitive abilities. Recent research has shown that these structures participate in learning and memory, and that their size increases with sensory experience rather than with age, indicating that these are plastic networks. Several authors have highlighted the functional similarities between the mushroom bodies and the vertebrate hippocampus, a brain structure also involved in learning and memory (see chapters 8 and 9). Likewise, a component of the arthropod brain called the central complex has been involved in foraging behavior in insects and displays similar functional, genetic, and developmental characteristics as the vertebrate basal ganglia, both being involved in motor control and controlled by the neurotransmitters GABA and dopamine. As mentioned in the previous chapter, reward-related and aversive circuits with a similar synaptic organization to those in vertebrates have been identified in insect brains. Another neural parallelism between arthropods and vertebrates concerns brain states like sleep with similar physiological characteristics to those of mammalian sleep. Finally, insects also show parental behavior that is apparently controlled by hormones homologous to the parental-related vasopressin/oxytocin of vertebrates.[14] Yet, despite these striking parallelisms, the critical question remains as to whether these similarities reflect a single origin of the respective organs in insects and vertebrates or the parallel recruitment of an ancestral genetic blueprint (personally I prefer the second alternative in this case).

Moreover, bees, ants, and termites are well-known for their complex social organizations, living in colonies where one or a few individuals can reproduce to start a new colony. Insect social behavior often results in strikingly coordinated actions that enable the colony to perform complex collective behaviors like finding the shortest routes connecting distinct places, moving large objects, or responding to temperature increases inside the colony. The communication of social insects has been considered to be largely instinctive, but recent work show that it displays important degrees of plasticity, like bees modifying the "waggle dance" according to the presence of predators.[15]

Insect Minds?

While usually depicted as displaying genetically fixed, stereotyped behaviors, recent research has shown that insects and other arthropods can have quite elaborate conducts, indicating a high degree of learning capacity and flexibility, displaying attentional capacities, navigational skills, and a good spatial memory. For instance, some species of wasps hide their nests in several places that they keep hidden and visit regularly to provide them with food. They apparently remember the quality of the supplies provided to each nest and the time when these were delivered in order to visit them again timely. This kind of memory (remembering the three wh's: what, where, and when something happened) was traditionally considered to be exclusive of higher vertebrates (see chapter 9) and is supposed to be a fundamental precursor of planning abilities and for the development of a complex representation of the world. In humans, this capacity has been termed episodic memory, but it critically differs from its nonhuman counterparts in being a form of explicit memory, that is, it can be verbally communicated to others (like when you tell your friend what happened last night at someone else's party).[16]

Bees are also able to learn from others. In a clever experiment, food was placed on a plate and covered by a transparent screen. However, the plate was attached to a string that could be pulled to bring the plate out and make food accessible. About one hundred bees from a colony were brought to the place, of which only two were able to solve the task by pulling the thread. However, the rest of the bees quickly learned the task after observing the others doing it. Also related to social cognition, a series of experiments have shown that social wasps are able to recognize individual wasp's faces and use this information for their benefit, similarly as social mammals and birds can do. For example, social wasps usually observe the fights of conspecifics and behave less aggressively toward individuals that have initiated or have won fights than to those who have lost fights or have been attacked. Insects may be able to do some math too, as they have a sense of numbers (one is less than two, and two is less than three; and zero is less than one!). This capacity is not exclusive of insects and has been reported in spiders and cephalopods as well. Finally, a recent report described bumblebees rolling over small balls just for the sake of it, which the authors considered as evidence of play, a capacity also considered to be exclusive of higher vertebrates. Similar findings have been obtained in fruit flies.[17]

Some studies also suggest that insects are able to display emotion-like states including pain (they have nociceptive-control mechanisms like those of vertebrates, although whether they do feel pain remains an open question), although it is highly debated whether nonhuman animals, and especially invertebrates, can have true emotions. Unlike Darwin, who proposed that human emotions derive from animal emotions, some authors claim that animal reactions may not be comparable with human feelings or emotions (see chapter 14). Thus, some scientists refer to these emotion-like states as "central states," driven by specific neural circuits in different species.[18]

An example of these states is provided by some innate behaviors like male courtship and aggression, each of them being triggered by the presence of a female or another male, respectively. The neural mechanisms involved in these behaviors show a striking parallelism in fruit flies and in mice but do not necessarily imply common ancestry. Another example, more related to human psychology, is the evidence of depressive-like behavior in insects. After experiencing a highly stressing event, flies tend to classify neutral stimuli as harmful and decrease mobility, resembling humans' "learned helplessness" that is usually linked to post-traumatic stress disorder. While scientists debate as to whether these internal states in insects can be called emotions, it is clear that they display complex mechanisms to control their behavior and maintain their internal homeostasis, which may be partly comparable to the physiological mechanisms underlying our own feelings.

Limb-Shaped Heads

Perhaps the most intriguing of invertebrate animals are the cephalopods. Although deriving from a bilaterian ancestor, cephalopods have quite unusual anatomies and nervous systems. Octopus brains are very complex and diverse, in relation to the species' ecology, visual capacities, and social behavior. However, they lack a central nerve cord proper, and a large proportion of their neurons are distributed as a complex interconnected network along their arms. An octopus nervous system has about 500 million neurons (about the same as a dog's brain), with some 60 percent of these neurons located in the tentacles. Interestingly, cephalopods display no color vision in their eyes, having usually only one color-sensitive photoreceptor (with a few exceptions) but have exploited polarized light vision which they use for increasing visual acuity.[19]

Octopuses have called the attention of both scientists and lay people with their behavioral and learning abilities, which have been popularized in different books, in movies, and in videos. Some of the more striking videos show octopuses using coconut shells as a shelter and opening jars with screw tops, something that chimpanzees have a hard time with. Yet, octopuses usually prey on bivalves and other mollusks encased in hard shells, and probably they have evolved an ability to pull apart and exert opposing forces to objects aided by their suckered arms. Octopuses have recently been found throwing objects to other octopuses or elsewhere, a behavior only seen before in birds and primates. Furthermore, cephalopods are impressive masters of disguise, being able to change their fluffy shape and their coloration according to the background, to mimic algae and unpalatable or nonthreatening animals to avoid predation or to approach their prey. (This contrasts with their apparent lack of color vision; perhaps the photoreceptors in their skin might be color sensitive.) In addition, octopuses' coloration patterns are highly unique for each individual, which may be used for subject recognition. Finally, and similarly to insects, cephalopods have demonstrated having a number sense and can remember the location, quality, and timing of a food source when foraging, displaying episodic-like memory. Likewise, they can learn from others like bees and vertebrates do. Cuttlefish have also shown the capacity to prefer larger rewards in the long term rather than smaller ones in the short term, which has been attributed to higher cognitive and planning capacities in vertebrates. Like the previously mentioned spiders, octopuses have also been found to engage in sleep stages similar to mammalians' REM sleep and slow-wave sleep. In the active, REM-like sleep, octopuses move their eyes like mammals do, and furthermore they undergo intense color changes similar to those observed in the waking state. Still, whether octopuses and spiders engage in dreamlike activity is an open question.[20]

Perhaps more impressive is the fact that octopuses have also shown apparent signs of emotional attachment to humans, something that is highly unusual for solitary animals, especially invertebrates. Lay people and scientists who have been close to octopuses have been quite impressed with them, some claiming that they display some kind of consciousness while others are more skeptical, warning that lay people and even scholars may be attributing human emotional and mental states to them like we usually do with our pets.

Unusual Evolution

Cephalopods have become highly successful since the early Cambrian and dominated the seas for quite a long time before the radiation of modern fishes, being probably the most intelligent organisms of earth during those times. Intriguingly, despite their diversity, cephalopods are not found in fresh waters or in terrestrial habitats, perhaps due to constraints related to osmotic balance in this group. Moreover, cephalopods may have evolved through quite special mechanisms. Not only several gene families like the protocadherins, involved in axonal guidance have massively expanded in this group, but in addition, cephalopod evolution is driven by diversification of RNA molecules that increase protein variability, a phenomenon that has also been found in human brain evolution.[21]

Having gained the reputation of being the most intelligent invertebrate animals, there is no clarity about the origins of cephalopods' cognitive skills. Their behavior defies the common notion that enhanced cognition results from complex social behavior (as it has been claimed in the case of human evolution) because, as said, they tend to live solitary lives, especially octopuses. Nonetheless, one proposal relates to the fact that cephalopods undergo fierce competition for mating, which is perhaps the most important instance of social behavior they have. In these instances, there might be intense selection for anticipating competitors' behaviors and deceiving them in order to gain access to reproduction. Some theories claim that human cognition evolved precisely by developing a "Machiavellian intelligence," the ability to develop appropriate social strategies in order to maximize profit, which includes the art of deceiving. Another hypothesis for cephalopod intelligence is that as they lost their protective shell (which is an ancestral feature of mollusks), they had to evolve increased behavioral versatility to survive. However, there are a number of fossilized and living cephalopods that have shells, including the ammonites. In this context, it would be highly interesting to know the behavioral capacities of the *Nautilus*, the only extant shelled cephalopod. Furthermore, many snails have lost their shells (like *Aplysia*, mentioned in the previous chapter), and beyond classical conditioning they are not known to have particularly sophisticated cognitive skills. Perhaps the evolution of cephalopod intelligence results from a complex combination of traits related to the jet-propulsion locomotion that expanded their habitat (like flight in birds), which is associated with complex sensory systems, and especially the manipulative abilities these animals acquired through their sucker-bearing tentacles. This is in line with human evolution again, as hand dexterity was probably a critical driver of brain evolution in our species (see chapter 11).

Perspective

Bilaterally symmetric animals are most of the animals we are commonly acquainted with. One of the basic characteristics of bilaterians relates to the differentiation of a head provided with a brain and sensory organs, perhaps derived from the sensory apical organ of larval jellyfish, plus a nerve cord along the body and a through-gut. The acquisition of a tubular gut separates most bilaterians into two groups, depending on whether the embryonic blastopore becomes the mouth (protostomes, most invertebrates) or the anus (deuterostomes, including starfish and vertebrates), although this distinction is not so clear-cut as was traditionally considered.

Bilaterians are organized according to the establishment of two axes of embryonic differentiation: in the dorso-ventral and in the rostro-caudal directions. These two axes are defined by the differentiation of specific morphogenetic centers at their extremes, which secrete morphogen molecules that diffuse in the different directions. A group of genes termed homeobox genes works downstream of these signals and establishes identity to different parts of the body, especially in segmented animals like arthropods and vertebrates. Homeobox genes become expressed at a specific embryonic stage, called the phylotypic stage, where embryos acquire the main characteristics of their taxonomical group (arthropods, vertebrates, mollusks, and so on). This provides a highly useful reference system for comparing different species.

Complex invertebrates have evolved brains and nerve cords, but cephalopods seem to have lost their ancestral nerve cord, developing instead a nerve plexus that innervates their different arms. The differentiation of sensory systems, particularly smell and image-forming vision, and extremities to facilitate locomotion, are important advances in the evolution of more complex behaviors and for the establishment of a rudimentary map or model of the outside world. Insects and cephalopods have been shown to display complex behaviors like face recognition, episodic-like memory, and a notion of quantity, besides being capable of social learning and perhaps emotion-like states, all of which point to some degree of intelligence and possibly sentience in these animals. While this chapter focused on the protostomes, in the next chapter, I will address the evolution of the deuterostomes and the origin of the vertebrates.

7 Tails, Bones, Jaws, and Limbs

Vertebrates are the animals with which we feel more familiar, as they have body plans, skeletons, and brains that we can recognize in our own bodies. This group arose as an early branch of the deuterostomes, originally as a lesser lineage compared with the major invertebrate branches. In fact, early deuterostomes are totally unimpressive regarding their ecology or their cognitive skills, but they evolved distinct characters and innovations that served as a scaffolding for constructing the body and brain organization of the vertebrates. Here, I will tell the story of the events that gave rise to a fishlike vertebrate starting from an early deuterostome ancestor, then I will describe the subsequent radiation of the vertebrates in the seas, and finally their colonization of land, yielding the major clades that compose this group. These processes lie at the root of our most fundamental human features like our voices and hand dexterity, not to mention our ability to walk and breathe air. I will deal more specifically with the evolution of the vertebrate brain in the next chapter.

Swimming Worms

Gill-Bearing Ancestors

Deuterostomes include quite dissimilar animals, divided into two main branches (figure 7.1). Morphology and recent fossil findings seem to indicate a single origin of this clade, but some genetic analyses have suggested that it is not a monophyletic group. One branch of the deuterostomes includes starfish and kin (echinoderms), which lost bilateral symmetry and acquired a radially symmetric body, and its sister group the hemichordates (acorn worms), which have in common with chordates the presence of gill openings. In addition, these animals display a diffuse neural network below the skin reminiscent of the jellyfish condition, plus two interconnected

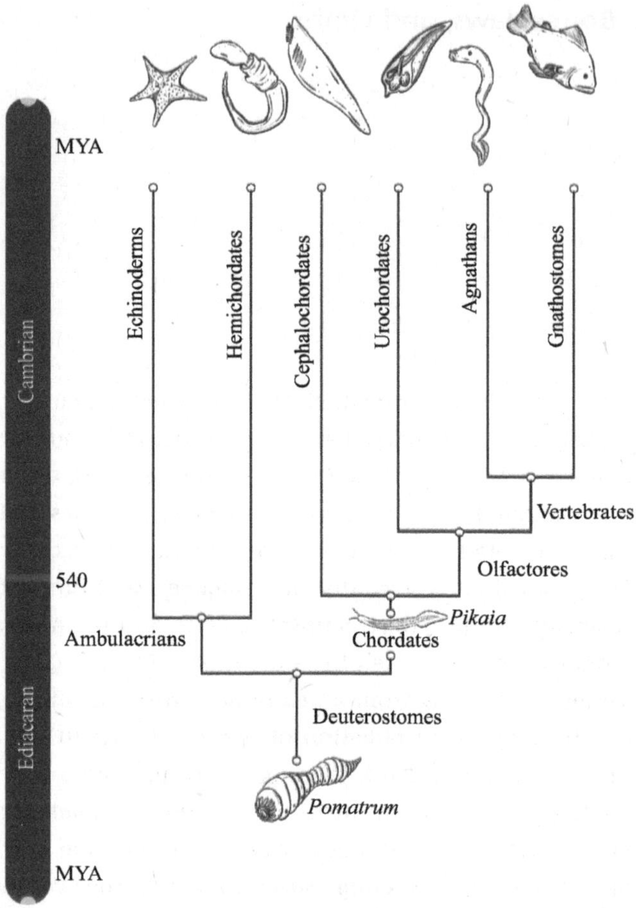

Figure 7.1
The phylogeny of deuterostomes. Urochordates are shown in the larval stage. *Pomatrum* is an early vetulicolian, while *Pikaia* is similar to extant cephalochordates.

nerve cords along the body, one dorsal and the other ventral, that fuse in the head. Furthermore, in the head region (the "collar"), the nerve cord has a small segment that cavitates on the inside, giving it a hose-like appearance. This is a possible forerunner of the tubular nervous system that is typical of chordates, which gives rise to the brain and spinal cord of vertebrates.[1]

A Tailed Larva

The second deuterostome branch includes the chordates, the first fishlike animals that appeared in the tree of life and, according to some views, represent the stem branch of this group. One of the principal characteristics of chordates is the presence of the notochord, a semi-rigid, rod-like structure that determines the animal's longitudinal axis and supports a swimming tail (figure 7.2). As opposed to the larvae of hemichordates and most other marine animals that use rows of beating cilia to propel themselves in water, larval chordates evolved a tadpole-like muscular tail to swim around. Putative chordates (or basal deuterostomes, depending on the interpretation) appear in the early Cambrian period, represented in the fossil record by the vetulicolians (exemplified by *Pomatrum*), provided with a mouth, gill slits along the trunk, and a swimming tail (figure 7.1). *Pikaia and Paleobranchiostoma* are confirmed chordates, similar to the vetulicolians but more laterally flattened and more fishlike in appearance. Living chordates are subdivided into the more basal cephalochordates (lancelets or amphioxus,

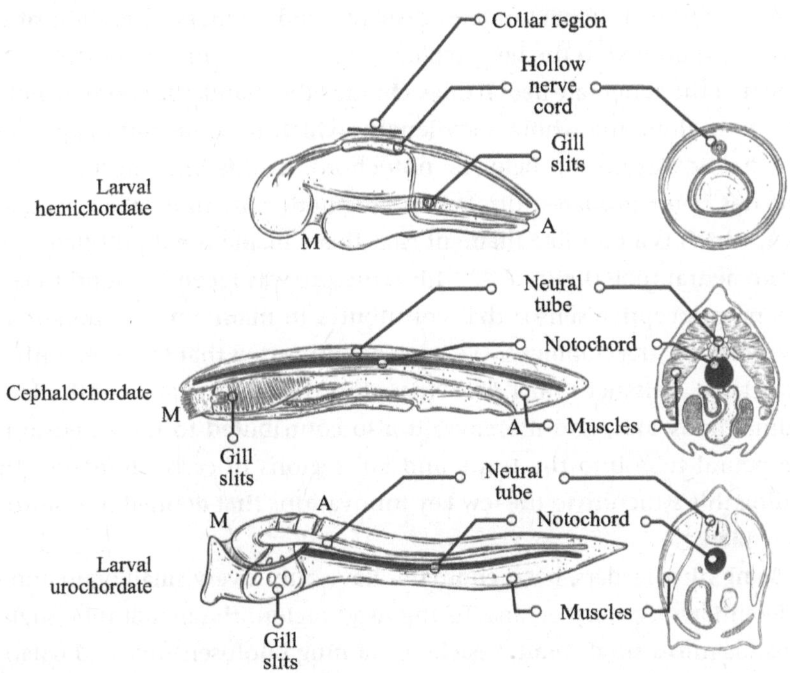

Figure 7.2
The body organization of different deuterostomes. The larval hemichordate worm has gill slits along its digestive tract and a hollow nerve cord restricted to the collar region in the head. In the chordates (cephalochordates and urochordates), beside the gill slits there is a hollow neural tube that runs into the animal's tail, and a stiff notochord below it. M: mouth; A: anus.

quite alike to *Pikaia* and *Paleobranchiostoma*), the urochordates (sea squirts and related forms, some of which are sessile as adults but have a swimming tadpole-like larva), and the vertebrates.[2]

Hollow Brains

In chordates, the notochord is fundamental not only for swimming but is also critical for the differentiation of the central nervous system, a process called neurulation (see figure 7.3). In the early embryo, chemical signals emerging from the notochord induce the formation of a neural plate in the dorsal epithelium. This plate folds into its sides, becoming a hose-like structure called the neural tube that will become the spinal cord and the brain of vertebrates. In the borders of the neural plate, there is a population of cells that will form the peripheral nervous system, which in vertebrates is termed the neural crest (see chapter 6).

As mentioned, hemichordates display a rudimentary cavitation of the nerve cord limited to the head (collar) region, which may represent a forerunner of the vertebrate neural tube. On the other hand, the chordate neural tube runs along the whole body length, which to some authors provides a hydraulic skeleton to help the notochord. In this line, most chordates (but not hemichordates) display an enigmatic structure called Reissner's fiber, which is a hair-like filament that floats inside the fluid-filled cavity of the neural tube (figure 7.4). This structure was recently found to serve as a proprioceptive sensor that contributes to maintain the straight axis in embryonic development. We have hypothesized that together with the notochord, Reissner's fiber participated in the maintenance of the longitudinal body axis, and moreover it also contributed to the extension of the neural tube into the trunk and tail regions of early chordates, thus adding this structure to the few key innovations that defined the chordate body plan.[3]

Being filter feeders, basal chordates have only a very small brain and no differentiated sensory organs. In the head region, the neural tube slightly expands into a single brain vesicle containing photosensitive and balance-sensing cells, and the spinal cord locates caudally in the trunk and tail, controlling muscular movements (figure 7.4). Sensory components connect with neurons that activate motor centers in more posterior brain regions (which has been compared to part of the brainstem of vertebrates), which in turn send axons to the spinal cord to elicit swimming movements.

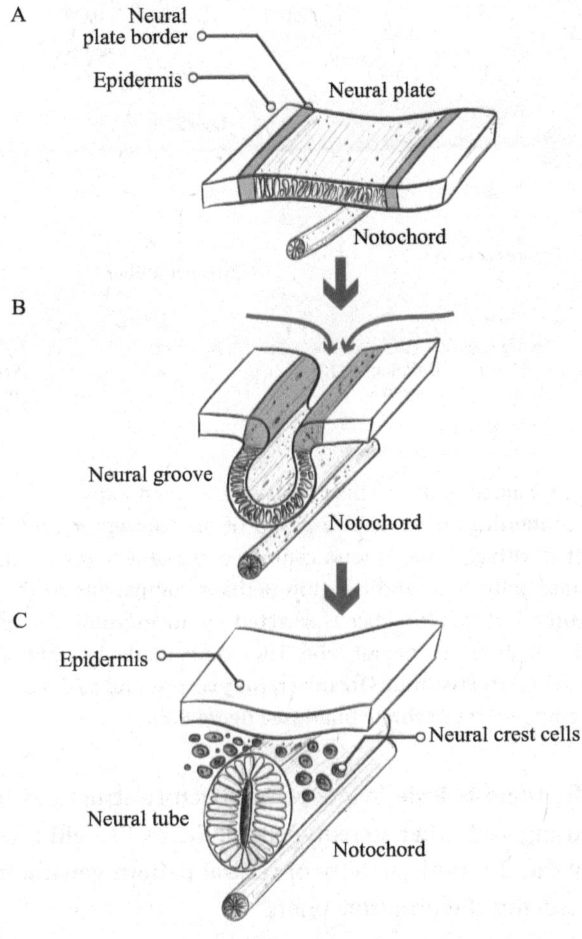

Figure 7.3
The process of neurulation. *A*: The notochord sends inducing signals to the neural plate (note the border of the neural plate in gray, which will become the neural crest). *B*: The neural plate folds on its side, forming a neural groove. *C*: The neural tube is fully formed, and the cells of the neural crest (marked in gray) start migrating in different directions to make the peripheral nervous system and other structures.

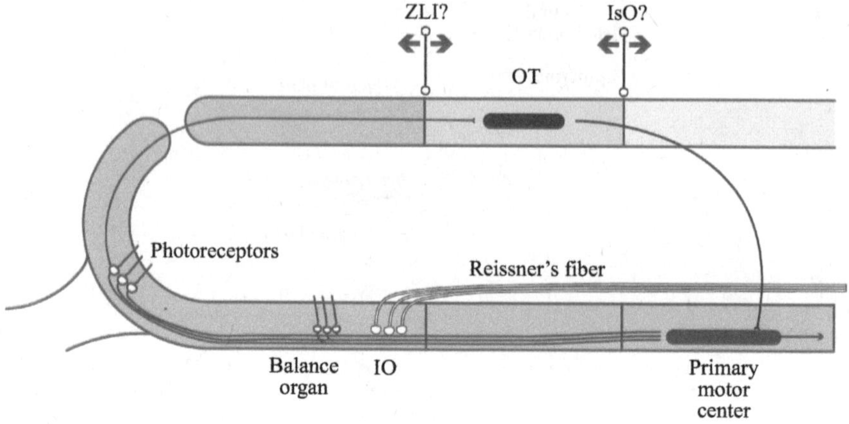

Figure 7.4
The brain of cephalochordates. This consists of a small expansion of the anterior neural tube, containing in its walls a group of photoreceptor cells and a balance organ connected with a primary motor center (comparable to the reticular formation of the vertebrate brainstem) and a region perhaps comparable to the optic tectum (OT) of vertebrates. Reissner's fiber is secreted by an infundibular organ (IO) and grows caudally all along the neural tube. The points marked by ZLI (Zona Limitans Intrathalamica) and IsO (Isthmic Organizer) may correspond to developmental landmarks of the embryonic vertebrate brain (see figure 8.2).

Paradoxically, there is little evidence of olfactory structures in these animals. Swimming and other stereotyped behaviors like gill movements are produced by the rhythmic activity of central pattern generators located in the spinal cord and the primitive brain.[4]

Turning Upside Down?

Finally, there is another striking peculiarity of chordates: they look as upside-down versions of other bilaterians. In most bilaterians, the nerve cord runs in the inferior (ventral) side of the body, the heart is located on the superior (dorsal) side, and the digestive tract is placed between both (as depicted in Urbilateria: figure 6.2; see also figure 7.5). On the other hand, in chordates the nerve cord (more specifically the neural tube) and the notochord run dorsally, the heart is ventral, and the gut is in between both systems. Molecular evidence has supported the notion that the genes determining the dorso-ventral differentiation of these organs are homologous between insects and vertebrates (see the previous chapter); it is only that they appear expressed in an upside-down pattern with respect to each

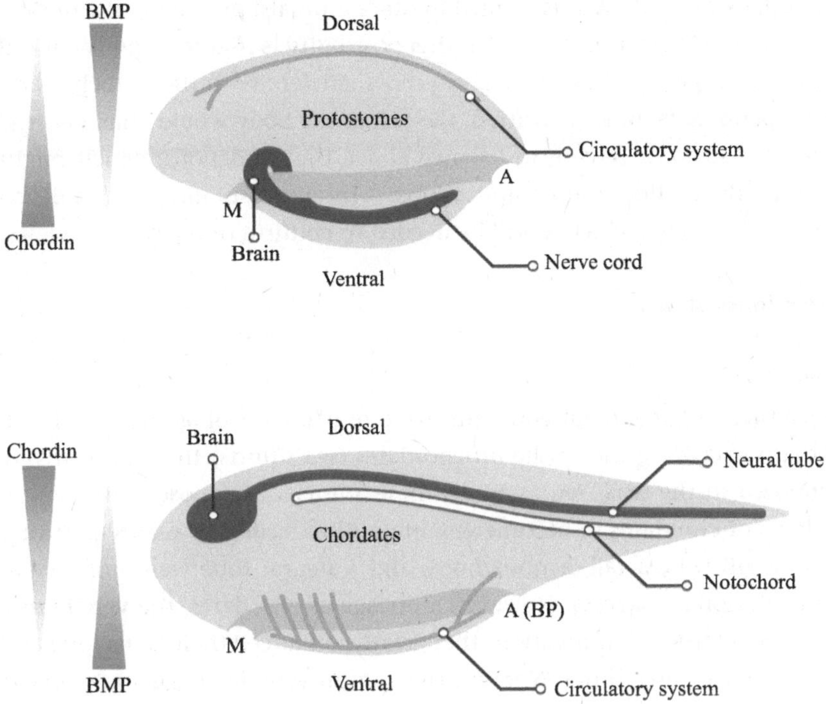

Figure 7.5
Putative dorsoventral inversion of chordates. Above, typical protostome, below, typical chordate. In most bilaterians, there are two opposed gradients of molecules (BMP and Chordin) determining the differentiation along the dorsoventral axis (see also figure 6.2). The BMP-rich side is related to the circulatory system (usually the dorsal side), and the Chordin-rich side is related to the central nervous system (usually the ventral side). These gradients are upside-down in chordates, yielding the nervous system in a dorsal position. The chordate mouth remains ventral and differentiates in the BMP-rich side, while in other animals it opens in the Chordin-rich side. Yet, some authors are recently questioning this view, claiming that chordates may be un-inverted.

other (although there may be dissenting views). Furthermore, in chordates the mouth apparently develops on the opposite side of the body as other animals, facing the now ventral side in order to feed from the bottom. More specifically, while in non-chordates the mouth forms at the same side as the anterior nerve cord, in vertebrates it appears to form on the other side, where the heart is placed. Attempting to explain this observation, some have argued that it relates to the expansion of the notochord in the ancestral tadpole larva, which forced the opening of the chordate mouth on the side opposite to where it is formed in other animals, producing the upside-down turn of the whole body. Another possibility is that for biomechanical reasons, tail-powered swimming was more efficient with the notochord on the superior side. In this position, the rest of the body would "hang down" from it instead of having to balance above the notochord, which might have produced flotation complications.[1,5] In any case, much more developmental genetic evidence will be needed to confirm or reject these views.

Great Innovations

A New Head

Since their origins, basal chordates have remained ecologically restricted, perhaps excepting the sessile urochordates (sea squirts) that are relatively abundant in the seas. Yet, as adults these animals lose most of the typical features of chordates to become sedentary filter feeders. It seems clear that having gill slits, a tail, a notochord, and a neural tube were not recipes for evolutionary success. However, a lineage of chordates, the vertebrates, acquired additional innovations that greatly enhanced their behavioral and evolutionary possibilities. Notably, these improvements largely rely on the diversification of a specific tissue, the embryonic neural crest that borders the neural plate in early development. This tissue amplified enormously in vertebrate origins, giving rise to a variety of structures that marked the emergence of this group. In different parts of the vertebrate body, the neural crest gives rise to a diversity of cell types, including the autonomic nervous system, endocrine organs, and pigmented cells, but in the head region they generate two of the most salient characters of vertebrates (see figure 7.6). Firstly, from the posterior brain, some of these cells migrate ventrally to the throat region, forming the cartilaginous branchial skeleton that enhances respiratory capacity and contributes to forming the vertebrate face (recent evidence indicates that in early vertebrates the neural crest contributed dermal skeletal elements, like scales, along the whole body, not only to the branchial system). Secondly, in the most anterior regions of the developing brain, where the cerebral hemispheres will form, the neural crest territory fragments in a series of cell clusters called placodes, which navigate like icebergs in the head surface in different directions, differentiating into or inducing the development of sense organs including the olfactory epithelium, the eyes, and the inner ear, as well as the neuroendocrine hypophysis. Recall that for some authors, the placodes represent an evolutionary

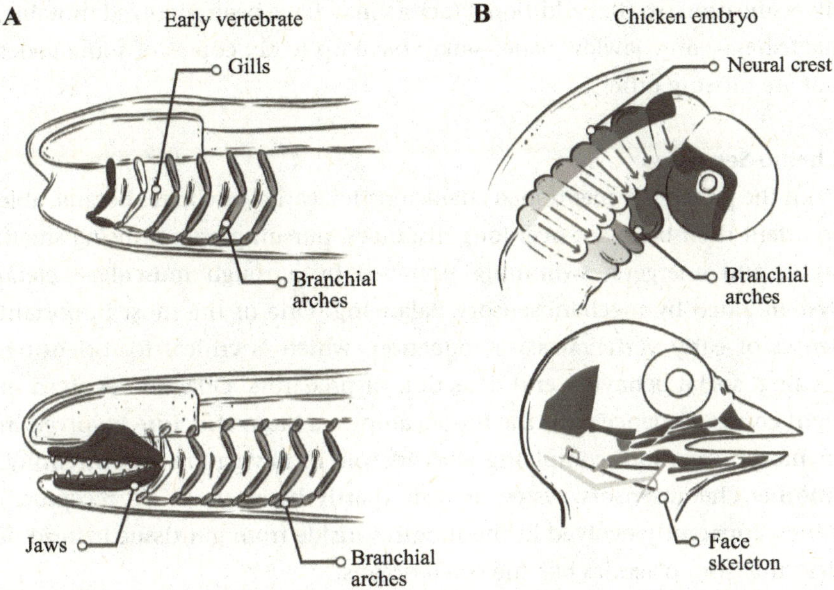

Figure 7.6
The branchial skeleton and neural crest in vertebrates. *A*: The gill slits in vertebrates became reinforced by a cartilage skeleton, the branchial arches (above). In jawed vertebrates, the anterior arches became transformed in the jaw (below). *B*: The branchial skeleton is produced by neural crest cells that migrate from the embryonic brainstem to the throat region (above). In higher vertebrates (a chicken is shown here), the embryonic branchial arches do not produce gills but give rise to much of the face and throat skeleton (below).

extension of the sensory-endocrine apical component of larval jellyfish (see chapters 4 and 6). The neural crest and placodes endowed the emerging vertebrates with a stronger respiratory system and keen sense organs, enabling them to navigate around searching for food and to colonize new ecological niches that supported their evolutionary success.[6]

Finally, these changes were associated with another major event in early vertebrate evolution, where the whole genome duplicated twice, making four copies of each ancestral gene. It is believed that the genetic redundancy produced by gene duplication released many genes to acquire new functions and opened a way to generate evolutionary novelties, like the amplification of the neural crest and all the cellular differentiation processes associated with it. However, evidence suggests that these gene duplications may not have inmmediately been followed by morphological diversification, so that additional factors may have been involved (notably, hagfishes—early, jawless fishes—may have up to six copies of some genes but are missing others).[7]

Chemo-Sensation

With the previously mentioned characteristics, early vertebrates became able to orient their behavior over long distances, pursuing prey by using smell, sight, and energetic swimming provided by a tough muscular-skeletal system aided by mechanosensory balancing. One of the most important senses of early vertebrates was olfaction, which is critical for orienting, feeding, social behavior, and detection of predators. Olfaction evolved in tight correspondence with the hippocampus, a brain structure involved in memory formation, complying with its role in navigation and orienting. Another chemosensory system is taste (partly based on GCPR receptors), which apparently evolved in the mouth's inside from gut tissue instead of deriving from placodes like the smell organs.[8]

Vision

Another crucial sense for vertebrates was vision. Vertebrates evolved paired eyes deriving from the single photoreceptor organ located in the anterior neural tube of early chordates. Cephalochordates display both rhabdomeric and ciliary photoreceptors (see previous chapter), but vertebrates have lost the rhabdomeric type and diversified the ciliary type into cones for bright, color vision, and the later emerging, highly sensitive rods for night vision. The vertebrate retina covers the internal surface of the eye, with a sheet-like structure composed of interconnected neurons and photoreceptor cells. A noticeable characteristic of the vertebrate retina is that the photoreceptor cells are buried inside the eye wall, while the neuronal layers are located in the retinal surface, interposed between the deep photoreceptors and the light coming through the eye cavity (figure 7.7). Therefore, light must traverse all the neuronal layers before impacting on the photoreceptors, which produces a degree of optical distortion. On the contrary, in the insect compound eyes and in the camera-shaped cephalopod eyes, photoreceptors are just in the retinal surface and light directly impinges on them. The reasons for this odd arrangement in vertebrates probably relates to their evolutionary history and the origin of the vertebrate eye as an outgrowth from the embryonic neural tube. In the semi-transparent basal chordates with no eyes, photoreceptors differentiate from ciliated cells lining the inner cavity

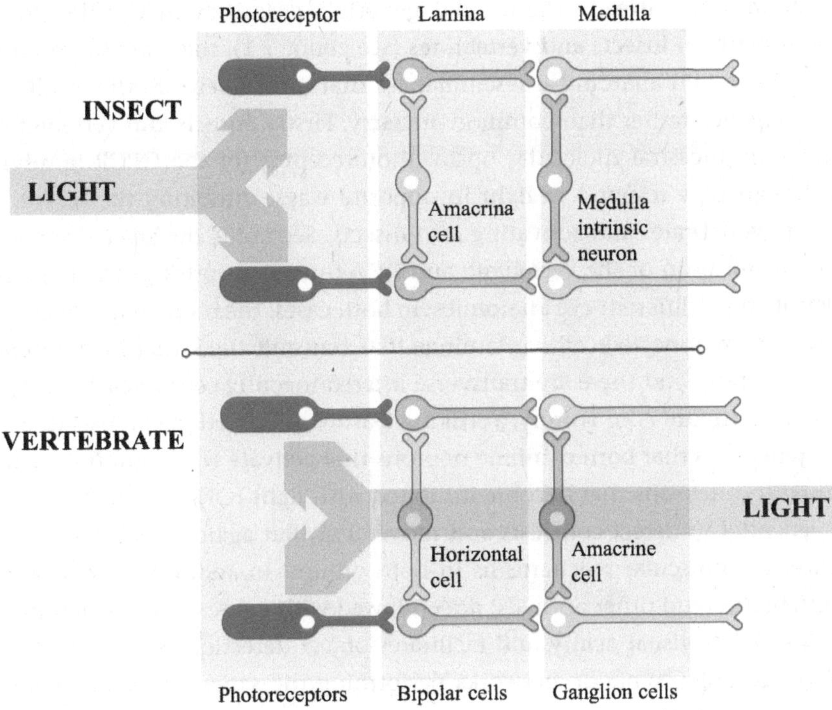

Figure 7.7
Synaptic organization of the retina in flies and vertebrates. In insects, the light, coming from the outside through the compound facets (not shown), directly hits the photoreceptors. In vertebrates, the light crosses the pupil and the eye cavity (not shown) before impacting on the retina, but must traverse the retinal network before reaching the photoreceptors. Also note the similar synaptic organization in flies and vertebrates. Medullar cells in insects, and ganglion cells in vertebrates, project their axons to the animal's visual centers of the brain.

(the ventricle) of the neural tube, while their axons are placed over these cells (more externally) in the walls of the tube (see figure. 7.4). Thus, light must cross the axon layer (and the organism's skin) before reaching the photoreceptors, which for a simple visual organ may not matter much. But as the eye increased in complexity, photoreceptors remained in the internal wall of the optic cup, and new neuronal layers were added making synapses with the photoreceptor's axons, which further interfered with light's trajectory. Moving the photoreceptor layer to the outside probably implied intermediate phenotypes that were not functional and were discarded by natural selection.

Besides the already mentioned genetic similarities in eye development between insects and vertebrates (see chapter 1), there are important functional and anatomical resemblances that most likely are the result of convergence rather than common ancestry. Firstly, insects and vertebrates share an ancestral molecular opsin photoreceptor (of the GPCR family), although they respond to light in opposite ways, inhibiting the receptor cell in vertebrates and activating it in insects. Secondly, the layered synaptic organization of the vertebrate and fly retinae is surprisingly analogous despite their different eye anatomies: in both cases, the retina is arranged in three interconnected cellular laminae that transmit the signal from receptors to brain, and there are transverse interconnecting cells in each neuronal layer (figure 7.7). Thirdly, a striking feature of the retinae of insects and vertebrates is that both combine neurons that activate with light (ON neurons) and neurons that become inhibited with light (OFF neurons), which is essential to detect contrasts and movement. But again, this is based on different molecular mechanisms in both groups. In addition, insects and vertebrates (and other animals) are endowed with color vision, which provides higher visual acuity and facilitates object detection (see chapter 6). Thus, complex synaptic networks performing the same function evolved strong similarities but at the same time were built according to different strategies, reflecting their separate evolution.[9]

As said, more than any other sensory system, the retina provides a first "window" to the world and provides the basic blueprint for the knowledge of our surroundings. A fundamental discovery in this line was made by Humberto Maturana and collaborators in the 1950s–1960s when analyzing the electrical responses of the frog's retinal neurons to light. While previous studies stimulated the neurons by directly illuminating a specific point in the frog's visual field, Maturana discovered that the simple movement of a shadow over the frog's eye generated a strong retinal response. Thus, it was not light itself but rather the changes in luminosity that produced a visual response. These findings led to the notion that the frog constructed its visual world starting from a few and very simple contrasts (like light-dark borders and moving spots—the latter perhaps related to the perception of a wandering fly to be caught), instead of making a direct image of the world like a photographic camera does. The finding that our basic perception of the world relates to changes in the environment rather than providing a "photography" of the surroundings has had profound consequences for cognitive neuroscience, and we will come back to them in the subsequent chapters.[10]

Balance and Hearing

Finally, in early vertebrates, balance became fundamental for proper swimming. In fishes, the elaboration of the balance sensors of the inner ear, provided by the semicircular canals, was probably key in developing a more active lifestyle. Fishes have also developed the lateral line, a system of channels in the skin that allow them to detect water movements, and many have developed electro-sensory systems that enable them to detect prey and communicate. The inner ear of fishes also contains an auditory component that detects water or air vibrations, which were later used by terrestrial vertebrates as a dominant sense.[11]

The Life of Fishes

Spreading through the Seas

Vertebrates emerged from Cambrian animals called yunnanozoans (for example, *Haikouella*), provided with pharyngeal slits, rudimentary eyes, a muscular body, and a tail with a dorsal fin that increased stroke during swimming. More complex anatomies were acquired by the conodonts that followed them, provided with well-developed eyes (see figure 7.8). Conodonts became quite abundant from the Cambrian to the Jurassic period, coexisting with more modern fishes, ammonites, and giant marine reptiles, being a common prey for many invertebrates and vertebrate animals.[12]

Together with conodonts, the extant eel-like lampreys (Cyclostomata) and hagfishes (Myxinoidea), which are not directly related to each other, represent the most basal branches of vertebrates. The latter two groups are provided with eyes and eye muscles to direct sight, gill slits supported by a cartilaginous skeleton instead of the collagenous branchial skeleton that is found in basal chordates, a tail provided with a caudal fin for swimming (but like the basal chordates they do not have paired fins), and oral denticles that they use to suck blood from other fishes (lampreys) or to eat carcasses in the sea bottom (hagfishes). Hagfish and lampreys also display vestigial cartilaginous vertebrae, probably representing a secondary reduction from a swimming ancestor. Fossil lampreys and hagfish date from about 350 MYA, having changed practically nothing since then.

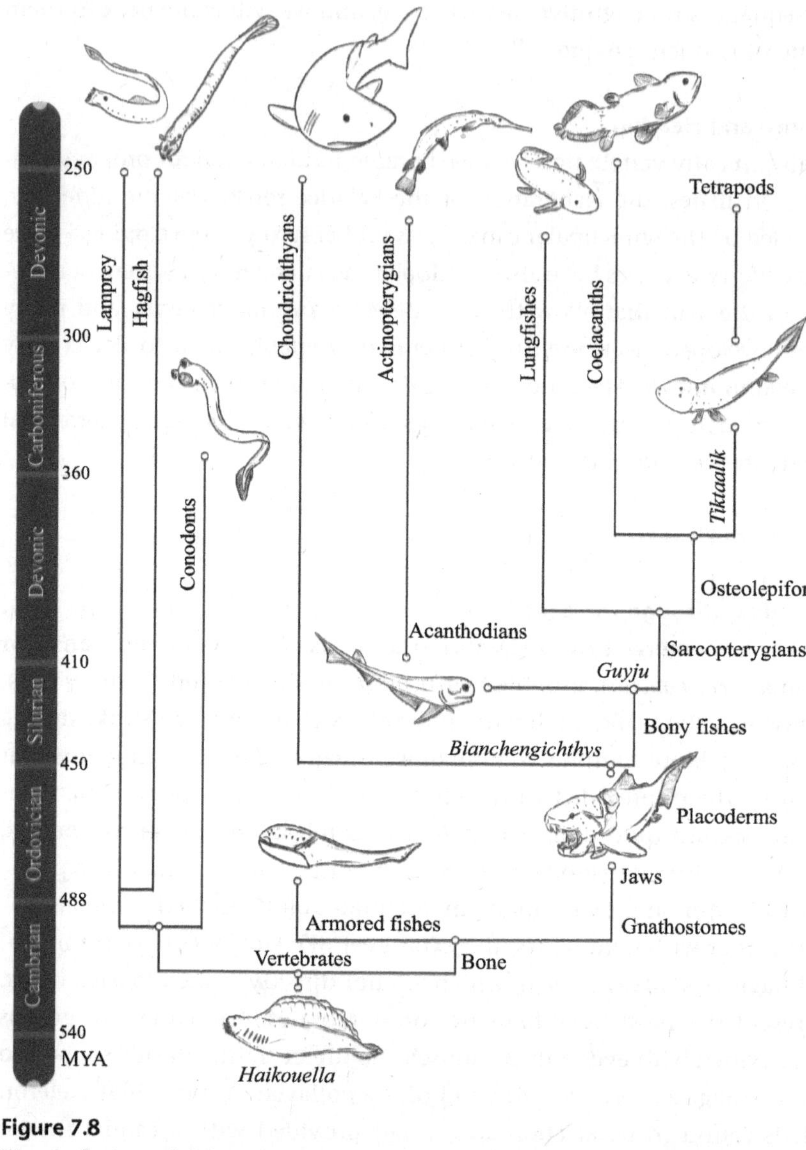

Figure 7.8

The evolution of fishes. Lampreys, hagfish, conodonts, *Haikouella*, and armored fishes are jawless vertebrates. Mineral bone appears in the armored fishes and jaws in the placoderms. Chondrichthyans lost the mineral bone, and the remaining jawed fishes are the bony fishes. *Bianchengichthys* and *Guyju* are early bony fishes, together with acanthodians. The lineage toward terrestrial vertebrates, or tetrapods, stems early in sarcopterygians with *Tiktaalik* being a key species.

Skeletons

Another fundamental acquisition of early vertebrates was the development of mineralized bone. Curiously, the earliest evidence of mineral bone, called dermal bone or acellular bone, is not the internal skeleton but can be found as external armors in the first fossil fishes dating from some 520 MYA. These external bony plates remain in all vertebrates until now, giving rise to most of the cranium and part of the face skeleton. Furthermore, external bone derived from small denticle-like plates called odontodes, which points to a common origin with teeth.

Armored fishes subdivided in two lineages: the older heterostracans and the more derived osteostracans. The former do not display evident signs of internal ossification, but some species (particularly *Tujiaaspis*) had lateral folds along the body that have been proposed as precursors of the paired, lateral fins of modern fishes. The more advanced osteostracans and kin display evidence of internal bone and developed paired pectoral fins that increased swimming skills. This was another great advancement because the internal skeleton provides support for the attachment of muscles, allowing a more vigorous swimming to chase prey and to escape predators.[13]

The origin of the paired fins has been a matter of controversy for evolutionary biologists, some proposing that the internal skeleton supporting them evolved as a transformation of the posterior branchial skeleton. Another hypothesis suggests that the lateral fins developed from anterior extensions of the caudal fin that reached the trunk region on both sides of the body, like the lateral folds of *Tujiaaspis*. Like other instances of evolutionary controversy, this one about fin origins may well remain for long. Anyhow, there is the possibility of a combination of the two hypotheses, where the lateral folds of heterostracans developed internal bone condensations by activating a set of genes partly related to gill skeleton development and recruited these later for pelvic fin formation.[14]

Jaws

The armored fishes gave rise to the placoderms, bearing a fundamental innovation that gave them a further similarity to modern fish: the presence of jaws (see figure 7.6). Interestingly, a recent study has shown that a specific homeobox gene (called Nkx3.2) is involved in the development of jaw articulation in most fishes but is absent in jawless fishes, and it may have played a role in jaw origins. The acquisition of jaws was associated with the origin of mandibular teeth and an increase in mouth size, enabling the early fish to feed on larger prey and to circulate more water through their gills and absorb more oxygen, promoting a more active lifestyle. Placoderms underwent a massive radiation starting in the Ordovician period about 485–445 MYA and became extinct by the Carboniferous period some 360 MYA, possibly due to competition with the emerging modern fishes. In their lifetime, they populated all the seas and became extremely diverse, including terrifying animals like the six- to seven-meter-long *Dunkelosteus* that became the largest predator ever at that time. Stemming from placoderms, early cartilaginous fishes or chondrichthyans made their appearance some 420 MYA or earlier, losing the external and internal bones that increased their buoyancy.

Bony fishes also originated quite early in vertebrate history and may have been related to the acanthodians, fossil fishes endowed with spines in their fins. Early bony fishes are *Bianchengichthys* and *Guyju* (ghost fish in Chinese), which show characters of both placoderms and bony fish. The osteichthyes or modern bony fishes are the largest and most diverse vertebrate group and divided very early into the less common lobe-finned fishes (Sarcopterygii; coelacanths and lungfishes, with fleshy fins), from which terrestrial vertebrates evolved, and the much more abundant ray-finned fishes (Actinopterygii). The latter are most of the bony fish we know, and are the most successful vertebrates ever, including some twenty-eight thousand extant species. Notably, early bony fishes acquired lung-like structures to capture air, as they may have lived in shallow waters. These evolved into the terrestrial vertebrates' lungs in the sarcopterygian lineage and into a swimming bladder to increase flotation in osteichthyes. Therefore, this tremendous innovation promoted the transition to land in the former group, and the dispersion into the seas and fresh waters in the latter group.[15]

Out of Water

The Great Leap Forward

In the Devonian period, a group of Sarcopterygians collectively known as Osteolepiforms, provided with robust fin skeletons, were the pioneers onto firm ground, using their primitive lungs to catch air and emitting sounds when exhaling air, which much later would allow our species to speak. Osteolepiforms were large, shallow water ambush predators living in the lakes and rivers of the Devonian period. A further step, closer to the earliest terrestrial vertebrates, is provided by the Elpistostegalians exemplified by *Tiktaalik*, provided with a flat, alligator-shaped skull. In general, these animals are quite difficult to distinguish from the earliest tetrapods (four-limbed terrestrial vertebrates), which are classified as amphibians together with the extant frogs and salamanders (living amphibians are called lissamphibians). The earliest amphibians resembled giant salamanders like the 370 MYA *Ichthyostega*, which was largely aquatic, spending little time on ground. Amphibians underwent a strong radiation during the Carboniferous and Permian periods (350–250 MYA), but they were confined to live near water in order to lay their eggs. Fossil amphibians practically disappeared in the late Permian extinction event, and their relations with present day amphibians are another enigma.[16]

The colonization of land implied surpassing a great physiological obstacle, which is the rapid loss of water in the body when living in a non-aqueous environment. This was partly resolved by making a tight integument that minimized water loss. Another adaptation to live on land consisted of one of the most spectacular skeletal innovations in vertebrate history: the transformation of fins into limbs. As expected, the development of vertebrate fins and limbs is driven by Homeobox genes and related ones. Perhaps the most dramatic modification in this process took place in the formation of the vertebrate hand and foot, in which a fleshy "palm" was formed together with digits. Digit formation took place concomitant with a dramatic reduction of the fin's rays present in sarcopterygians, whose vestiges may correspond to our nails. The earliest amphibians already had well-formed digits in their four limbs, but their numbers were highly variable, usually more than five (up to ten), and only in later stages of tetrapod evolution did the number became fixed to five digits. Notably, the genes involved in fin–or limb–differentiation may have been coopted in terrestrial vertebrates to develop the external genitalia. Another important but less known acquisition was the development of a muscular tongue that enabled early tetrapods to swallow prey on land, since aquatic fishes usually swallow their food by creating a water current to suck prey inside. The muscularization of the tongue is another great innovation that later amplified the feeding and communicative abilities of terrestrial vertebrates, pollen-feeding in birds and bats, mastication in mammals, and complex vocalizations as in bats and human speech.[17]

Laying Eggs on Land

Another breakthrough took place when some small tetrapods became able to lay their eggs on ground by covering them with a shell and a membrane containing a fluid-filled cavity, the amnios (the chicken egg's white), which protects the embryo from desiccation.[13,18] These animals also had a slender but well ossified skeleton provided with mobile ribs to inhale air, resulting in a smaller mouth and a modified jaw and dentition that enabled them to catch prey more efficiently and feed on plants. Amniotes, as they are called, radiated quite early in two main branches. On the one hand, there were the synapsids or so-called mammal-like reptiles (strictly they were not reptiles), characterized by a single opening in the side of the skull to make space for jaw musculature and by the appearance of incipient canine teeth, possibly both features related to more powerful biting. On the other hand, the ancestors of modern reptiles and birds emerged (diapsids or sauropsids) that had two skull openings (see figure 7.9). Among the synapsids, carnivore and herbivore pelycosaurs featuring the sail-bearing species *Dimetrodon* and *Edaphosaurus* among many other forms dominated the earth during the Carboniferous. In the Permian, pelycosaurs gave rise to the therapsids, more mammal-like in shape and behavior, who also evolved into carnivorous forms like the sabertoothed gorgonopsians and herbivores like the dicynodonts, some of which achieved rhinoceros-like sizes. These animals modified their gate by rotating their limbs, placing them under the body unlike most reptiles and amphibians, which have the limbs oriented sideward from the body. Some authors suggest that therapsids could have had hair, skin glands, and an intermediate degree of homeothermy (warm-bloodedness), presaging these conditions in the early mammals. Close to the end of the Permian period, the major biological catastrophe in earth's history took place due to intense volcanism and global warming. This event eliminated most therapsid lineages excepting a few like the pig-like dicynodonts and the smaller sized cynodonts. The former became soon extinct, but the cynodonts survived longer into the Mesozoic period and gave rise to the early mammals.

The other branch of early amniotes, the diapsids (or sauropsids), emerged by the late Carboniferous, diverging into two main branches. The first one gave rise to lizards, snakes, terrifying marine mosasaurs, and the lizard-resembling rhincocephalians. In the second branch are the crocodiles, pterosaurs, dinosaurs (including birds), and possibly turtles. Diapsids diversified

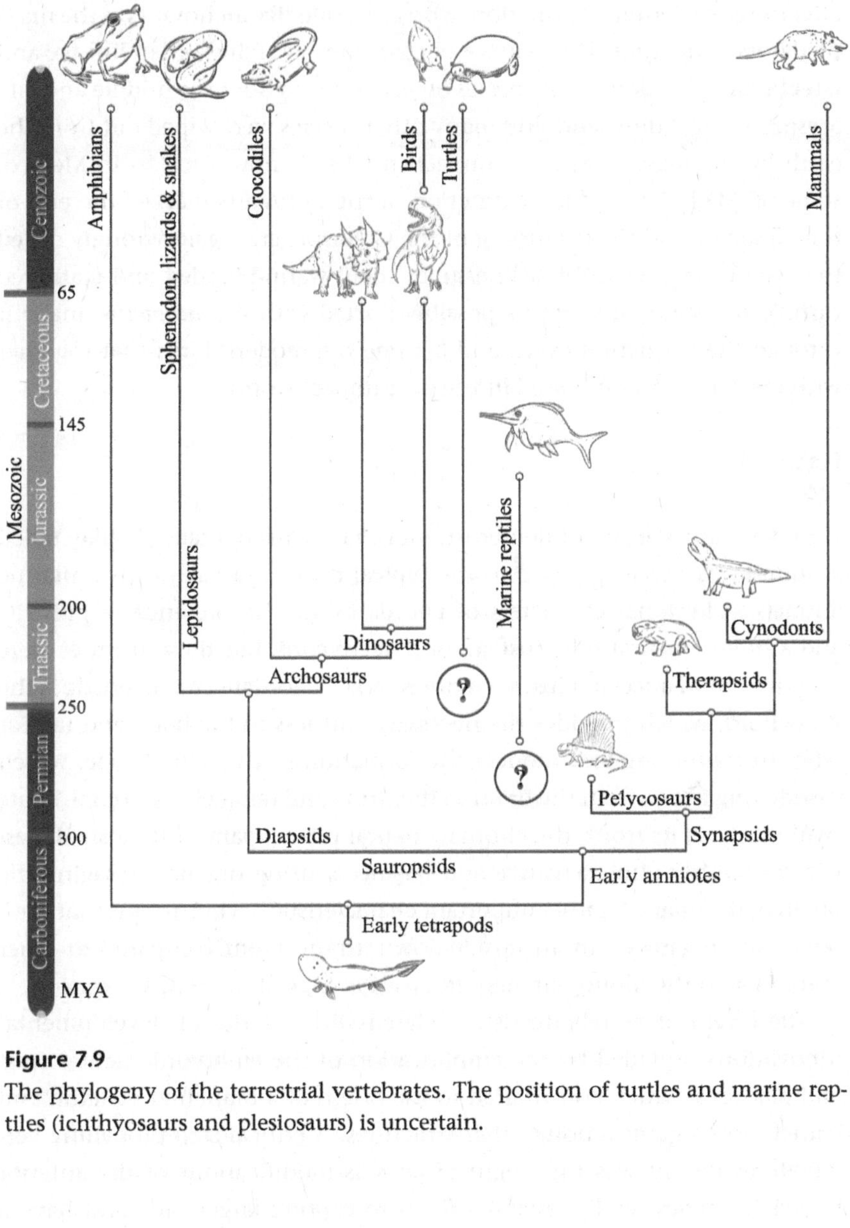

Figure 7.9
The phylogeny of the terrestrial vertebrates. The position of turtles and marine reptiles (ichthyosaurs and plesiosaurs) is uncertain.

after the great Permian extinction, with crocodile-like archosaurs as the main predators of the time. Dinosaurs emerged from an archosaurian lineage and later became the dominant species on earth throughout the middle and late Mesozoic. Most dinosaurs and many other species were wiped out from the earth by the massive asteroid impact in what is now Yucatán in Mexico, some 65 MYA. The sudden extinction of the dinosaurs marked the end of the Mesozoic and the beginning of the Cenozoic era, again wrongly called the age of mammals. Yet, a lineage of small warm-blooded and feathered carnivorous dinosaurs, which possibly started with *Archaeopteryx*, made it through the extinction event and became the modern birds that together with mammals disseminated in the post-impact world.

Perspective

Chordates are a group of deuterostomes that evolved a tadpole-like, tailed swimming larva, as opposed to the typical ciliated larvae of most marine animals. Additional characters of chordates are the presence of gill slits and a hollow neural tube instead of a nerve cord, but these features were inherited from ancestral deuterostomes. A key chordate characteristic is the notochord, which provides the necessary stiffness to the body and tail for efficient swimming and induces the formation of the neural tube, which carries long axons from the brain to the trunk and tail regions to coordinate swimming. At its front, the chordate neural tube expands into a small vesicle containing photosensitive and balance-sensing organs, presaging the origin of the brain. Another important characteristic of chordates is that their bodies are organized in an upside-down arrangement compared to other animals, and the mouth changes in position according to this inversion.

The origin of vertebrates is associated with a series of developmental innovations provided by the amplification of the embryonic neural crest and placodes, which contributed to the origin of a branchial skeleton and paired sense organs among other structures. A critical step propelling vertebrate evolution was the origin of jaws as modifications of the anterior branchial arches, which enabled them to capture larger prey and have a more active life. Bony fishes split into the ray-finned fishes, and the lobe-finned fishes. The latter became able to colonize land by taking advantage of a primitive lung and by transforming their fins into limbs. Another great innovation was the origin of the amniote egg, which enabled a branch of tetrapods to live their full lives outside the water unlike amphibians. Early amniotes divided again into two branches: the synapsids, which dominated the Permian period and eventually gave rise to mammals, and the diapsids that evolved into all living reptiles and birds. The latter became more abundant in the Mesozoic period, which ended after a massive asteroid impact to give rise to the Cenozoic or the age of mammals. The next chapter tells the story of the expansion of the vertebrate brain, its development, and its sensory and motor organization in different lineages.

8 A Masterpiece of Evolution

In this chapter, I will discuss the evolution of the vertebrate brain and its diversification among the different vertebrate classes, emphasizing the brains of birds and mammals. I will first refer to the phylogenetically conserved aspects of brain structure and development, and in the second part of the chapter I will address the morphological divergence of the vertebrate brains, which is associated to the expansion of "higher" brain regions involved in sensory perception and the organization of behavior. This second part is the most controversial, as for more than a century there have been agitated discussions among neuroanatomists to establish homologies between different vertebrate brains without reaching a consensus.

Before we begin, a brief sketch of vertebrate brain neuroanatomy may be most helpful (see figure 8.1). The vertebrate brain is subdivided in two main regions: the forebrain and the brainstem. The forebrain includes the cerebral hemispheres and some deep brain structures like the thalamus that relays sensory inputs to the cerebral hemispheres. In addition, the forebrain includes the hypothalamus, a small but critical complex of nuclei that controls neuroendocrine functions and general body physiology. Behind the forebrain is the brainstem (divided into the midbrain and the hindbrain), containing sensory and motor nuclei that control sensory-oriented body reflexes and basic behaviors like locomotion, breathing, and swallowing. The brainstem also modulates the general activity of upstream forebrain centers and downstream spinal cord circuits. In the brainstem lies the cerebellum or "little brain" that coordinates motor behaviors and will be briefly discussed.

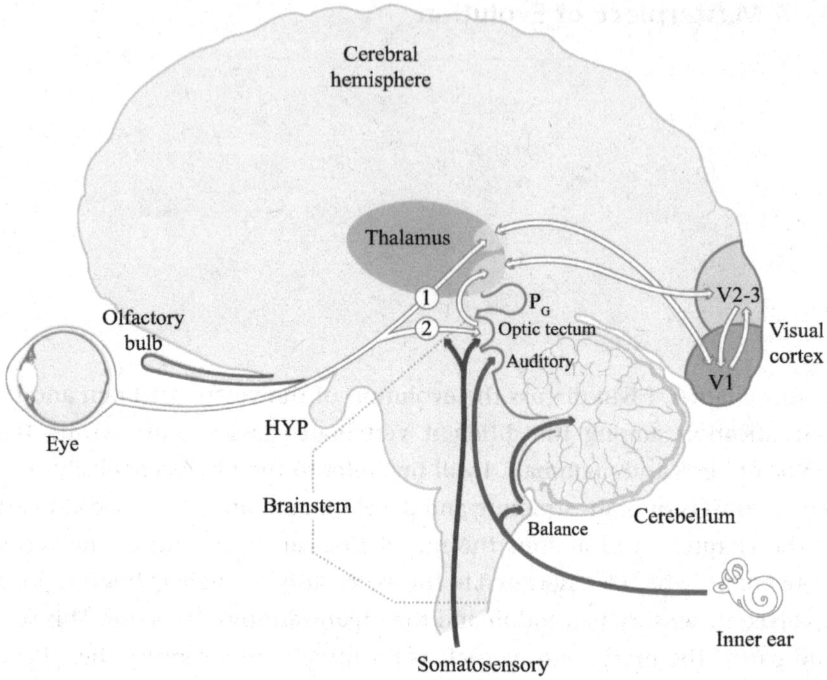

Figure 8.1
The main components of the brain and sensory inputs, exemplified in a human brain. The cerebral hemispheres comprise the cerebral cortex, thalamus, and hypothalamus (HYP) among other structures. The brainstem includes several brain nuclei involved in reflex actions and the cerebellum. The main sensory pathways (excluding olfaction and taste) are also shown: vision (from the eye), somatosensory, and audition (inner ear), all of which end up in the thalamus (only the visual thalamic input is shown for clarity). Note that the visual pathway is divided into two components: (1) a "direct" pathway that goes directly from the retina to the thalamus, which ends in the primary visual area (V1); and (2) a pathway that passes through the brainstem, which ends in the secondary visual areas of mammals (V2-3).

The Stem

Axes and Segments (2)

Like the body organization of most bilaterians, brain development is organized according to two main axes of differentiation: a rostro-caudal (longitudinally along the neural tube) one and a dorso-ventral (back to belly) one, which are determined by the presence of distinct molecular concentration gradients in all four directions (see chapters 6 and 7 and figure 8.2). In this

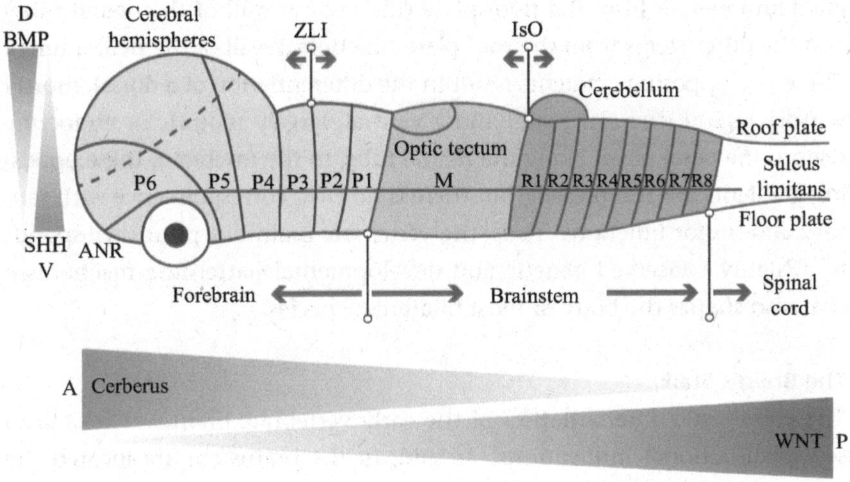

Figure 8.2
Differentiation of the embryonic brain. Antero-posteriorly, a complementary gradient of morphogens (Cerberus and Wnt), specify an axis over which a segmented structure differentiates, with rhombomeres in the posterior brainstem or hindbrain (R1-8), the mesencephalon or midbrain (M) in the anterior brainstem, and prosomeres in the forebrain (P1-6). The dorso-ventral axis is driven by complementary gradients (BMP and Shh), and in the brainstem and spinal cord specifies a dorsal or alar sensory plate from a ventral or basal motor plate, separated by a sulcus limitans. The zona limitans intrathalamica (ZLI) and the isthmic organizer (IsO) are morphogenetic centers that pattern structures anterior and posterior to them and may also be present in cephalochordates (see figure 7.4).

process, some genes involved in the formation of the embryo's head (one of these is eloquently called *Cerberus*) induce the formation of the brain and brainstem. In the hindbrain (posterior brainstem), the classical homeobox genes contribute to the differentiation of specific segments (called rhombomeres), which are associated with distinct cranial nerves and embryonic branchial arches (see chapter 7). On the other hand, the forebrain becomes patterned by a mosaic of more complex homeobox-containing genes that subdivide the organ into a series of segments called prosomeres. As it is strongly conserved across species, this segmental organization has been a useful tool to establish homologies between different vertebrate groups.[1]

Together with the rostro-caudal axis of differentiation, the vertebrate brain (and the whole embryonic neural tube for that matter) differentiates in the dorsoventral dimension, driven again by two opposing gradients. One gradient emerges from the floor plate (the ventral wall of the neural tube) and the other stems from the roof plate (the dorsal wall of the neural tube). These two opposing gradients result in the differentiation of a dorsal, mostly sensory region (the alar plate) and a ventral, largely motoric or endocrine region (the basal plate) along the neural tube. In the forebrain, these opposing gradients are also present, but there is no clear correspondence with sensory and motor functions. Thus, the vertebrate brain blueprint corresponds to a highly conserved genetic and developmental patterning mechanism that also shapes the body of most bilaterian species.

The Brain's Stalk

The embryonic differentiation of the early vertebrate brain was related to several functional innovations. As said, in the brainstem are located the motor and sensory circuits that control rhythmic movements for feeding, respiration, locomotion, swallowing, chewing, and eye movements. Another crucial brainstem innovation was provided by sensory systems and complex sensory-guided reflexes. For example, the eyes remained heavily connected to a brainstem structure called the optic tectum that drives visuomotor orientation reflexes (see figure 8.1, pathway 2). Early chordates have a rudimentary forerunner of the optic tectum (see figure 7.4), but the vertebrate tectum becomes organized similarly to the retina (or the mammalian cerebral cortex), in a series of cellular laminae that repeat the topographic organization of the visual scene so that different positions in space correspond to different retinal positions and to different tectal positions. The optic tectum also sends visual input up to the cerebral hemispheres. However, this ascending projection loses the topographic organization of the visual field. Consequently, in most vertebrates, spatial vision and orienting visual reflexes remain largely restricted to the brainstem. An exception are the mammals and a few other vertebrates in which a second visual pathway that goes directly to the cerebral hemispheres has gained predominance (figure 8.1, pathway 1).[2]

The Little Brain

Another key feature of vertebrates is the cerebellum, a prominence in the back of the brain that coordinates complex movements but also contains an important but understudied sensory and cognitive component. The cerebellum seems to have emerged in jawed vertebrates, possibly related to a more active lifestyle (cartilaginous fishes have relatively large cerebellums). In terrestrial vertebrates, there is a close coevolution between the cerebellum and the cerebral hemispheres where both increase in size together, amplifying their reciprocal connectivity as they grow. Notably, the cerebellum contains an astounding number of densely packed neurons, which in mammals add up to more than five times the neurons in the cerebral cortex.[3]

The Emergence of the Cerebral Hemispheres

Smell and the Brain

In the first vertebrates, olfaction was a critical element for exploratory behavior and orientation. Very early in their history, they acquired a pair of large olfactory bulbs and incipient cerebral hemispheres, as small evaginations in the front of the neural tube. The olfactory bulbs relayed chemosensory input to the emerging hemispheres, reaching a rudimentary olfactory cortex. This was closely connected to the hippocampus, another early cerebral hemisphere structure that participates in learning and memory. An associated chemosensory system corresponds to the accessory olfactory bulb, receiving pheromone inputs that drive social and sexual behavior, which are largely conveyed to the amygdala, a deep complex of nuclei in the lateral hemispheres that is involved in emotional behavior, and to the neuroendocrine hypothalamus. Thus, the cerebral hemispheres of early vertebrates may have received abundant chemosensory projections and may have been used to make different kinds of memories, spatial and emotional.

In agreement with this, the hemispheres of the jawless hagfishes are largely dominated by olfactory input, while other sensory modalities like vision have a more restricted projection. Nonetheless, another jawless fish, the lamprey, has a more active lifestyle than the hagfish and its cerebral hemispheres receive a strong visual input. It is thus likely that the progressive invasion of the telencephalon by non-olfactory inputs was an early event, or appeared more than once in vertebrate brain evolution.[4]

Ballooning Vesicles

Subsequently, the cerebral hemispheres underwent a tremendous expansion in most vertebrate lineages, reaching in some species up to billions of neurons. This was achieved by a highly conserved embryonic process where stem brain cells (the radial glia mentioned in chapter 5) underwent rapid proliferation and produced large quantities of neurons. As said in chapter 5, these cells begin dividing symmetrically, generating two equal progenitor cells in each round, which increases their numbers exponentially. In later developmental stages, progenitors undergo neuron-generating, asymmetric divisions where one of the daughter cells becomes an immature neuron and the other remains as a stem cell. Some genes, like ASPM and the already mentioned Pax6, control the transition from symmetric to asymmetric cell divisions, perhaps contributing to the expansion of the vertebrate brain including the mammalian cerebral cortex and the human brain (figure 8.3).

However, these mechanisms may be insufficient when it comes to producing very large brains like those of birds and mammals because developmental time is short and the internal ventricular zone, where neuronal progenitors divide, is limited in how much it can expand. These vertebrate groups have taken advantage of another strategy to increase neuronal

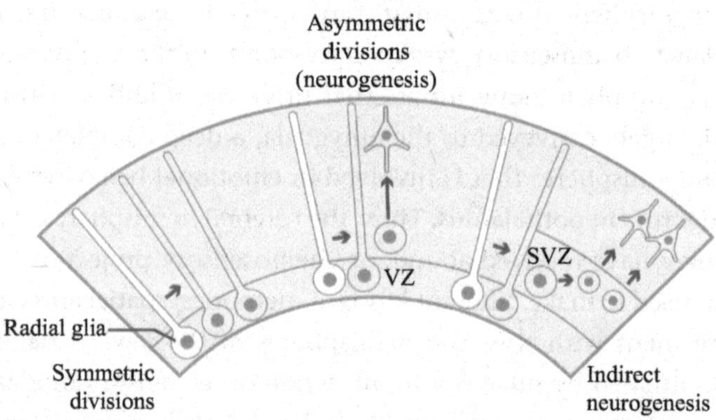

Figure 8.3
Neurogenesis in the cerebral hemispheres. For clarity, only a small region of the embryonic cerebral hemispheres is shown. Note that cell division takes place in the ventricular zone (VZ), located in the internal surface of the developing hemispheres. Neuronal progenitors (radial glia, shown in white) first divide symmetrically (left) generating two identical daughter cells (shaded in gray). In later stages, progenitors undergo asymmetrical divisions (middle) where one daughter cell differentiates as a neuron, detaches from the ventricular zone (VZ), and migrates to its adult position. In the cerebral hemispheres of birds and mammals, indirect neurogenesis (right) produces intermediate progenitors in the subventricular zone (SVZ) that keep dividing before they start migrating.

production, called indirect neurogenesis. In indirect neurogenesis, instead of producing an immature neuron, one of the daughter cells becomes an intermediate neural progenitor that keeps dividing but detaches from the internal surface of the cerebral hemisphere (the other daughter remains as a standard progenitor). In mammals and birds, intermediate progenitors accumulate to form a distinct layer (called the subventricular zone, SVZ, just above the ventricular zone) where they increase neuronal production without expanding the ventricular surface. It looks like indirect neurogenesis became amplified separately in mammals and birds, in a case of evolutionary convergence. Furthermore, primates go a step further from other mammals by generating a special kind of radial glia-like cells (called basal radial glia) in the SVZ, further augmenting neuronal numbers.[5]

A Long Journey

Especially in vertebrate brains, newborn neurons face another big challenge, which is to migrate from the inner ventricles to outer regions to coalesce in specific nuclei or laminae. In the cerebral hemispheres, neuronal migration follows a rather unexpected pattern, partly provided by a striking differentiation between the upper (dorsal) and lower (ventral) parts of these structures. The dorsal region is just a thin sheet in the early embryo and is termed the pallium, while the ventral region is more bulbous and is called the sub-pallium. In the adult, the pallium receives most sensory inputs and generates motor outputs, and gives rise to the cerebral cortex of mammals and other structures (see figure 8.4). On the other hand, the sub-pallium differentiates into the basal ganglia and other structures that are involved in coordinating motor and regulatory functions. Intriguingly, in embryogenesis the pallium gives rise mostly to excitatory neurons, while inhibitory neurons are largely generated in the sub-pallium.[6] Thus, the migration of excitatory neurons into the pallium is relatively simple: they just move outward to the surface, in a process called radial migration, aided by the radial glia. But since the pallium also contains inhibitory neurons, it must "import" the latter from subpallial regions, in a process termed tangential migration where cells move in a direction parallel to the hemisphere surface. This mysterious process has been documented in all vertebrates studied, perhaps resulting from early established regional differences in neuronal type production that were acquired by the common ancestor of the group.

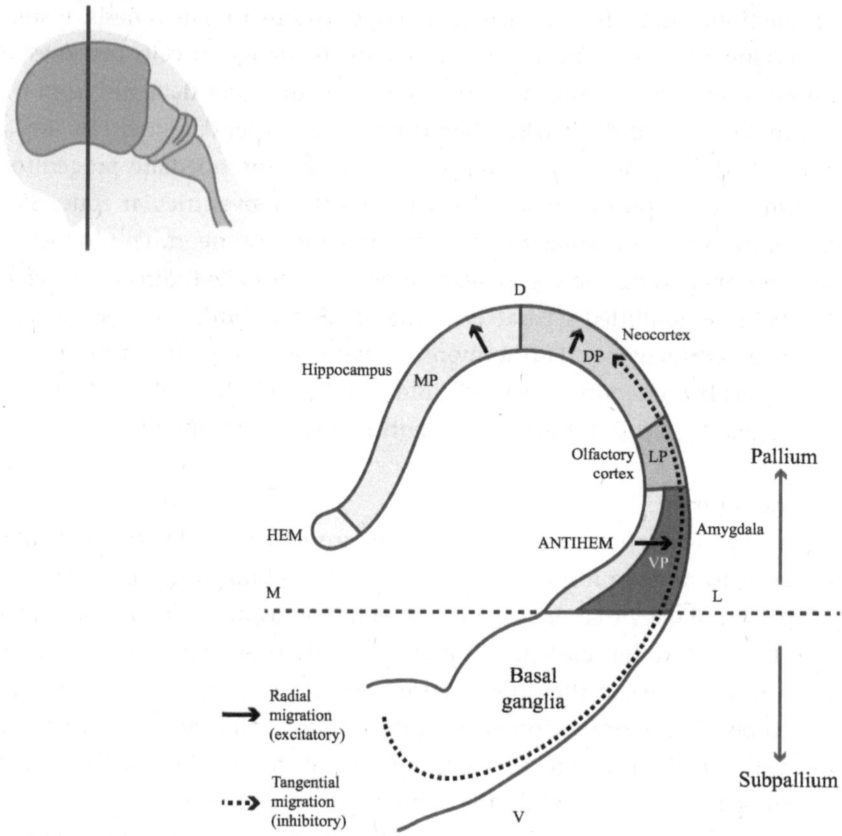

Figure 8.4
The main components of the embryonic cerebral hemispheres in mammals. This is a transverse section of one hemisphere's pallium, subdivided into medial (MP), dorsal (DP), lateral (LP), and ventral (VP) parts, which roughly give rise to the hippocampus, neocortex (mostly), olfactory cortex, and the amygdalar complex, respectively. The cortical hem and the antihem are embryonic morphogenetic centers that drive the differentiation of these components. The sub-pallium differentiates into the basal ganglia involved in motor functions, among other structures. Excitatory neurons are produced in the deep part of the pallium and migrate radially, outward (short arrows). Conversely, inhibitory neurons are produced in the subpallium and migrate tangentially to reach the pallium (long, segmented arrow). D: dorsal, L: lateral, M: medial, V: ventral.

Hemispheric Memories

Memento

I have reviewed the cellular development of the vertebrate brain, which is highly conserved across species. Now I will focus on specific components of the cerebral hemispheres that can be easily recognized across vertebrate groups and some of which participate in one of the most relevant functions of the vertebrate brain: memory formation. Of special interest in this context is a pallial region called the hippocampus (see figure 8.4), which differentiates near the dorsal midline of the embryonic hemisphere. Most sensory modalities directly or indirectly converge into it, especially olfaction and vision. Many studies have shown that the hippocampus participates in integrating multisensory information and in memory formation (particularly, spatial and episodic-like memory) in all vertebrates studied. I will further discuss hippocampal functions in the next chapter.[7]

Emotions and Routines

Another fundamental component of the cerebral hemispheres is the amygdala, a complex set of nuclei lying on the border between the pallium and the basal ganglia. This structure receives multisensory inputs from the brainstem and thalamus and has connections with hypothalamic areas involved in emotional or instinctive responses, evoking generalized physiological responses to frightening and pleasant stimuli, but it also contributes to motivation and drives purposeful behavior.

The third memory-related component of the cerebral hemispheres is the subpallial basal ganglia, involved in motor or procedural memories (like bike riding). In fishes and amphibians, the basal ganglia are largely connected with the thalamus and the brainstem, and they participate in the control of stereotyped behaviors like feeding, locomotion, posture, and eye movements. However, in birds and mammals, the basal ganglia increasingly engage in a reciprocal circuit with the pallium that allows more complex and versatile behaviors based on motor learning. The basal ganglia are strongly modulated by the neurotransmitter dopamine (see chapter 5), secreted by the axons of neurons that have their cell bodies in the brainstem. Furthermore, in a region called the ventral striatum, dopamine participates in motivational mechanisms, driving behavior toward or away from emotionally relevant stimuli, that have been fundamental for vertebrate success and evolution.[8]

A Neuroanatomical Puzzle

Diverging Shapes

Up to here, I have highlighted the conserved components of the vertebrate cerebral hemispheres and their general functions in behavior and memory. I will now come to a more controversial issue, which is the evolution of some specific areas of the pallium that receive sensory afferences and participate in behavioral organization. These regions have enlarged enormously in different species, yielding an intricate rainbow of morphologies. Neuroanatomists have struggled to compare the different vertebrate brains to find homologies among them and to unveil their evolutionary origins, engaging in highly intense and emotional controversies for more than a century.[5,9]

Interestingly, the simplest cerebral organization is found in the amphibians and the lobe-finned fishes (like the lungfishes), resembling the balloon-like embryonic condition of other vertebrates (figure 8.4). On the other hand, ray-finned and cartilaginous fishes have undergone significant increases in brain size, strongly distorting the ancestral pattern and making a nightmare for species comparisons. Despite usually lacking elaborate or grasping appendages like insects or mollusks, fishes display sophisticated cognitive skills and evolved brains more complex than those of most invertebrates (with the possible exception of cephalopods), largely associated with exploratory behavior. Some fishes can even consider who is observing them when performing an action: cleaner fish, which remove dead skin and other products from the skin of other fishes, improve their cleaning behavior if there is a third fish nearby so that the observer will allow it to clean her later. Moreover, some fish can recognize themselves in a mirror, a trait that was initially believed to be privative of large-brained mammals and birds and has been considered by some as a test for consciousness in animals.[10]

Brains on Land: A Thorny Dispute

In the amniote lineage, brain structure dramatically diverged after their separation into two lineages (sauropsids and mammals). On the one hand, reptiles developed an internal neuronal mass called the dorsal ventricular ridge (DVR), which receives most of the auditory and visual sensory inputs ascending from the brainstem (in the case of the visual system, pathway 2 in figures 8.1 and 8.5). More dorsally, reptile brains display a very small cortex containing the hippocampus, the olfactory cortex, and a small dorsal cortex interposed between them, which corresponds to the embryonic dorsal pallium in figure 8.4. The large brains of birds hypertrophy into a complex group of nuclei but retain the basic organization of the reptilian hemispheres. On the other hand, mammals developed the neocortex (or cerebral cortex) in the dorsal aspect of the pallium; it is a six-layered structure that receives most of the sensory inputs and sends downstream motor commands.

Despite this great morphological divergence, there are notable similarities in connectivity between the sauropsid and mammalian brains. Both

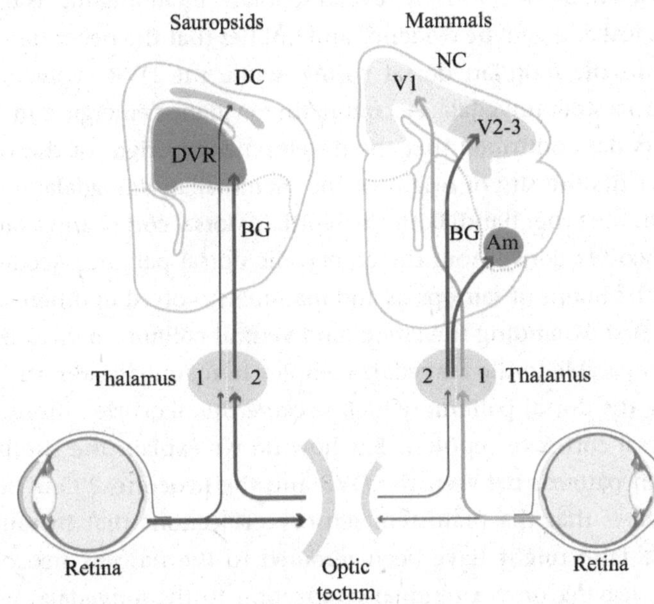

Figure 8.5
The visual pathways to the brain in sauropsids and mammals. The direct pathway (1) reaches directly the thalamus and ends in the dorsal cortex of reptiles (DC) and in the primary visual area (V1) of the mammalian cerebral cortex. The brainstem-relayed pathway (2) ends in the DVR of sauropsids, while in mammals this pathway reaches the secondary and tertiary visual areas (V2-3), and the mammalian amygdala (AM). The brainstem-relayed pathway (2) is more prominent in sauropsids and nonamniotes, while the direct pathway (1) is more developed in mammals.

the sauropsid DVR and a part of the mammalian neocortex receive major visual and auditory inputs that are relayed in the brainstem (figure 8.5). Conversely, visual and somatosensory inputs that do not relay in the brainstem but ascend directly to the brain reach the reptilian dorsal cortex (or its avian equivalent) and the mammalian neocortex (see figures 8.1 and 8.5; visual pathway 1). Based on this evidence, some authors proposed that part of the DVR is homologous to the sensory areas of the mammalian neocortex that receive the brainstem-relayed projections (exemplified as areas V2-3 in figures 8.1 and 8.5, which receive input from pathway 2). As a consequence, the mammalian neocortex would have had two separate origins, one deriving from the reptilian DVR and the other deriving from the reptilian dorsal cortex.[11]

A different perspective, first proposed in the early twentieth-century and promoted again in the 1990s by several scholars, including me, is based on embryological and genetic evidence and implies that the neocortex derives largely from the reptilian dorsal cortex, while the DVR evolved mostly from the ancestral amygdala. Accumulating evidence emerging in the last thirty years has confirmed that the developmental origin of the reptilian DVR better fits the site of origin of the mammalian amygdala and other regions. On the other hand, both the reptilian dorsal cortex and a large part of the neocortex derive from the embryonic dorsal pallium. According to this view, the brains of sauropsids and mammals evolved in different directions, the first expanding the lateral and ventral pallium to yield the DVR (partly comparable to the amygdala complex in mammals) and the second expanding the dorsal pallium, which became the neocortex (homologous to the dorsal cortex of reptiles). But how do we explain the similarity of innervation patterns between the DVR and the neocortex? One plausible explanation is that the brainstem sensory projections that in sauropsids end in the DVR might have been rerouted to the nascent neocortex in mammals, leaving only a residual connection to the amygdala, which is involved in reflex behaviors.[12]

A Way Out?

Attempting to account for these transformations, Juan Montiel and I have proposed a hypothesis based on the modulation of different morphogenetic centers or "hotspots" during the development of the cerebral hemispheres, comparable to the gradients that specify longitudinal and dorso-ventral patterns in the rest of the brain (see figures 8.4 and 8.6). Studies have

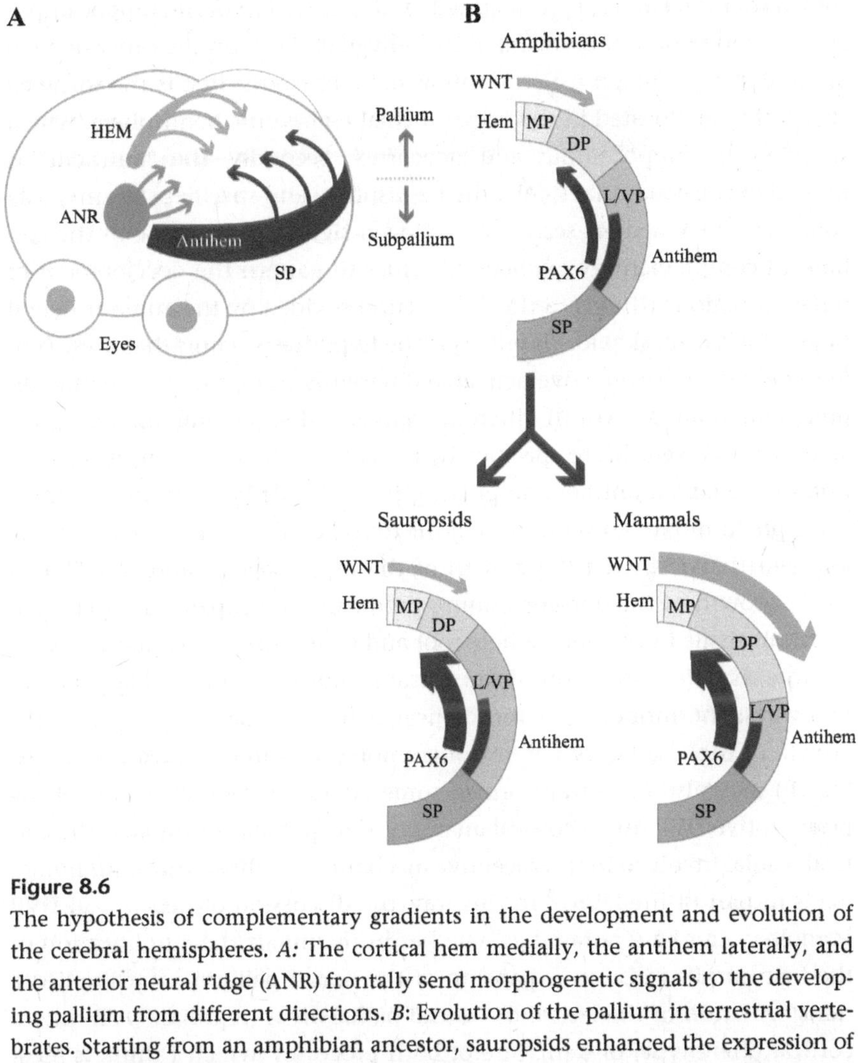

Figure 8.6
The hypothesis of complementary gradients in the development and evolution of the cerebral hemispheres. *A:* The cortical hem medially, the antihem laterally, and the anterior neural ridge (ANR) frontally send morphogenetic signals to the developing pallium from different directions. *B*: Evolution of the pallium in terrestrial vertebrates. Starting from an amphibian ancestor, sauropsids enhanced the expression of signals deriving from the antihem (Pax6), while mammals enhanced signals deriving from the cortical hem (Wnt), possibly also augmenting antihem signaling. This led to the development of the DVR in sauropsids in the ventral pallium (VP) and of the cerebral cortex of mammals in the dorsal (and parts of the lateral) pallium. MP, DP, LP, VP, medial, dorsal, lateral, and ventral pallium, respectively. SP: sub-pallium.

shown that the embryological growth and maturation of the hippocampus and neocortex on one side and of the amygdala/DVR on the other depend on at least two antagonistic morphogenetic hotspots: first is the so-called "cortical hem" located in the dorsal-medial embryonic hemisphere, which specifies the hippocampus and neocortex. Secondly, the "antihem" is located in the equatorial side of the hemisphere and specifies the amygdala complex. These centers secrete morphogen signals that diffuse in the pallium in complementary gradients in order to pattern the development of different regions (there is a third gradient, provided by the anterior neural ridge which I will skip for simplicity). The hypothesis claims that these centers could have become overactivated differently in reptiles and mammals, producing brain growth in different regions and specifying different gross anatomies as well. More specifically, in reptiles, there was an overactivation of the lateral antihem amplifying the DVR, while in mammals there was a predominant amplification of the dorsal cortical hem (but also of the antihem that restricted the growth of the hippocampus and contributed to the growth of the dorsal pallium), producing the expansion of laminar (cortical) tissue in the hemisphere roof and giving rise to the neocortex.[13]

Note that the expansion of the dorsal pallium could have happened in two ways: the influence of the cortical hem may have (i) increased the growth of the small dorsal cortex of amniotes into an extended neocortex, but (ii) may also have transformed some lateral or ventral regions of the presumptive DVR into a cortical anatomy. One possible example is the cortical insula, involved in interoceptive mechanisms, whose sauropsid homolog is in part of the DVR.[14] In this way, the discussion of neocortical-DVR homology can be rephrased as whether in mammals the amplification of the hem implied just an expansion of the dorsal pallium, or it implied a transformation of the lateral or ventral pallial areas (reptilian DVR) into a cortical phenotype, or a mixture of both processes (which I think is most likely). However, this question becomes secondary to the more fundamental issue of the identification of the developmental mechanisms driving vertebrate brain evolution and diversification.

Is Bigger Better?

Brain and Body

What are big brains—especially in our own species—good for? Brain size has been long associated with increasing intelligence and has been viewed as key for our evolutionary success. However, comparing intelligence across species can be a very complicated issue, since each species faces a different set of challenges to survive, and scientists usually focus on human-like capacities to assess cognitive skills. Nonetheless, some features like learning capacity, problem-solving abilities, memory, or social complexity are many times used as indexes of intelligence across species, especially when studying mammals or birds.

At odds with the above assumption, the main determinant of brain size across species is not intelligence (however you may define it) but overall body size so that larger species tend to have larger brains than smaller species. A likely explanation is simply that the brain follows the general rules of embryonic development and if the body grows larger, the brain grows larger as well. Nonetheless, smaller animals tend to have larger brains relative to their body size than larger animals (similarly to children having larger brains–or heads–for their bodies than adults) so that the body expands in size faster than the brain in a so-called allometric relation where both variables increase at different rates.[15]

Notably, the brain-body relation is not the same for every vertebrate group, with homeothermic birds and mammals having larger brains for a given body size than other vertebrates. Homeothermy may provide the energy needed to grow such an expensive organ (neural tissue consumes about ten times more energy than the rest of the body). Still, there are many further limitations including the amount of maternal investment needed for growing a larger brain, or the tradeoff with other expensive tissues like the gut or the immune system, which might become reduced as the brain increases in relative size. On the other hand, a larger brain must also be of selective advantage; otherwise, the energy required to build it might well be invested in other processes. It is commonly assumed that intelligence will always be a favorable trait, but most animals do very well without sophisticated cognitive capacities.[16]

Moreover, within each vertebrate group, there are important differences in relative brain size. For instance, primate brains nearly double the volume of brains from other size-comparable mammals. The encephalization quotient (EQ) refers to the difference between the actual brain size of a species and its expected brain size given its body volume if it followed the allometric curve for a certain animal group. High EQs indicate larger than expected brains and are consistently found in primates (especially humans), elephants, dolphins, and in some birds like parrots and crows, all of which are widely considered to be "smart" animals, with many of them showing strong parental care of their young.[15]

Size versus Numbers

The previously mentioned studies have largely worked on the assumption that larger brains have more neurons, and that more neurons imply more processing capacity. However, new studies have reported a high variability in the number of neurons per unit volume, where species with comparable brain sizes may significantly differ in total neuronal numbers. For example, birds tend to have more neurons in their brains than mammals with brains of similar size. Moreover, allegedly clever animals like primates, parrots, and crows have higher neuron numbers per unit of brain volume than related species with similar brain sizes. Some authors claim that instead of EQ, absolute neuronal numbers are the best proxy for animal intelligence. However, the number of neurons in human brains is exceeded by the number of neurons in large cetaceans like the pilot whale, and possibly by the even larger baleen whales. Besides, the cerebellum contains about five times the number of neurons of the cerebral cortex, yet the evidence shows that cognitive and conscious mechanisms rely more heavily on the latter.[17]

Perhaps the missing link in the relation between the brain and cognition is the capacity for brain remodeling. In my view, encephalization and neuronal numbers may be good proxies for cognitive abilities across species, but these perspectives tend to ignore the role of brain network reorganization and plasticity. As mentioned, to increase brain power, there must be evolutionary selection for enhancing cognitive capacity, and brain circuits may remodel to achieve this result by decreasing some connections and increasing others. Selection for new circuitry, enhanced neuronal plasticity, and a more robust connectivity focuses on the properties of neurons, synapses, and network organization, which may be a less evident but perhaps equally important driver of cognitive evolution than increasing neuron production.[18]

Whole or Mosaic?

Finally, an additional issue is that the brain is not a homogeneous network of neurons, but rather it consists of many different parts, subdivided into other components and so on, that are involved in different functions. An important line of research has established that these components may expand at distinct rates in evolution, but they mostly follow a concerted pattern (there is allometric growth of the distinct parts). Another group of researchers claims that there is a significant degree of independent variation in the growth of different parts, which reflects sensory and cognitive specializations. There are examples favoring both views, and probably the answer lies somewhere in the middle, with a general pattern of brain growth but a residual variability that may account for the behavioral and cognitive differences that are observed.[19]

Perspective

The vertebrate brain evolved in correspondence with the differentiation of sensory systems, leading to the differentiation of the paired cerebral hemispheres, the brainstem and the cerebellum. The regional development of the vertebrate brain is driven by a conserved genetic pattern, probably inherited from early bilaterians. This pattern determines early differentiation gradients, establishing morphogenetic "hotspots" that specify the main brain components, which are eventually parcellated into different segments. The expanding cerebral hemispheres are also patterned by a shared genetic gradient that specifies dorsal and ventral components (pallium and subpallium, respectively). In addition, the mechanisms of brain cell production are also conserved, depending on the early differentiation of proliferating radial glia that works as a stem cell for neurons and glia in the developing brain. Large-brained species like birds and mammals have evolved so-called indirect neurogenesis to increase neuronal production in a small proliferative compartment.

Three main components of the hemispheres are highly conserved across vertebrates, all involved in memory mechanisms: the hippocampus (spatial memory), the amygdala (emotional memory), and the basal ganglia (motor memory). The cerebral hemisphere components that expand the most are those involved in processing sensory input and organizing behavior; they underwent a notorious morphological divergence in the different vertebrate classes. Particularly, the brains of reptiles and birds (sauropsids) on one hand and those of mammals on the other evolved quite different morphologies, which has led to long-standing disputes about their homology. While the former underwent an amplification of nuclei related to the amygdala, the latter largely expanded a small region adjacent to the hippocampus, giving rise to the mammalian neocortex. We have proposed a developmental mechanism based on the differential modulation of morphogenetic gradients that might account for brain diversification in terrestrial vertebrates and might help solve the homology controversies.

Finally, while body size is one of the major determinants of brain size and neuron numbers across species, there can be selective conditions that favor the production of larger brains and more neurons than expected for a given body size, provided there is a sufficient energy supply to grow and maintain these extra neurons. Furthermore, selection may also favor genetic variants increasing plasticity and refinements of connectivity that enhance neural processing capacity. In the next chapter, we will delve into the origin of mammals and their cerebral cortex, a character that is as definitory of this group as the milk glands are.

9 Mesozoic Eden

After the great Permian extinction, a surviving branch of small therapsids, the cynodonts, gave rise to the earliest mammals, which coexisted with the rising dinosaurs. In this chapter, I will briefly narrate the history of early mammals during the Mesozoic, the formation of their body plan and especially their brains, including the origin of the laminar neocortex, a quite unique character among vertebrates. Although this structure has long been considered a pinnacle of brain evolution, it does not represent by itself a higher level of neuronal complexity than exhibited by other animals (without a neocortex, birds rival mammals in cognitive performance). In this chapter, I will emphasize that the origin of the neocortex resulted from specific contingencies to which early mammals had to adapt, more than as an innovation to increase cognitive capacity.

Living with Dinosaurs

Flurry Beginnings

Cynodonts, the Mesozoic ancestors of mammals, have been depicted as small, hairy, and possibly warm-blooded animals. They also had sensitive whiskers around the mouth as evidenced by a mammal-like expansion of the trigeminal nerve. Moreover, the vertebral spine became divided into four main regions (cervical, thoracic, lumbar, and sacral, which was regulated by Hox genes), and the ribs formed a ribcage associated with the development of a muscular diaphragm, providing inspiration force during respiration. Modern mammals, also called Mesozoic mammals, emerged in the late Jurassic period, being mostly about the size of a house mouse or smaller, but there were also badger-sized species. Mesozoic mammals underwent a notable radiation, diversifying into multiple branches. One of these gave rise to the modern egg-laying monotremes (the echidnas and the platypus). Another branch included the successful multituberculates, rodent-like animals that shortly survived the Cretaceous mass extinction that killed the dinosaurs.[1]

Another group of early Cretaceous mammals, the therians, became viviparous and soon divided into the marsupials (metatherians) and the placental mammals (eutherians). Placental mammals branched into four main clades during the Mesozoic, all of which thrived and diversified after the impact. The most basal of these groups are represented by the African aardvarks and elephants on one side (Afrotheria), and by the South American sloths, anteaters and armadillos on the other. Another branch split into the present-day insectivores, ungulates, and carnivores on one hand, and rodents and kin, as well as primates, on the other (see figure 9.1).

Soft Tissue Innovations

Most of the fossils mentioned earlier are represented by highly fragmentary remains of minute animals, largely based on details of the teeth and jaws, which are more likely to resist disruption. However, these characters do not reflect the whole set of changes that took place along the cynodont-mammalian lineage, including soft-bodied and physiological innovations that leave no fossil traces. Early cynodonts already foreshadowed some key mammalian characters, but there are many additional features that may have been evolving in this emerging group. For example, milk production is one of the key characters shared by all living mammals and may have provided the nutrients required to grow large brains. Notably, there is one homeobox gene called *Msx2* that is involved in the development of hairy skin, the mammary glands, the cerebellum, and craniofacial morphogenesis. Mutations in this and other genes probably became quite relevant for the acquisition of diverse features that were appearing in the early mammals.[2]

Mammals also underwent major improvements in body physiology that had consequences for their behavior and brain evolution. Together with the diaphragm, they acquired a four-chambered heart, which provided more efficient blood oxygenation. In relation to this, the oral cavity developed an internal palate separating the nostrils, which are involved in

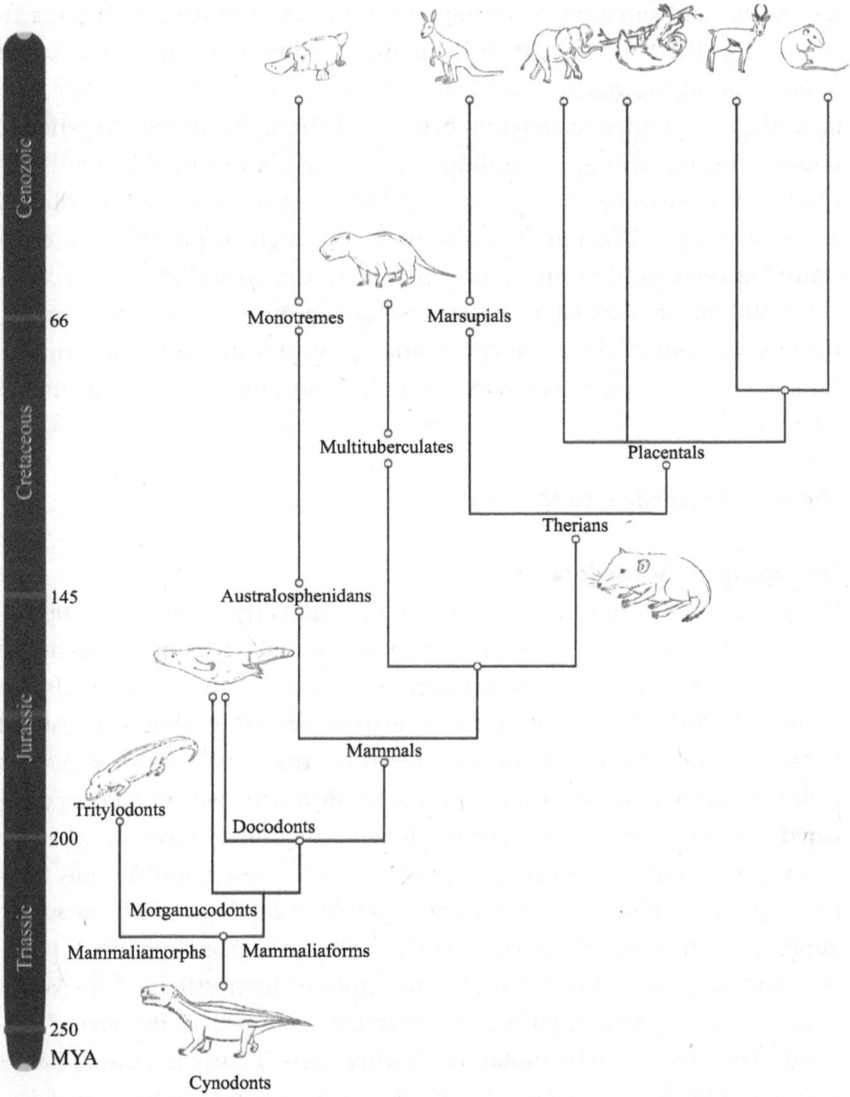

Figure 9.1
The phylogeny of mammals.

respiration and smelling, from the mouth involved in feeding and mastication. Together with this, the tongue increased mobility and contributed to lactation in the young and to mastication in the adults (the palate may have also contributed to lactation behavior, helping the tongue to generate pressure during feeding). In addition, the throat became muscular and cartilaginous to enhance suckling capacity. These innovations also contributed to enhanced vocal communication, including high frequency calls, especially between mother and pups, driven by the social hormone oxytocin. (Millions of years later, speech emerged in our species.) Furthermore, the enlargement of the nasal cavity and the expansion of the labyrinthic turbinal bones helped to conserve body heat and moisture and facilitated olfaction.

The World According to Mammals

The Nocturnal Bottleneck

Perhaps hiding from dinosaurs, cynodonts and early mammals were predominantly nocturnal and semi-burrowing animals, which is reflected in the eye anatomy and visual perception mechanisms of living mammals. For example, mammals (except primates) usually have large quantities of rod photoreceptor cells, which are specialized for nocturnal vision, while the cones, involved in daylight and color vision, are much lower in number compared to other vertebrates. Likewise, while non-mammals have four or more types of opsin molecular pigments for color vision, most mammals only have two pigments ("blue" and "red"). Later, primates developed trichromacy by duplicating the "red" gene, yielding an additional "green" pigment, possibly as an adaptation to daily fruit consumption. Interestingly, New World primates have an additional "yellow" pigment, from which the "green" pigment may have arisen by mutation. Further signs of early nocturnal habits are present in the enhancement of binocularity, which increases visual acuity and depth perception (as in the frontal eyes of owls). There are other visual changes, as the ON-OFF visual systems apparently segregated from other visual features like color (see chapter 7), generating distinct channels for ON-OFF and color vision. In association to these adaptations, mammals also amplified the direct visual pathway to the brain (bypassing the brainstem relay, see the previous chapter), which became their main visual processing system.

Jaws and Ears

Compensating for their relatively poor vision, early mammals made audition, smell, and touch their principal senses. Like I said, the jaws are among the more useful diagnostic characters for mammal paleontologists, as they underwent notable changes that have been preserved in the fossil record. The mandible of Mesozoic mammals and their kin reorganized with a specialized dentition that facilitated the grinding of food and increased the absorption of nutrients. Furthermore, the jaw itself was profoundly modified, enlarging a teeth-containing bone (the dentary) that gradually established a new articulation with the cranium, providing stronger masticatory force and lateral mobility. In this process, the ancient reptilian and synapsid jaw articulation, consisting of two small bones (the quadrate and the articular) became detached from the jaw and was incorporated as two tiny ossicles (the incus and the malleus, respectively) that joined the tympanic bone, making up the chain of ossicles of the mammalian middle ear. The middle ear bones increased hearing sensitivity, especially for high frequencies, which much later enabled our ancestors to acquire speech and music. The jaw articulation is considered by paleontologists to be the diagnostic character to define a true mammal, something that is useful for classification but obscures the whole set of changes that were taking place. Furthermore, it seems that the transition from a therapsid-like articulation to a mammalian one was a long process with many transitional and diverging forms, that possibly took place separately in more than one lineage of early mammals (a Mesozoic gliding species displayed five auditory bones instead of the typical three).[3]

Smell and Touch

In addition, smell became greatly enhanced in the mammalian ancestors. Early mammals were able to breathe at high frequencies, enabling them to sniff around and capture the surrounding smells. Olfaction works as a major reference sense in mammals, contributing to spatial navigation by recognizing places and driving the animal toward food sources, mates, or even just home. Concomitantly, early mammals enlarged their olfactory bulbs and cerebral hemispheres, associated with the expansion of the nascent neocortex.[4]

Furthermore, early mammals acquired a soft, cartilaginous, and movable nose and muscular lips covered with sensitive whiskers, probably connected to a rudimentary somatosensory cortex. Thus, hearing, sniffing, and whisking, more than eyesight, became the main drivers of exploratory behavior in the first mammals. Concomitant with increasing somatosensory perception, early mammals also evolved a corticospinal tract from the cerebral cortex to the motor centers in the spinal cord, providing fine manual skills and enabling them to grasp small objects and bring them to the mouth, a behavior that had important consequences for primate and human evolution.[5]

Multimodal Memories

Replaying Experiences

After briefly reviewing the bodily and sensory transformations of early mammals, we will now enter their neural characters that are of most interest to this book. Next I will propose a hypothesis of neocortical origins that is strongly based on the intimate relationship of olfaction with the hippocampus (remember that the neocortex is proposed to emerge largely through amplification of the hippocampus-related cortical "hem" in the early embryo). Therefore, in order to understand the emergence of the neocortex itself, it may be important to discuss first the role of the mammalian hippocampus in memory and spatial cognition and how its functions may have contributed to the expansion of the former. Below I will review findings on visual learning, but it must be remembered that the hippocampus of most mammals receives a powerful olfactory input that intersects with visual inputs and is key for spatial learning and orientation.

The hippocampus has long been studied in humans and mammals for its relevance to spatial and episodic-like memory (remembering what, where, and when something happened; see chapters 6 and 8). This brain region is heavily and reciprocally connected with the olfactory cortex and with sensory areas of the neocortex, generating a multimodal integration of these inputs to make enduring memories of space. Basically, the sensory input coming to the hippocampus from cortical areas becomes amplified and is later returned to the cortices to consolidate these experiences as solid memories.

A crucial finding in the study of hippocampal memory was the discovery of "place cells," which are neurons that activate when an animal is in a specific place when running in a lab-made labyrinth. The sequential activation of place cells mimics the animal's trajectory in the labyrinth. Notably, at rest or during sleep, high-frequency neuronal oscillations called sharp-wave ripples activate the different place cells in sequence, replaying the animal's experience when learning the task. Furthermore, a notable finding is that when having to choose two distinct routes to find a similar goal, the replayed sequence at rest predicts which route the animal will later choose, as if it had the selected pathway "in mind" before executing it. A more recent report shows that the animal can replay the sequence while standing still if trained to do so, perhaps indicating that they can "mentally access" the places they previously visited.

Another important discovery was the "grid cells" in a region closely related with the hippocampus called the entorhinal cortex. Instead of activating at specific spots, grid cells fire in multiple places of the labyrinth following an extremely regular pattern across space (hence their name). Together, place cells and grid cells provide a kind of Cartesian reference system an animal can use to locate itself as it explores the environment. Finally, there are head-direction cells that signal the direction of movement of a mouse while running through the labyrinth. Notably, during REM sleep, eye movements align with the activity of head-direction cells, which is reminiscent of the coordination of head and eye movements when exploring the surroundings or escaping a predator, suggesting that mice might have dream-like experiences as they sleep.

In addition, the hippocampus participates in the nonspatial memory of sequences of events during complex behaviors, concatenating distinct events into a definite historical sequence. In other words, the hippocampus contributes to bind the successive events that the animal experiences during exploration using different sensory modalities (like vision, olfaction, and touch), aligning them in a sequence that provides a reconstruction of the events leading up to the present, as well as the immediate future. In our human experience, these mechanisms give rise to subjective time and the notion of an external world around us.[6]

The Emergence of the Neocortex

Considering the evidence in the previous section, we have postulated that in early mammals, the hippocampus was key for establishing olfactory, visual, and tactile memories of the space surrounding them, including places like the animal's den and routes to follow. While the olfactory cortex is directly connected with the entorhinal cortex and hippocampus, the neocortex may have emerged as an expansion of the dorsal cortex of early amniotes, contributing visual and somatosensory (whisker-driven) inputs to the hippocampus that were used for exploration and place recognition. In this process, the entorhinal cortex may have served as an interface between the hippocampus and the growing neocortex. Subsequently, the emerging neocortex also received invading auditory and visual sensory afferents from the brainstem that as said are directed into the DVR in reptiles (see the previous chapter) and in mammals may have deviated their trajectory toward the cerebral cortex. In addition, the neocortex gave rise to motor areas that send descending projections to execute voluntary behavior. Thus, the neocortex became the repository of multimodal events that were long-term consolidated in it by the hippocampus, generating episodic-like memories that became integrated with motor planning circuits.[7]

Navigation

Some authors have proposed that the formation of complex episodic-like memories in hippocampal-neocortical circuits may have transformed instrumental or operant conditioning, based on the association between specific behaviors and rewards, into goal-directed behavior by the elaboration of cognitive maps. Furthermore, mammals are considered to use these representations to plan future behavior and anticipate events.

Adding one more step, in humans the hippocampus participates in the generation of the so-called semantic memory, which is the long-term memory of world facts of any kind that are common knowledge to all of us (like knowing who the president of your country is). This semantic network allows us to organize our knowledge of the world and navigate through it. However, recall that the notion of a cognitive map or representation may imply that there is someone observing the map or the representation, but there is no inner spectator in the brain. Instead, perhaps what the brain does is to bridge separate experiences and to establish links between them that are independent of the animal's position (i.e., allocentric) to drive behavior in different contexts, be it in physical space-time or in the social domain of shared knowledge as is the case in humans. Moreover, the point where a "representation" of the world appears in animal evolution is rather blurry, not the least because there is little agreement on the details of what a representation is. We have seen that invertebrates and fishes, not to mention birds, may display goal-directed behaviors and elaborate cognitive skills, which to some may account as representations of the external world.[8]

Assembling the Neocortex

Columns and Layers

The next stage involved the expansion of the neocortex to cover the whole brain. Emerging from a small structure like the reptilian dorsal cortex and neighboring cell populations, the neocortex grew in two different domains: increasing in depth by adding cellular layers in the vertical dimension and increasing in surface as a large sheet that ended up covering most of the brain's surface.

We will start with cortical lamination. As opposed to the single- or double-cell layered cortex of reptiles, the neocortex is made up of five horizontal cellular layers plus a superficial cell-free marginal zone (see figure 9.2A). In addition, neocortical neurons are arrayed vertically in a series of parallel columns, associated to the radial organization of glial progenitors (for simplicity we will skip the tangential migration of inhibitory neurons).

Inside-Out Migration

In mammalian origins, successive neuronal layers were added to the ancestral one- or two-layered amniote cortex through sequential waves of radial migration from the deep ventricular zone. However, the way neuronal layers were added to the neocortex is highly unusual among vertebrates, and in my view reflects the evolutionary history of this organ. In most brain regions and in the reptilian brain, the early-produced neurons typically migrate longer distances than late-produced neurons, which migrate shorter distances and usually end up locating beneath the older neurons. This pattern is known as the outside-in gradient of neurogenesis, where the outside or more superficially located neurons are those that were born earlier and the deeper cells are produced in later stages. In the neocortex, the early-born neurons migrate up to the external surface of the cerebral hemisphere and stop just before penetrating into the superficial marginal zone. However, as opposed to other brain regions, later-born neurons migrate further outward than the early neurons, traversing the layers of previously migrated neurons to locate more superficial to them (see the 1-to-5 neurogenetic sequence in figure 9.2A). At the end, the latest-produced neuronal layers end up in more superficial (outward) positions, while those that were produced earlier end up in deeper positions. This inverted pattern is termed the inside-out gradient of neurogenesis and is characteristic of the mammalian neocortex.

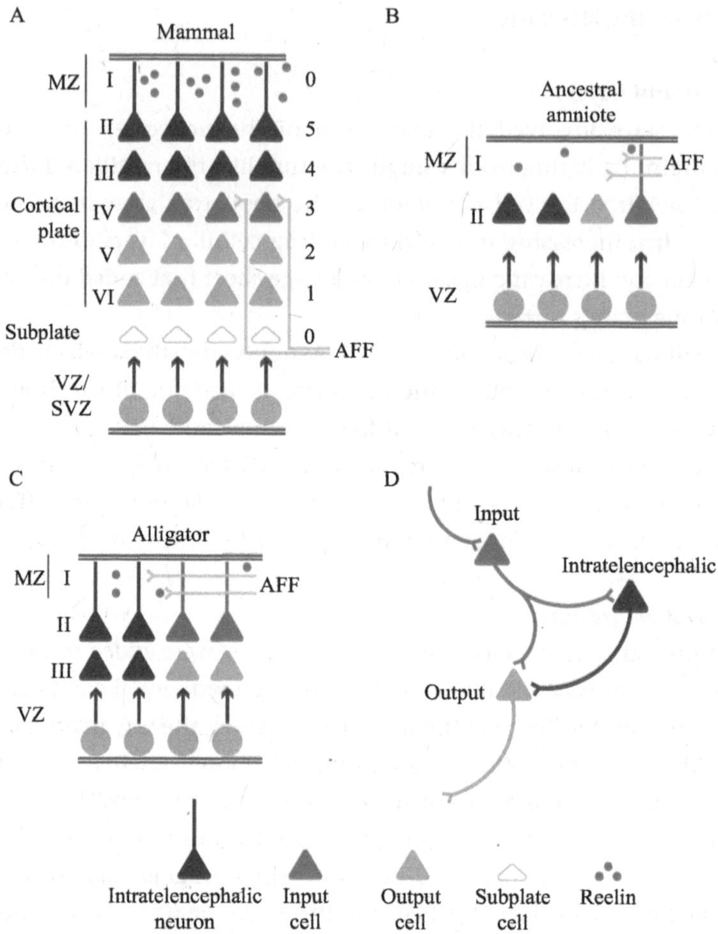

Figure 9.2
The cellular organization of the mammalian neocortex. For clarity, only excitatory neurons (triangles) are shown. *A*: Anatomical organization of the neocortex in six horizontal or tangential layers (I-VI) and vertical or radial columns, which derive from progenitor cells (gray circles) in the internal ventricular and subventricular zones (VZ/SVZ). Note that layer I (also called the Marginal Zone, MZ), is largely cell free in the adult and contains large amounts of reelin (gray spots), secreted in the embryo by the Cajal-Retzius cells (not shown). Afferents (AFF) mostly enter the neocortex from the underlying white matter, where the embryonic Subplate is located (white triangles). During development, the earliest formed structures (labeled as "0" to the right) are the MZ and the Subplate. In later stages, the neurons that will form the mature cerebral cortex (the Cortical plate; layers II-VI) migrate from the VZ/SVZ and interpose between the Subplate and the MZ, separating these two early layers. In

Notably, a protein called reelin, which is secreted by very early migrating neurons in the marginal zone, was found to have a critical role in the generation of the inside-out gradient, working as a stop signal for migrating cells. Normally, when the first cohort of migrating neurons touches the reelin-rich marginal zone, they stop there. However, neurons produced later move across the layer of previously migrated neurons, stopping only once to make contact with the reelin in the marginal zone, and become located above the older neurons. This process is repeated in the migration of all five cellular layers, yielding the inside-out gradient. Reelin expression has greatly increased in mammalian evolution, making it an effective stop signal during neocortical development, besides other developmental functions.[9]

Why did the neocortex acquire this odd arrangement? We have proposed a hypothesis based on the organization of axon inputs to the so-called "primitive cortices" like the reptilian cortex and the mammalian hippocampus and olfactory cortex. Neuronal afferents to these regions run parallel to the brain surface in the uppermost marginal zone, contacting the ascending apical dendrites of the excitatory cells (see figure 9.2B). In this way, one incoming axon contacts many dendrites along its path, stimulating a stripe of the cortex instead of a spot of it. We suggested that the acquisition of the mammalian inside-out gradient originated as, together with increasing neuronal production, the late-born neurons differentiated as input-receiving cells, and crossed the layers of early-produced neurons (that differentiated as output neurons), gaining access to the marginal zone where the afferent axons were located. A similar but more rudimentary arrangement may be

---

Figure 9.2 (continued)
this process, the deep output layers VI and V are the first to arrive (light gray; 1, 2), followed by the input-recipient layer IV (mid-gray; 3) and finally the intratelencephalic neurons of layers III and II (dark gray; 4, 5) that establish associative connections. This sequence from 1 to 5 constitutes the inside-out neurogenetic gradient that is characteristic of the neocortex. In many species, cells from the MZ and the subplate are largely eliminated around birth, leaving only the cortical plate neurons from layers II-VI in adulthood. *B* and *C*: Cortical organization of a hypothetical ancestral amniote and an alligator, respectively, showing one or two neuronal layers, little reelin expression and a predominantly superficial distribution of the afferent axons (AFF). Note that in the alligator, the input cells appear to be located more superficially, close to the afferents (Briscoe et al., 2018). *D*: The canonical intracortical circuit, possibly shared by all amniotes, including input cells, intratelencephalic neurons, and output cells.

seen in the alligator dorsal cortex, where thalamic-receiving neurons seem to locate above the output neurons, forming contacts with the superficial axons (see figure 9.2C). If this is correct, it remains to be assessed whether these input neurons are born earlier or later than the output neurons.[10]

However, in the mature neocortex, the association or "intratelencephalic" neurons are those that are produced later in development, locating more superficially than the input-receiving layers. Perhaps the reason for this had to do with a subsequent major change in neocortical development: the displacement of afferents from the superficial marginal zone to the white matter below the cortex, as I will explain next.

From Tangential to Radial Entry

As mentioned, there is another unusual characteristic of the neocortex: thalamic afferents to the reptilian cortex run tangentially in the most superficial cortical layer. However, in the mammalian neocortex, afferents predominantly enter from the underlying white matter, penetrating vertically upward (see figure 9.2). Our hypothesis is that this change of afferent orientation was related to the tangential expansion of the neocortex, after the inside-out gradient had established. As cortical surface increased (most likely by increasing the number of cortical columns), incoming axons had to run over longer and longer distances over the expanding cortical sheet in order to reach their targets. Furthermore, they could also make contact with too many apical dendrites along their path, losing their synaptic specificity. An alternative for these axons was to run a much shorter distance from below the developing cortex to finally ascend radially, or vertically, into it. This change of route was possible by the development of a new cellular layer, the subplate, that is located in the deepest part of the embryonic neocortex but in many species disappears after birth. The subplate contributes to direct the subcortical thalamic axons to their cortical targets from below the developing neocortex, and participates in the development of the first cortical circuits. In addition, the subplate may have provided guidance for the midbrain-ascending sensory pathways (that in reptiles end up in the DVR) into the nascent mammalian neocortex (see chapter 8).[11] Therefore, once the afferents started entering from below the developing cortex, there was no pressure for the input cells to be located superficially. As the neocortex kept increasing its thickness, it retained the inside-out gradient and the late-produced intratelencephalic neurons could be arranged in the superficial most layers, without interfering with the input-receiving middle layers. Note that a small set of afferences from hippocampus, amygdala and some thalamic nuclei maintain an input to layer I in mammals, which has been proposed to promote plasticity and provide multimodal inputs about bodily and sensory context to the neocortex.

Cortical Maps

A critical consequence of this new arrangement is that the radially oriented incoming axons ended up innervating a more restricted region of the neocortex (more specifically, a vertical column), instead of a tangential stripe of it as they do in the reptilian cortex and the hippocampus or olfactory cortex. This new arrangement enabled this structure to elaborate point-to-point maps of the sensory surfaces, where each cortical column received input from a specific spot of the retina, the body surface or the auditory cochlea (whose topography represents different sound pitches), depending on the respective cortical area. In this way, the mammalian neocortex acquired a fine topographic organization of its connections that could be used for increasing sensitivity (audition) and spatial behavior (visual and somatosensory inputs), as opposed to the cerebral hemispheres of nonmammals that contain only blurred spatial maps (recall that in these species visual topographic information is largely processed in the brainstem's optic tectum; chapter 8).

A Canonical Circuit

Associated with the fine-grained columnar input, neurons within each cortical column are highly interconnected, making up a modular input-output processing unit spanning neurons of different layers within the column. This microcircuit, or canonical circuit, is repeated along the whole neocortex, serving as the basic building block to assemble large-scale neural networks involved in motor programming, perception, and cognition (see figure 9.2D). Likewise, genes that label input and output elements of the canonical circuit are also expressed in the amygdalar complex of mammals, indicating that at least its basic cellular components may be present there. Notably, a strikingly similar input-output circuit organization is present in the brains of reptiles and birds, both in the dorsal cortex and in the DVR, although its components are distributed in different anatomical compartments instead of being compacted into a column as in the mammalian neocortex. One possibility is that this modular input-output organization is ancestral at least to amniotes, and its main elements have been preserved in different pallial regions of mammals and sauropsids.[12]

The Sheet Expands

Areas in the Cortex

The previous sections have referred to the origins of the microstructure of the neocortex like its lamination and columnar organization, but these features assemble into large-scale networks that process and integrate perception and behavior. If we take a higher-level aerial view of the cerebral cortex, different regions performing distinct functions will be evidenced: visual, auditory, somatosensory, and motor areas to name a few. The ancestral neocortex of all living mammals was probably based on a shared common plan, likely consisting of some twenty cortical areas that are present in all species. This contained at least distinct primary and secondary visual areas, an auditory area, two somatosensory areas, and a motor area (in small mammals, the somatosensory and motor areas overlap in a "sensorimotor amalgam," perhaps indicative of an ancestral condition). As in the evolution of vertebrate brains, different lineages tended to expand their neocortices independently, starting from this basic plan. As brains grew in size, the number of cortical areas increased in numbers (about two hundred cortical areas have been reported in humans). In general, the growth of the cortical sheet in different species is associated with an increase of so-called higher-order and association areas interspersed between the sensorimotor areas, which in large-brained mammals results in a dwarfing of the primary sensory and motor regions compared with the expanding rest of the cortex (see figure 9.3). Higher-order areas are usually unimodal, that is, they respond to just one sensory modality and are heavily connected with their respective primary and secondary sensory areas, while association areas are multimodal, receiving input from more than one sensory modality.[13]

Parcellation

A proposed mechanism for the proliferation of cortical areas is the parcellation theory, in which different kinds of inputs that converge in a cortical region become gradually segregated as the cortical surface increases, resulting in separate areas. The process of input segregation and areal parcellation may

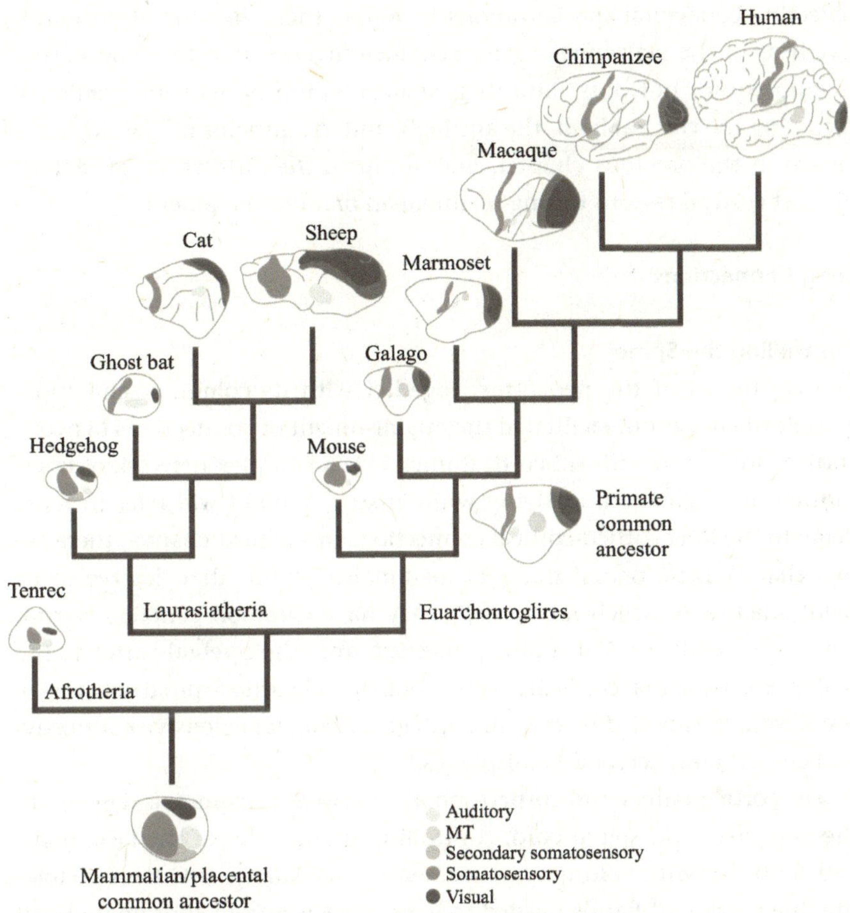

Figure 9.3
Lateral views of the brains of different mammals. These show the disposition of primary and secondary sensory and motor areas (different shades of gray). As mammals increase their brain size, higher-order and association areas expand (white regions), dwarfing the relative area of sensory and motor regions. Data from Kaas (2019).

be based on simple mechanisms of activity-dependent synaptic plasticity, where the inputs remodel their positions according to congruent patterns of activity.[14] While segregation is probably concomitant to increases in cortical area, this need not be a process of homogeneous expansion across the neocortex. Some areas have expanded more than others in certain species, like the association areas in large brains and the sensorimotor areas reflecting behavioral specializations in some others (like the electrosensitive beak of the platypus). Again, evidence suggests that the modulation of embryonic "hotspots" and their associated morphogenetic gradients, including the cortical hem, the antihem, and the anterior neural ridge discussed in the previous chapter, may result in the differential growth of distinct cortical regions during mammalian brain development.

Long Connections

Controlling the Spine

The expansion of the neocortex, together with its columnar and topographic organization, facilitated the establishment of connections between cortical areas, and with subcortical nuclei via a complex network of tracts running through the underlying white matter. While I will refer in more detail to the long cortico-cortical connections in the next chapter, there are two characteristic neural tracts in mammalian brains that deserve to be mentioned here, which incidentally may have common genetic determinants: the first is the descending projection from the cerebral cortex to the brainstem and spinal cord (the cortico-bulbar and cortico-spinal pathways); the second is provided by commissures generating an extensive connectivity between both cerebral hemispheres.[15]

The cortico-bulbar and cortico-spinal pathways in mammals descend to the brainstem and spinal cord, controlling the muscles of the face, body, and limbs, driving voluntary movements. Particularly, this tract enables the fine control of hands needed to grasp objects and to coordinate hand and mouth for feeding. Particularly in primates, the corticospinal tract directly innervates the cervical spinal motor neurons that control the hand muscles. In most other mammals, this tract innervates nearby interneurons that themselves contact the motor neurons but does not contact the motor neurons directly. The development of corticospinal terminations in the spinal cord is under tight developmental control, with the production of an excess of projections in early development that later become pruned to achieve more precise motor tuning. Recent studies have identified a few genes regulating the extension and retraction of these projections. Whether mutations in these genes or related ones were involved in the evolution of dexterity in primates (especially humans) is an open and intriguing question.[16]

Across Hemispheres

In addition to the descending tracts to the brainstem and spinal cord, mammals have unique, profuse connections between both cerebral hemispheres through the anterior commissure and the corpus callosum.[17] In previous works, we claimed that the establishment of interhemispheric fibers in mammals took place associated with the development of topographic maps of the sensory and motor surfaces in the cerebral cortex. Along this line, it is important to remember, first, that topographic maps are restricted to the brainstem's optic tectum in nonmammals, and that in mammals, each hemisphere contains the projection of the opposite side of the body and external world.[18]

According to this view, interhemispheric connections join the two halves of the sensorimotor map to gain a unified perception of the left and the right visual or sensorimotor fields. Likewise, nonmammals have commissural connections between the two sides of the optic tectum that can help fuse the two half-maps of their midbrain's sensory fields. Experimental studies in rodents have supported this hypothesis, showing that callosal connections are required for the development of visual acuity and depth perception in the frontal visual field, and that they provide a mechanism to predict the entrance of a moving stimulus from one visual field into the other across the midline. In somewhat later stages, interhemispheric fibers may have provided a circuitry for bimanual coordination as well, allowing the manipulation of objects to bring them to the mouth. Finally, interhemispheric connections may have helped to coordinate the activity of large-scale neuronal ensembles in both hemispheres, contributing to higher cognitive functions. Yet, birds like crows and parrots rival most mammals in cognitive abilities despite having only minimal interhemispheric connections, which casts doubts on the relevance of these fibers for higher cognition (see the next chapter).

Perspective

Mammals originated from a cynodont lineage during the Mesozoic period, exploiting a niche provided by the emerging flowering plants and the diversification of insects. While their vision was reduced due to nocturnal habits, their olfaction, touch, and audition became greatly enhanced, firstly through the expansion and elaboration of the nasal cavity and the mouth including the tactile whiskers, and secondly through the incorporation of the ancestral jaw articulation into the ossicle chain of the middle ear. The olfactory system and its connections with the hippocampus and neighboring regions became fundamental for the expansion of the nascent cerebral cortex. The hippocampus is critical for consolidating short-term into long-term memories that become stored in the neocortex, facilitating the establishment of episodic-like memories and the development of "maps" of the animal's surroundings. In this context, the emerging neocortex expanded to include different kinds of sensory inputs for the generation of multimodal, enduring memories of the world.

The neocortex was formed by two processes, firstly an increase in depth, adding new neuronal layers to the primitive monolayer cortex of early reptiles, and secondly increasing in surface, which enabled it to receive more sensory inputs. Associated with the six layers that characterize the neocortex, neurons organize in radial columns that receive most of their thalamic input from the underlying white matter as opposed to reptiles in which the sensory input is superficial. The radial columns represent a basic canonical microcircuit that is repeated across the entire neocortex, and provides a scaffolding for the formation of large-scale neuronal networks in the brain.

The neocortex has evolved from having some twenty areas shared by all extant mammals into a highly complex mosaic of regions as seen in large-brained species. The other two characteristics of the neocortex are its long projections to the brainstem and spinal cord that control voluntary and skilled learned movements and the long-distance connections between both cerebral hemispheres. In the next chapter, I will delve into the characteristics of the neocortex that make up the basic framework of connectivity of the mammalian brain and have been related to distinct cognitive and behavioral capacities.

10 The Symphony of Cognition

In this chapter, I will provide a brief account of the functional organization of the neocortex and its relation to behaviors comparable to human cognition. I will put emphasis on primates and especially on the human brain, which in the last instance is one of the main subjects of this book. Yet, the general framework of cortical organization is quite conserved across species, and most of what I will speak about in this chapter we share with other mammals (the next section of the book deals with what may be unique to our species).

Cognition refers to a set of processes including thought, memory, attention, language, decision-making, and planning among others, which are usually ascribed to the human mind and are largely (but not exclusively) dependent on the neocortex. Furthermore, cognitive processes can be assessed by specific behavioral tasks, and their neural correlates can be studied by analyzing brain activity during these tasks or in the case of brain lesions. Similar tasks (excepting language which is uniquely human) can be applied to nonhuman animals, where we can also study their neural correlates and propose neuronal-behavioral homologies or analogies with the human brain. Nonetheless, as I have discussed in previous chapters, we cannot access the mental processes of other animals if they do have any, while we usually take for granted that other humans possess a rich mind inside them just like each of us. This is the matter of heated controversies among behavioral scientists, neuroscientists, and psychologists, which I will attempt to address in the last part of the book.

Global versus Local

Charting the Brain

Early human neuroanatomists soon found that the cerebral cortex could be charted in a mosaic pattern associated to different mental functions. Perhaps the most successful of these was Korbinian Brodmann, who parcellated the neocortex into a series of numbered areas (see figure 10.1). Yet, these areas are highly interconnected, first in reciprocal loops with other brain components that continuously modulate its activity, and secondly between themselves via specific axonal tracts arranged in the underlying white matter. However, this is not a random mesh but is organized in a combination of very long tracts connecting distant areas, and more abundant short-cut connections between neighboring regions, emulating the arrangement of a city with small streets connecting nearby neighborhoods, and long highways to travel across the city. Some areas act as sites of convergence of many connections, working as hubs in the network, and areas more interconnected have been termed the "rich clubs." This pattern of interconnectivity fits what is called a "small world" organization where local and global processing combine and are kept in appropriate balance. A new view proposes a fractal-like organization of the brain, in which different levels of neural activity, from microcircuit to large-scale ensembles, behave in scale-invariant patterns that somehow mimic each other. Therefore, rather than a strict localization of functions or a diffuse meshwork, the cerebral cortex is a highly interconnected but heterogeneous organ, where some regions concentrate distinct kinds of input but maintain strong dynamic interactions with other systems.[1]

An Overall View

Before we get into the more specific functions of the cerebral cortex, it may be useful to recall its basic anatomical arrangement as depicted in figure 10.1. The cortical mosaic is divided into a posterior region that is predominantly sensory, and includes the parietal, occipital, and temporal lobes (roughly for touch, vision, and hearing, respectively). In these regions, there are primary and secondary sensory areas (directly receiving the sensory stimulus from the thalamus), higher-order sensory areas surrounding them, and finally associative, multimodal areas that integrate different sensory modalities. For instance, two regions called the temporoparietal junction

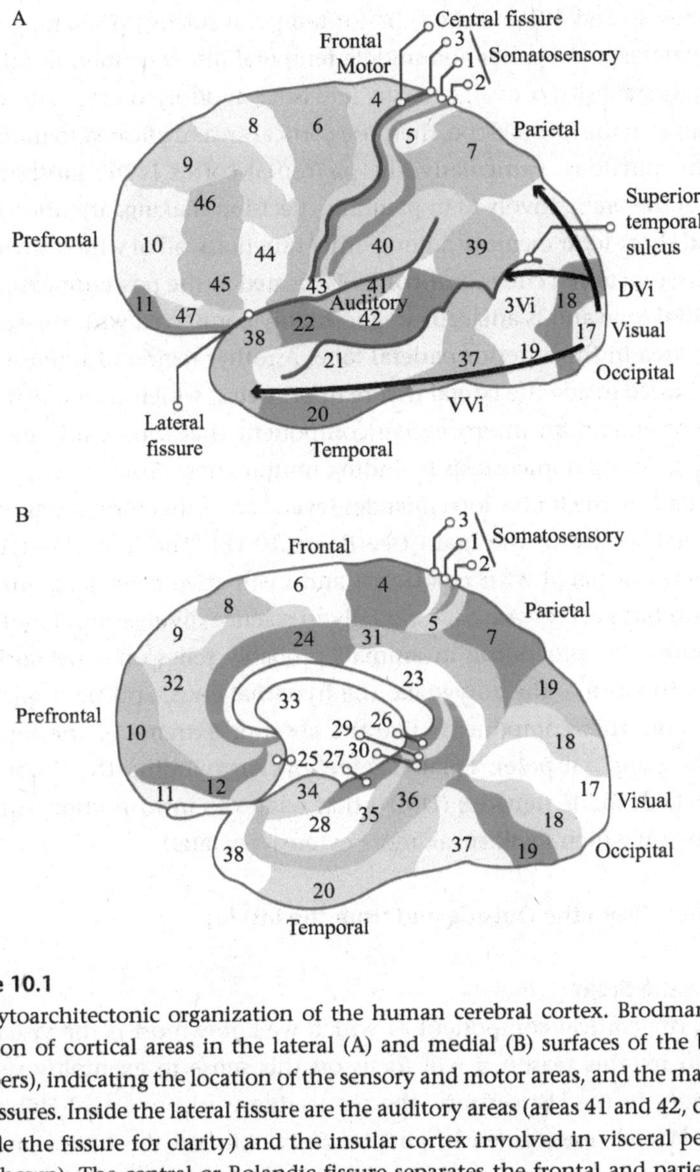

Figure 10.1

The cytoarchitectonic organization of the human cerebral cortex. Brodmann's parcellation of cortical areas in the lateral (A) and medial (B) surfaces of the brain (in numbers), indicating the location of the sensory and motor areas, and the major lobes and fissures. Inside the lateral fissure are the auditory areas (areas 41 and 42, displayed outside the fissure for clarity) and the insular cortex involved in visceral perception (not shown). The central or Rolandic fissure separates the frontal and parietal lobes and contains the motor (area 4) and somatosensory areas (3, 1, and 2), respectively. The primary, secondary, and tertiary visual cortices (areas 17 or V1, 18 or V2, and 19 or V3) are located in the occipital lobe. Finally, the dorsal, ventral, and third visual streams (DVi, VVi, and 3Vi) are shown as arrows originating from the occipital lobe.

(TPJ, areas 39 and 40) and the superior temporal sulcus (STS, a long fissure that separates the superior and middle temporal lobes) contain multimodal areas that establish a convergence site for visual, auditory, and somatosensory modalities. In the frontal lobe, there are cortical areas dedicated to motor and foresight functions. Particularly, the prefrontal cortex is the most anterior region of the brain, involved in planning, decision-making, attentional control, and short-term memory among other functions, all of which are termed executive functions. The motor cortex is located at the posterior extreme of the frontal lobe and is adjacent to and highly connected with the somatosensory area in the anterior parietal lobe. Another region of interest is the insula, buried inside the lateral fissure in primates, which contains the gustatory cortex and an interoceptive component that senses internal body states, regulating homeostasis including immune responses.

In addition, the limbic lobe includes several areas that form a ring around the medial surface of the brain (see figure 10.1B). The limbic system has long been associated with emotional and instinctive behaviors, although its notion has been overused, especially in science divulgation. Emotional, or emotion-like, processing in animals probably relies on a network that involves the insula, the amygdala, the hypothalamus, and the hippocampus among other components like the anterior extreme of the temporal lobe (the temporal pole). Finally, somehow surrounding the limbic lobe is the default mode network (DMN) that relates to introspection and daydreaming, of which I will speak more extensively later.

Perception (from the Outside and from the Inside)

Vision in the Brain

Perhaps the cortical component of which we know most is the visual system, and for this reason, I will focus on this sense to exemplify sensory perception. In the last century, the pioneering works of David Hubel and Torsten Wiesel (mentioned in chapter 5) established the basics of visual processing in the cerebral cortex, and their work has had an enduring influence not only in vision research but also in cognitive and computational science. Hubel and Wiesel had observed that retinal and thalamic neurons of the mammalian visual pathway were maximally excited or inhibited by a contrast stimulus consisting of a spot of light surrounded by a dark ring, or vice-versa, a dark spot surrounded by a lighting ring. On the other hand, neurons in the thalamic-recipient layer (layer IV) of the primary visual cortical area (also called V1, or Brodmann's area 17 in figure 10.1) responded maximally to light bars (as in a fluorescent tube) oriented in specific directions. In more superficial layers, cells had more complex responses, and outside V1 the responses were classified as "hypercomplex." Hubel and Wiesel proposed a serial and convergent arrangement of these cell types, where several aligned "spot" cells in the thalamus converged into one "simple" cell in the primary visual cortex generating a response to bar orientation; then, several simple cells converged into a complex cell, and finally several complex cells converged into a hypercomplex cell in the secondary and tertiary visual areas. This perspective implies not only a convergent or hierarchical array but also a sequential organization where each successive step processes more abstract aspects of the stimulus. However, more current views point to visual perception taking place in large-scale associative networks that link previous memories with current perceptions (see the next section).[2]

Streams of Perception (and Action)

Besides the primary and secondary visual areas, vision entails a complex array of higher-order and multimodal areas in the posterior part of the brain, which are particularly well developed in primates. Although these visual areas receive modulatory afferences from the brainstem-tectal visual pathway (see chapter 8), their main input comes from the primary visual area that as said, is innervated by the "direct" visual pathway bypassing the midbrain. Outside V1 (area 17), two main processing streams emerge: the dorsal and ventral streams (see figures 10.1 and 10.2B). The dorsal stream is directed to the parietal lobe, which projects into frontal regions coding for body movement. A characteristic area of this stream in primates is V5 or MT, which is particularly responsive to moving objects. MT is highly interconnected with additional areas like areas MST and LIP, which coordinate spatial and movement-related actions such as grasping. The dorsal stream has been also dubbed the "where" stream since it codes for stimuli location, although some authors refer to it as the "how" stream since it processes visuomotor transformations in orientation and manipulation behaviors. (I will speak further of manipulative behavior in the next chapter.)

On the other hand, the ventral visual stream runs along the inferior temporal lobe toward the anterior temporal lobe, including the amygdala, but also has strong connections with inferior frontal regions. The ventral stream is involved in the recognition of visual stimuli like objects, hands, faces, and even food and has an important emotional component, for which reason it has been also termed the "what" or "who" stream. An important component of this stream is the color-responsive area V4, particularly in primates. Color perception is another phenomenon that provides important insights about nervous system function. Recall from chapter 6 that rather than responding directly to distinct wavelengths, color vision results from the integration of the activities of different chromatic receptors. For instance, color illusions can be produced when ambient luminosity is changed, and subjects report seeing different colors than the specific wavelength that is presented to the eye. Apparently, the brain makes a balance of the ambient illumination, and its responses are related to the luminosity relations between the object that is in focus and its surroundings, rather than to the physical wavelengths that the object reflects. Thus, variations in the composition of the ambient light can yield paradoxical effects like observing colors that are physically not present. Notably, studies in monkeys show that in early visual areas (V1 and V2) there are neurons that are specifically responsive to wavelength, regardless of behavioral color reports. On the other hand, "color-sensitive cells," that activate together with the color reported by the subject, tend to be found in V4 (although some recent works found these cells in early visual areas as well). This phenomenon shows again the fact that the nervous system perceives the world in terms of contrasts rather than directly recording the physical stimuli it receives.[3]

A second area of interest in the ventral stream is the face-responsive area, located in the inferior regions of the temporal lobe. Lesions in this area usually produce a condition termed prosopagnosia, where subjects cannot recognize the faces of known people. Face perception is a fundamental and highly adaptive process, firstly because your life may depend on detecting a predator's or an enemy's face. Nonetheless, face-responding neurons have also been found in several other regions including the hippocampus, suggesting that there is a widespread network involved in face perception. Furthermore, face-responding neurons are not strictly face-specific, responding to other stimuli as well. This indicates that these neurons may participate in other networks responding to different kinds of patterns. In this line, there is a reading-selective cortical region in the human brain that is in close association with the face region, termed the visual word form area (VWFA; see chapter 13).[4]

Finally, a third visual stream has been depicted running just above the inferior temporal lobe, in the previously mentioned STS. This is claimed to relate to the dorsal stream, receiving strong input from the movement-sensitive area V5/MT, and contains areas responding to moving faces and bodies of humans and different animals, a feature that has been termed biological motion. I will come back to the VWFA and the STS in chapter 13, when discussing the neural underpinnings of language.[5]

Binding and Predicting

The separation of vision in two or more cortical streams, each processing different aspects of the visual scene, poses the binding problem that asks: How do these separate attributes get joined in a coherent perception? Very importantly, these routes converge in regions of the prefrontal cortex, the anterior temporal cortex, and in the hippocampus, the dorsal route being more involved in spatial memory and the ventral route being more related to object memory. Perhaps the convergence of these routes in regions like the hippocampal formation provides grounds for forming multimodal episodic and episodic-like memories that I have referred to in previous chapters.

Another issue is that while the previously mentioned streams have usually been depicted as highly directional from the primary visual area to the motor or memory regions, there are two critical features challenging this perspective. The first is that, as said, these projections are strongly bidirectional, that is, they are largely formed by the successive reciprocal connections between neighboring areas, with bottom-up projections transmitting sensory input and top-down projections modulating the activity of the downstream areas. Top-down control depends on widespread brain regions involved in memory, contextual processing, and motor execution. This array has been conceived as a predictive and error-signaling mechanism, where the expectations or predictions of the imminent sensory input are provided by the top-down signal (from higher-order areas to sensory areas) in relation to context, while the bottom-up signal (from sensory areas to other regions) conveys sensory novelty. That is, the brain prepares for what the next stimulus should be. Interestingly, if the sensory signal adjusts to the top-down expectations, it becomes suppressed and is only activated if there is a mismatch between the sensory input and the top-down prediction, that is, when there is an error in the prediction. In other words, top-down signals are generating a sensory scene that fits the general context, based on memory, ongoing behavior, planning mechanisms, and multimodal sensory integration. Some authors like Anil Seth claim that we are constantly hallucinating reality, that is, constructing the world according to our memory traces and expectations about it. This is what makes our everyday present. Only when the expectations are broken, the processing stream proceeds to generate appropriate responses and new memories, updating the context and continuing the construction of reality. Moreover, what has been termed spontaneous activity in the brain and has been typically interpreted as just noise produced by the ever-active networks, has been recently proposed to participate in this predictive top-down mechanism.[6]

Feelings

Finally, in this section, I will say a few words about a different kind of perception, which concerns the internal conditions of our body. Our brains not only receive external sensory inputs like the classical five senses but they also receive critical information about our inner states that drives our behavior to feed, drink, mate, and so on. The insular cortex, buried inside the lateral fissure, receives visceral stimuli including taste and gut sensations as well as bodily sensations including pain. Furthermore, it serves as a sensor and regulator of the internal states and participates in some basic functions like fear responses and respiration. The insula balances these internal inputs, making predictions of bodily physiology in order to anticipate future events and modulate them. Therefore, the insular cortex provides the brain with an update of the physiological conditions of the body at every time, which is used to assist behavioral decision-making and social behavior by prefrontal areas, and is fundamental for the conscious perception and processing of emotions (called feelings).[7]

Action and Thought

Organizing Behavior

Earlier I argued that perceptual circuits are not encapsulated systems that passively view reality but are strongly and reciprocally connected with motor and planning networks. Now I will refer to the frontal cortex that is involved in voluntary control of behavior, containing the motor and premotor areas in its posterior aspect, and the prefrontal regions anteriorly. These areas, robustly connecting to sensory temporoparietal regions, are highly developed in the large-brained primates, but these networks have expanded in other lineages as well, like for example monotremes. Prefrontal regions participate in the executive functions of cognition, like attention, working memory, and decision-making. These functions are carried in the context of a hierarchy involved in goal-selection and action planning at different levels, from general orienting to direct execution and sequential behavior at the fine detail. A critical feature of the prefrontal cortex is that, like the basal ganglia, it receives a profuse dopaminergic innervation from the brainstem. Thus, dopamine strongly supports the workings of prefrontal cortex neurons (and other cortical regions as well) and is required for its appropriate functions, modulating large-scale network activity in which the prefrontal cortex is a critical hub.[8] The prefrontal cortex is subdivided into several functional and anatomical domains (dorsolateral, ventrolateral, and polar on one side and dorsomedial and ventral on the other), all of which participate in organizing behavior according to the sensory or bodily contingencies.

Attention and Memory

As its name says, the dorsolateral prefrontal cortex is located in the lateral sides of the frontal lobe (areas 9, 46 and neighboring regions) and contributes to some so-called executive functions of the brain (see figure 10.2). The dorsolateral prefrontal cortex participates in the dorsal attentional network (DAN), which is related to sustained attention and task engagement during the execution of cognitive tasks, in which eye movements become tightly controlled. On the other hand, there is a ventral attentional network (VAN), located in the ventrolateral prefrontal cortex, involved in bottom-up switching attention mechanisms in response to sensory signals, driving eye and head movements away from fixation toward the source of salient stimuli. In a way, the DAN and the VAN are complementary partners that coordinate the focusing of attention with the attentional shifts during a cognitive or behavioral task. A good example of this is driving. The DAN keeps you focused on the road, but the VAN makes you look into the mirrors and to the sides to update your contextual information. The appropriate balance between both networks is needed for successful driving.[9]

Another highly studied function of the dorsolateral and ventrolateral prefrontal cortices is working memory, a kind of short-term memory that enables us to keep a few sensory items "on mind" for a short while, before

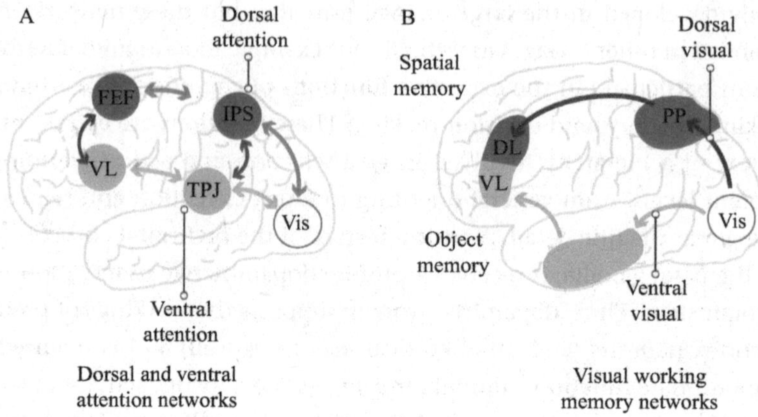

Figure 10.2
Major cortical networks involved in attention and working memory. *A*: The dorsal and ventral attention networks for sustained attention and attentional shifts, respectively. *B*: Networks for visual working memory that use the ventral visual stream for object recognition and the dorsal visual stream for spatial orientation and project to different regions of the prefrontal cortex. DL: dorsolateral prefrontal cortex; FEF: frontal eye fields; IPS: intraparietal sulcus; PP: posterior parietal lobe; TPJ: temporoparietal junction; VL: ventrolateral prefrontal cortex; Vis: visual area.

using this information to perform some task. Examples of this are mentally recalling the position of an object that is not present anymore in order to signal its previous location, or in the case of audition, remembering a telephone number to be dialed soon. Visual working memory is shared by other mammals, but auditory-based working memory has evolved especially in humans and will be discussed in chapter 12. As in the organization of visual perception, visual working memory has two main components, spatial memory (Where was it?) and the memory of objects (What was it?), closely related to the prefrontal projections of the dorsal and the ventral visual streams, respectively. Working memory is highly dependent on attention, and the networks involved in both processes are not easy to disentangle. In fact, attention is such a fundamental component of working memory that it may not make much sense to separate both phenomena. According to many authors, beside these circuits working memory relies on the activation of widespread cortical networks including the hippocampus, all involved in coordinating perception and action. In these networks, neurons specific to the remembered stimulus would remain persistently active during the time delay in which the relevant items are maintained in memory. Other authors argue that this "persistent" memory activity may rather correspond to a fluctuating process that goes up and down, associated with the oscillatory functioning of the neural networks that participate in this process.[10]

Where the Outside Meets the Inside

Other prefrontal components are the dorsomedial prefrontal, the orbitofrontal, and the ventromedial prefrontal cortices, which participate in weighing the internal and external conditions and using them to plan adaptive behavior, considering its possible consequences. These regions are intimately connected with the insula, linking emotional and executive systems, and are related to the detection of incongruences and to decision-making mechanisms according to these inputs, including social behavior.[11] The role of these areas was highlighted by Antonio Damasio when he referred to the now famous patient Phineas Gage, who suffered a major lesion in these regions (and other areas of the frontal lobe), rendering him a highly impulsive, violent, blasphemous person without control of his impulses.

Another prefrontal area that has grasped recent attention is the frontal pole (area 10), after several studies that reported a role of this area in complex behaviors, particularly social conduct. The frontal pole has been related to a function called "cognitive branching," which is to select among different goals and decide over one, keeping the second goal in working memory for a later event, or making subgoals in order to achieve a final goal. Area 10 has expanded more in our species than in other primates, but like other amplified regions of the human brain, this is most probably an effect of allometric enlargement of the cerebral cortex where different components grow at different rates, rather than a localized expansion of a set of areas (see the next chapter).[12]

The Idle Brain

All the previously described cortical functions are related to a specific general process: sensorimotor coordination, or what has been termed the perception-action cycle (behavior generates a feedback response via the sense organs or by internal mechanisms that again changes sensory activity, continuing a perception-motor-perception cycle during the execution of a task). However, most of the time we are not engaged in cognitively or physically demanding tasks, and much of it we spend daydreaming, imagining scenes, or remembering past events. Our minds are in a constant flux that runs parallel to the external events that we perceive and the things we routinely do. Taking again the example of driving home, you can automatically drive your car while your mind engages in what you did during the day, whether you will see your partner or not this evening, and so many other things.

Some twenty years ago, Marcus Raichle and collaborators performed a seminal experiment in which, in contrast to all cognitive experiments that had been performed to that date, brain activity was recorded while the individuals were at rest (without falling asleep), instead of performing say, a memory or an attention task as previous studies had done. Next, he contrasted the average brain activity in these subjects with an average of activity patterns observed during various cognitive tasks. As expected, he determined that the diverse executive tasks largely activated frontal and parietal areas (particularly the DAN and the VAN), and these decreased their activity when the subjects were at rest. But in contrast, there was a group of areas in the medial aspect of the hemispheres and a small region in the inferior parietal lobe that increased their activity during rest. Raichle called this cluster of areas the default mode network (DMN) and proposed that these were involved in internal thoughts like daydreaming, introspection, inner speech, and other processes including social behavior, empathy, and the ability to infer mental states in others (see figure 10.3). Yet, no brain region "shuts off" either during rest or during task execution, and the DMN interacts with the different brain systems, both during rest and when focused on a specific task. In this line, some authors use the term "resting state networks" to refer to a constellation of networks that remain active and interact during rest, including the DMN.

Interestingly, a DMN has been characterized in fetuses, rodents and in nonhuman primates, suggesting that it may contribute to brain homeostasis by regulating the balances between different types of activity. A highly intriguing hypothesis is that the DMN represents an axis to replay patterns of activity associated with perception and mnemonization of episodic events, in a sequence that starts in the hippocampus and propagates to the DMN and to the rest of the cortex (see the previous chapter). This view not only provides a phylogenetic interpretation for the functions of the DMN in nonhuman animals but also provides a substrate to explain deep conscious mechanisms like daydreaming in the context of memory

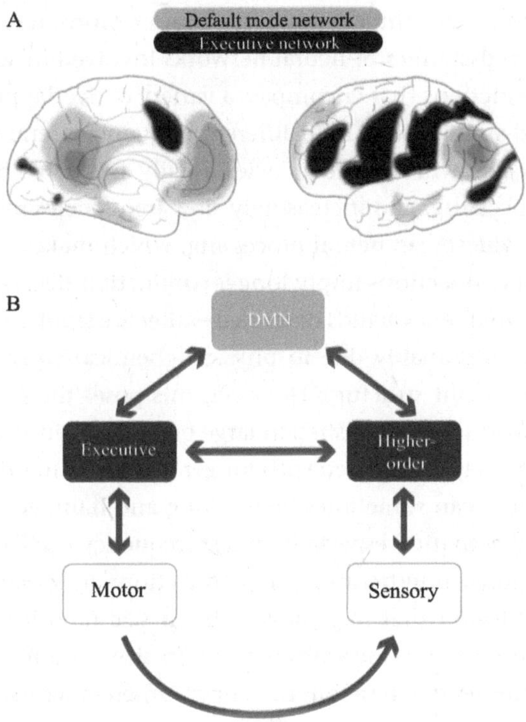

Figure 10.3
The default mode network (DMN). *A*: The DMN is more active at rest (gray) and is involved in introspection and mind-wandering, while the executive network (black) activates more while performing different cognitive tasks. *B*: A diagram showing the hierarchical organization of cortical regions, with the DMN at the top.

consolidation. In fact, DMN replaying activity may give rise to some of the fundamental characteristics of the human mind: mind wandering, inner speech, and the construction of an internal narrative. [13]

Brain Pulses

Ongoing Cycles

But how do all these different cortical and subcortical regions interact and generate the highly dynamic neural networks producing cognition? In the human brain, oscillatory activity was first recorded by the use of electroencephalography (EEG) more than a century ago. The reciprocal arrangement of cortical circuits at all scales, from the columnar microcircuit to the large-scale networks encompassing several brain regions, results in a highly complex cyclical dynamics of neural networks involved in sensory, motor, and cognitive functions that encompass a variety of overlapping oscillatory frequencies and amplitudes. These different frequencies tend to work at different spatial scales, with fast frequencies usually associated to short-range, local neural processing, and increasingly slow frequencies relating more to large-distance, widespread neural processing, which makes sense considering that distant connections imply longer conduction delays.

Notably, the frequency bands described earlier are similar for most mammalian species, presumably due to physical-chemical constraints of neuronal and microcircuit structure. However, this poses the problem of how large-scale networks are established in large brains, considering the increasing conduction delays associated with longer distances. In other words, the conduction delays can sometimes be too long and hamper the synchronization of neural activities, especially in fast frequency oscillations. We and others have studied interhemispheric connections in species of different brain sizes and found that regardless of brain size (and interhemispheric distance, which naturally increases with brain size), the fibers connecting both hemispheres tend to be about the same diameters across species, which implies an increase in conduction delays as brains get larger. However, there is a small population of fibers that rapidly increase their diameter (and conduction velocity) concomitant with increasing interhemispheric distance across species. These fibers may support fast synchronic ensembles in both hemispheres, providing a powerful drive to recruit distant neuronal populations into organized circuits.[14]

Sleep Waves

A classical instance of oscillatory mechanisms is seen in sleep. Sleep is related to bodily resting and repair requirements in most animals including jellyfish, and as we have seen, is also associated with learning and memory mechanisms. Moreover, the autonomic nervous system has been found to support memory formation during sleep, establishing a link between learning mechanisms and body physiology. The sleep-wake cycle is controlled by a deep brain network involving the hypothalamus, the brainstem, and the pineal gland, and is neurochemically mediated by the hormones orexin and melatonin, which are found in many bilaterian animals and have similar functions across them. As seen in the mammalian EEG, there are basically three referential brain states through our circadian cycle: wakefulness, characterized by a predominance of fast oscillatory waves; deep sleep with slow EEG oscillations together with hippocampal-related high frequency bursts (spindles) that participate in memory consolidation (see chapter 9); and REM sleep displaying fast oscillatory activity, rapid eye movements (what REM stands for), and muscular paralysis. Dreams take place mostly during REM sleep, although they can also occur in slow sleep. An intriguing relation is that of the DMN and sleep stages. Some studies suggest that both the DMN and the executive networks tend to decrease their functional connectivity during sleep, while other reports have likened REM activity with mind-wandering and have evidenced short episodes of REM sleep concurrent with DMN activation during wakefulness.[15]

Cognition without a Cortex

Clever Birds

While I have referred to cognition as largely seated in the neocortex (and its connections with other systems), these processes are by no means dependent on having such structure. Birds, particularly some lineages like the songbird group (oscines, order Passeriformes, especially corvids) and parrots (order Psittaciformes), are well-known for their behavioral and memory skills. Anecdotes and scientific observations about these animals doing clever things and imitating the human voice and other sounds abound. For example, in the lab, birds have been shown to make new tools to solve a food retrieval task, something that apes are quite slow to do in captivity. Perhaps one reason why birds are particularly good at these manipulative tasks has to do with their nesting behavior. Making a nest may be largely instinctive, although increasing evidence indicates an important role of learning in this behavior. Furthermore, birds need to assess the physical resistance, length, and flexibility of the twigs and materials they use to make their nest, which implies knowledge of the physical properties of objects.

Likewise, corvids have shown a tremendous capacity to remember the three classical components of episodic-like memory (what, when, and where something happened), an ability that is highly adaptive as they store food in several distinct places during the winter, recalling which items were stored in which locations and when they were cached. Thus, avian hippocampal neurons display similar features as those in the mammalian hippocampus, exhibiting place cell-like and entorhinal-like activities, particularly in food-caching birds. In addition, birds have developed an equivalent to the prefrontal cortex of mammals, involved in behavioral control and planning. As in the latter, this structure displays a strong dopaminergic innervation, which again is much stronger in songbirds and corvids than in the quail or the chicken. Still, determining whether birds rely on similar or different large-scale network organizations to achieve similar cognitive goals as mammals represents another challenge for evolutionary neuroscience.[16]

Perspective

The neocortex or cerebral cortex is considered an essential component for mammalian cognition, including human consciousness. It is subdivided into a series of areas, some having sensory and motor functions, and others working as nodes in a network that drives cognition and behavior. Likewise, the cerebral cortex engages in complex dialogues with subcortical systems to process sensory inputs and execute behavior, and it relies on tight loops with the hippocampal formation and the amygdala to assemble sensory-based and emotional memories, respectively. Finally, deep brain nuclei involved in arousal, reward processing, and motivation exert a strong modulation of cortical activity and drive behavior.

Sensory processing has been well characterized in the visual system. While the earliest studies depicted a sequential and convergent organization for visual processing, more recent studies suggest a more dynamic mechanism involving top-down and bottom-up influences along the cortical system. Yet, the visual cortex appears segregated in two main domains: a dorsal stream involved in visuomotor transformations for body orientation and hand grasping, and a ventral stream involved in visual recognition and memory. A third stream relates to moving socially relevant stimuli along the STS. These streams concur in different areas of the frontal cortex, where behavior is organized according to current goals, and in the hippocampus where episodic-like memories are formed. The networks connecting temporal, parietal, and prefrontal areas participate in the so-called executive functions including attention, working memory, and more complex ones like decision-making and behavioral planning. An intriguing cortical network that works in an ongoing balance with the executive networks is the DMN, which relates to introspection, empathy, and daydreaming.

The neocortex displays an intense oscillatory activity at very different frequencies that are highly conserved across species. Oscillations also participate in memory formation during sleep, where extended networks including the neocortex, the hippocampus, and the thalamus engage. REM sleep is a particular state, similar to the awake condition, where the EEG signals show fast activity together with eye movements.

Finally, without a neocortex, birds can display behavioral and cognitive capacities not unlike or better than those seen in monkeys or apes in some cases. Up to here in this book, I have attempted to provide a background on the evolution of body and brain organization in animals, vertebrates, and mammals. In the next part of the book, I will delve into the origin of our own species and the evolution of our brain and cognitive capacities that has led us to where we are now.

IV A Singular Ape

Possibly soon after the dinosaurs' extinction, primates became relatively common animals and diversified in many branches. One of these was a tailless lineage that gave rise to the last common ancestor of chimpanzees and humans. Our immediate ancestors colonized the African savanna, evolving an obligate erect posture and acquiring a rudimentary tool technology together with increasing brain size. Modern humans arose some 300,000 years ago and rapidly spread across the planet, developing increasing technological skills that eventually led to the formation of modern civilization. Toolmaking culture evolved together with increasing hand-coordination capacities, and probably took place in a context of intense social interactions where vocalizations played a significant role providing group and mother-child bonding. A key feature of human evolution was the acquisition of language, starting as multimodal vocal-gestural communication, and the eventual recruitment of an auditory-vocal brain network that gave rise to modern speech. Language provided the possibility to share the world and develop a common semantics, depicting not only objects and events but also describing complex actions through the emergence of grammar. Language also enabled us to reflect about our own minds, and about others' minds, contributing to human consciousness. However, consciousness may be the most difficult problem for modern neuroscience, and there is not even a clear agreement on how to define it. Furthermore, the more complex issue of explaining the subjective experience of say, seeing colors, remains an inscrutable problem. Finally, there is the concern about our species' future. While we may keep slowly evolving biologically, the explosive rise of technology in the last years may generate profound changes in our well-being while at the same time it may be the source of deep inequalities that can affect our future as a species. We are rapidly changing the face of the earth, and it depends on us to decide what world we want to live in.

11 Homo

The word primate comes from the Latin meaning "first rank," in a clear allusion to our anthropocentric view of life. While not being hierarchically superior, primates represent an unusual clade displaying adaptations to arboreal life, strong frontal vision initially associated with nocturnal habits, and particularly large brains. Most primates, especially the anthropoids, live in groups and develop complex social lives. Our species emerged as a highly social species, in a unique series of events that have changed the face of the world in a few hundred thousand years. This chapter refers to the origin of humans starting from our primate ancestors, and to some special adaptations that characterized our lineage, like a very large brain, the manipulative and toolmaking skills, and our social behavior.

From Climbing to Walking

Tree-Dwellers

The earliest primates recorded in the fossil record are the arboreal shrew-like purgatoriids, probably originating shortly before the asteroid terrestrial impact that killed the dinosaurs (see figure 11.1). Primates soon developed several characters like opposable thumbs in hands and feet and a sophisticated neural control of them, enhanced hand-mouth coordination for feeding on fruits and leaves, and finger pads and nails instead of claws facilitating grasping. Other features of primates are that they are usually highly social species, relying on extended use of gestural, vocal, and facial signals to communicate.[1]

Modern primates split into two main lineages: the "wet-nosed" primates (strepsirrhines: lemurs, galagos, and lorises) and the "dry-nosed" primates

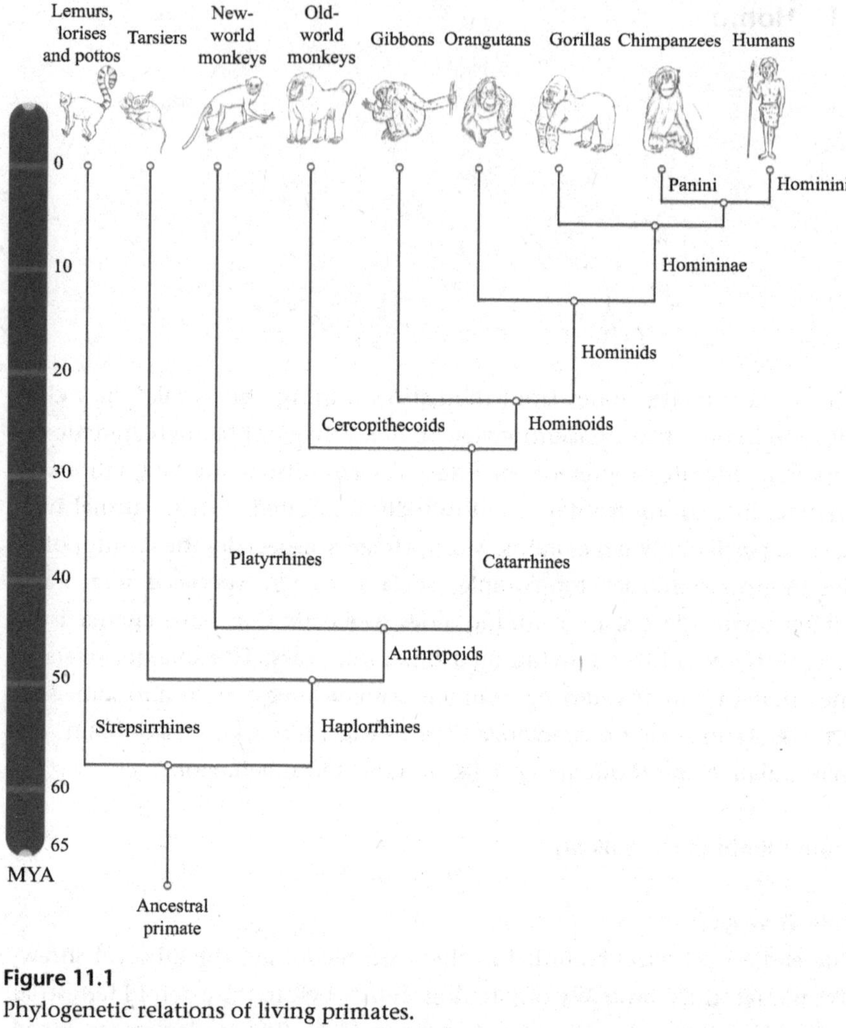

Figure 11.1
Phylogenetic relations of living primates.

(haplorrhines: monkeys—including us—and the tarsier, a small, large-eyed creature from Indonesia). Wet-nosed primates have an upper lip that curves in the midline and reaches the nose as in cats, dogs, and rodents. In dry-nosed primates, the lip forms a continuous band of skin below the nose that gives it more flexibility (as in our upper lips). Monkeys use the upper lip for socializing and produce a series of signals and voiceless sounds like "clicks," "kisses," and "whistles" that represent an early scaffolding for the origin of speech in our species.

Tail Loss

Anthropoids (monkeys, apes, and us) emerged in Asia some 58 MYA, separating from tarsier ancestors as their sister group. Anthropoids became diurnal and regained color vision, associated with fruit feeding, that had been partially lost in early mammals. Anthropoids further subdivided between platyrrhines (new-world monkeys) and catarrhines (old World monkeys). Subsequently, monkeys that remained in Africa and Asia split into two branches, the Cercopithecidae (colobus, macaques, baboons, and their kin) and the Hominoidea or apes: flat-faced, long-armed, and tailless primates. Early apes were distributed in Asia, Europe, and Africa during the Miocene period, including the macaque-like *Proconsul* and other species. As mentioned in chapter 1, tail loss in this lineage was probably caused by the insertion of a mobile DNA sequence (a "jumping gene") into a gene called TBXT. However, there was a cost to this innovation as the mutation in TBXT is related to increasing incidence of neural tube defects. Tail loss may have been associated with a new mode of locomotion in the canopy called brachiation, which consists of balancing on the branches hanging by the arms like a trapeze artist, presaging a vertical body posture. Notably, the early fossil ape *Danuvius* seems to have walked upright in trees about 11 MYA, long before bipedality arose in the hominins. Other evidence suggests that bipedality arose in seasonal woodlands rather than on deep forests where apes had to ascend and descend from trees. In this line, a recent proposal suggests that ape forelimb morphology relates more specifically to down-climbing from the trees, which apes perform differently from monkeys.[2]

Gibbons and Siamangs from South Asia represent the earliest branch of living apes, usually living in the high trees and maintaining stable couples during their lives. Importantly, gibbons are highly vocal species and have elaborated complex songs. Despite their complexity, gibbon songs are quite stereotyped and species-specific. Perhaps more important for the context of human evolution is the fact that gibbon couples sing in highly coordinated duets led by the female, where both sexes alternate their calls like we do during a conversation. A more flexible mechanism of "conversation" is observed in the South American marmosets, who display complex communal lives. I will come back to these animals in the next chapter.

The rest of apes include the living orangutan, the gorilla, the chimpanzee, the bonobo, and our own lineage. The chimpanzees and bonobos (Panini) are our closest living relatives, while Hominini refers to the human lineage after the split with the chimpanzee ancestors, that is, excluding Panini. Despite our anatomic, cognitive, and cultural differences, estimates of genetic sequence divergences between these apes and humans range between 1 to 3 percent, although large chromosomal rearrangements are much more common in both groups and may have been relevant in the divergence of both lineages. While chimps tend to be more aggressive and more carnivorous, bonobos are gentler and have a juvenilized appearance. Some authors have argued that the life of bonobos better resembles that of human ancestors, with a trend to maintain juvenile features, decrease aggression, and increase cooperation. Likewise, early fossil hominins like *Orrorin*, *Sahelanthropus*, and *Ardipithecus* had small canines and large molars compared to other primates, pointing to a less aggressive way of life. Yet, the brains of these species were similar in size to those of other apes. Another critical character of these fossils is their incipient bipedality. Still, the social life, posture, and locomotion of early hominins is a matter of heated disputes among paleoanthropologists, and the last word has not been said yet.[3]

Featherless Bipeds

A Walk in the Savanna

Our own lineage (Hominini) probably starts with *Ardipithecus* and the closely related *Orrorin* and *Sahelanthropus,* all showing a mixture of arboreal and pedestrian features (see figure 11.2). The later *Kenyanthropus* and the Australopithecines displayed more adaptations to bipedal locomotion, with a more human-like vertebrate column to support the standing weight. The erect posture may have been useful for losing heat in the open savanna, exposing the body to the wind and minimizing sun irradiation, which was accompanied with the increase of sweat glands and probably loss of hair (except in the head, which protected the brain from overheating). At the same time, the straight posture offered a wider visual perspective to detect danger or food. But perhaps the more radical consequence of the obligate bipedal posture was that it freed the hands to carry food, babies, sticks, and stones for defense and for crushing carcasses.[4]

Hominins probably had an omnivorous diet similar to that of modern chimpanzees, including fruits, leaves, seeds, insects, and occasional meat. Some 3–2.4 MYA, butchery marks made on bones of large animals start

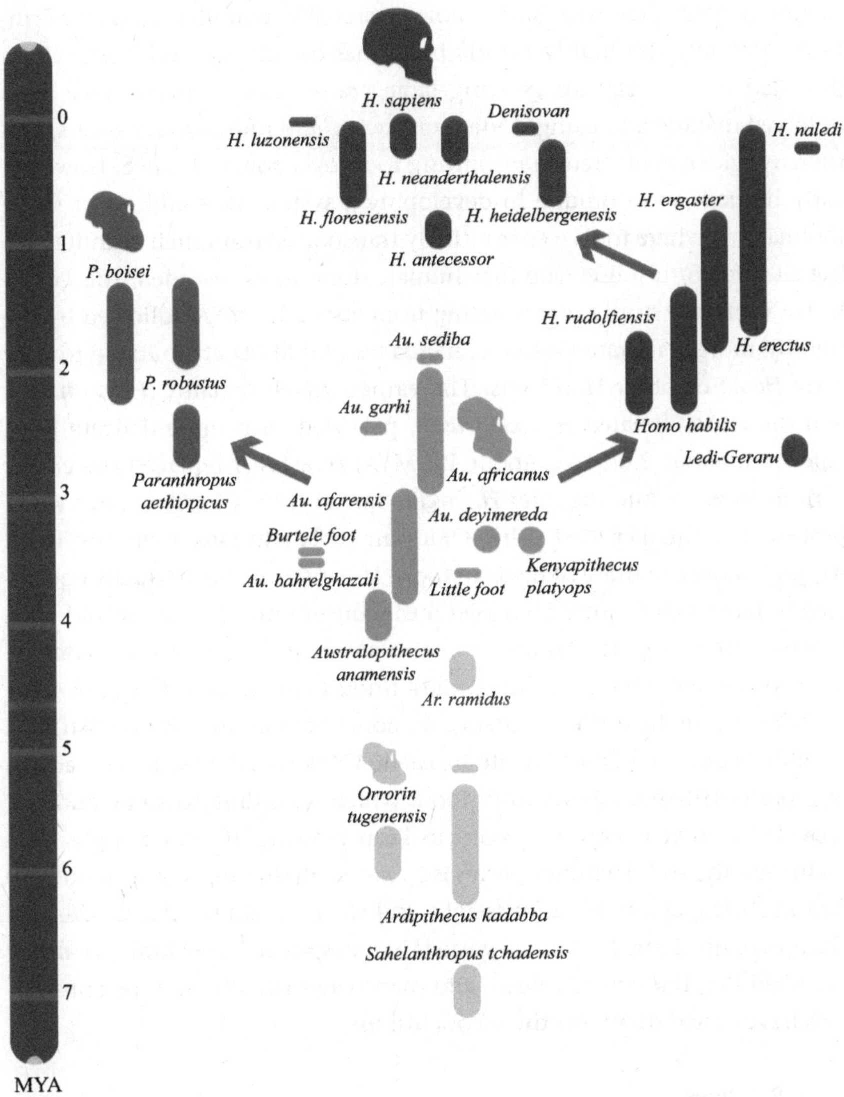

Figure 11.2
The evolution of Hominins.

to appear, associated with stone tools presumably manufactured by hominins to extract the highly caloric bone marrow inside the bones, which provided new energy for growing large brains. Chimpanzees have been observed making and using stone tools, twigs, and rudimentary spears, and there is evidence of them transporting tools over some distance. However, early humans were unique in developing a systematic tradition of stone toolmaking, where tools were routinely transported from their manufacturing sites for future use. The first human stone tools recorded, the Lomekwian tools, are sturdy pieces dating from about 3.3 MYA, followed by the slightly more elaborate Oldowan handaxes (2.9 MYA) attributable to very early *Homo* or other Hominins. The earliest *Homo* remains (*Homo habilis* and the closely related *H. rudolfensis*), provided with more delicate jaws, date from about 2.8 MYA. About 1.8 MYA, *H. ergaster* replaced the earlier African forms, while the later *H. erectus* spread into Asia. The latter was a proficient toolmaker, first using Oldowan tools and later replacing them with the more elaborate Acheulean tools. *H. erectus* has been usually considered to have significantly increased meat consumption, which would have propelled brain growth. Another early human achievement was fire control. Data for use of fire is elusive, the earliest findings pointing at *H. erectus* some 1.5 MYA, while the earliest evidence for cooking food (in this case fish) was recently found in a 780-KYA site in Israel. Cooking allowed for extraction of more nutritive contents from food, which according to some authors, provided further energy necessary to keep growing an increasingly large brain. Finally, an intriguing species is *H. naledi*, dating from only about 230 KYA in Africa, about 45 kg of weight and with a brain similar in size to a chimp's brain. Some reports suggested that, despite its small brain, *H. naledi* mastered fire, buried their dead, and made cave art, although recent findings have casted doubt on these conclusions.[5]

Close Relatives

The recent human lineage appeared about 1.2 MYA, with *H. antecessor* and later *H. heidelbergensis* from Europe, the stout Neanderthal man in Europe and Asia (*H. Neanderthalensis* or *H. sapiens neanderthalensis*), and the Denisovan man in Asia (*H. denisova* or *H. sapiens denisova*). Neanderthals (and possibly Denisovans) had a complex culture, with cave paintings, hunting large game, cooking their food, and making bone ornaments, indicating they displayed manual and cognitive expertise, including symbolic thought. Two side branches of late *Homo* were the dwarfed island dwellers Flores man (*H. floresiensis*) from Indonesia and *H. luzonensis* from the Philippines. According to recent findings, early *H. sapiens* arose some 300 KYA and spread into diverse populations in different parts of Africa, all showing a mixture of modern and archaic anatomical traits and possibly exchanging cultural and genetic traits among them. These populations gradually acquired modern physical characters and a more sophisticated stone technology than their ancestors. Recent studies indicate that early *Homo sapiens* reached the Arabian peninsula, southern Europe, and Asia several times in history, carrying with them a sophisticated culture perhaps similar to that of Neanderthals. Yet, all non-African living humans apparently descend from a great migratory wave that took place some 60 KYA.[6]

As modern *H. sapiens* reached Eurasia, they rapidly encountered other early human populations in the middle east like the Neanderthals and the Denisovans. There is mounting evidence that these three groups regularly interacted and transmitted cultural and genetic traits among them. Neanderthals and Denisovans became extinct 40 or 25 KYA, possibly killed directly or indirectly by modern humans, although there was a degree of interbreeding between these groups. About 1–4 percent of the modern European *sapiens* genes were inherited by the Neanderthals, and modern Tibetan and Oceania populations have inherited a significant proportion of Denisovan genes, particularly related to immune function.[7]

Runaway Culture

After spreading into Eurasia and the other continents, the history of *Homo sapiens* went runaway. While for more than 2 MY technical innovations were quite slow, some 100 KYA our African *Homo* ancestors started changing their diets by including shellfish and fishing, making ornaments, elaborating spear heads and other innovations that took place concomitantly with increasing population size and long-distance trading of materials. The first confirmed abstract artwork dates from 73 KYA, marking the beginnings of figurative art. The earliest musical instruments known, flutes made from vulture and swan bones, date from about 30 KYA. Many authors claim that the cultural explosion is directly related to the development of abstract thoughts and especially with the capacity to plan future actions.

Cultural innovations rapidly accumulated through time, accompanied by another great invention of humankind: the domestication of other species by controlling their reproduction, and generating our own food supplies, which in a way is an additional step in the self-maintenance mechanisms. By the Neolithic, some humans transitioned from a hunter-gatherer lifestyle to a pastoral-farmer way of living, inventing agriculture in several places and at different times around the globe. The onset of agriculture provided a new dimension to human history, further increasing population numbers and trading. The increase in population density led to the concentration of people in large, collective settlements called megasites like those found in Ukraine about 6.2 KYA, which later became the earliest cities in Mesopotamia. In these conditions, population density rose highly and individual aggression too, making life much harder. Together with cities, higher-order organizations began to emerge to regulate social behavior, controlling aggression but also increasing inequality (hunger and famine became increasingly common).[8]

The evolution of culture and technology in our species has been referred to by some authors as an instance of "niche construction," that is, we have been able to generate and modify our own environment in order to satisfy our needs. In my view, a more biologically grounded notion for this phenomenon is Richard Dawkins's "extended phenotype," which is found in many other species, like the beaver's dam and the spider's net. These behaviors are likely coded in the beaver's or spider's genes, while in our species, genes have coded for large and highly plastic brains that enabled us to develop our culture. I will come back to this point in chapter 15.

Eighty Billion Neurons

Brain Explosion

All these spectacular innovations were associated with one of the most notable features of our evolution: a dramatic increase in brain size. Starting from an ape-like brain of about 500 cc., the *Homo* brain reached 1,200 cc volume in Asian *H. erectus* some 200 KYA. More recent humans and Neanderthals (possibly Denisovans too) achieved some 1,300–1,600 cc of brain capacity, but there is a big dispersion and overlap in these data.

Intriguingly, there is a discordance between cultural development (evidenced by stone tool technology) and brain evolution in our lineage, the former increasing quite slowly at the beginning and speeding up only in the last 100–200 TY when our brains had already achieved their modern size. One possibility is that increasing brain volume in our early *Homo* ancestors was related to social and behavioral skills that have left no traces in the archaeological record.

Diet, Birth, and Heat

Increasing brain size in Hominins was not without drawbacks. First of all, the brain is an energetically expensive organ and needs an extra caloric increase for its maintenance. Some authors have proposed that the increase of meat and bone marrow in the diet provided this extra energy. This also required tighter cooperation between individuals, promoting prosocial behavior. It has been also argued that the use of fire and the invention of cooking provided the energy for brain growth as it enabled a more efficient extraction of nutrients. Nonetheless, archaeological evidence does not indicate a correspondence between fire control and human brain expansion. Furthermore, fire control and cooking require significant cognitive abilities to begin with, implying that selection for enhanced cognition (if not brain size itself) was already on its way when fire was domesticated.[9]

Secondly, there was an obstetric problem: as babies were born with larger brains, the passage through the pelvis during birth became increasingly difficult. Increasing the size of the pelvic canal had its limits as it anatomically interfered with walking efficiency. A complementary strategy was that our ancestors started delivering increasingly premature babies, which brought together increasing obstetric problems compared to other primates. Thus, care of the newborns may have required assistance from others, putting an additional pressure on the evolution of our social life. A third problem was that the head (and the brain) was maximally exposed to the savanna's sun. Together with keeping hair on the head, the venous blood return from the brain was modified in early *Homo* to increase heat dissipation. Thus, making large brains entailed a very complex set of anatomical compromises and functional tradeoffs, which implies that this probably provided benefits that so far outweighed the costs.[10]

Larger but Conservative

We have the largest brains relative to body size of all primates and of all animals as well, and possibly we have the largest number of brain neurons in relation to body size of all animals (whales may have more brain neurons but their bodies are humongous). As in other large-brained mammals, the human brain has not grown uniformly, but some regions like the prefrontal cortex and higher-order areas (particularly the region surrounding the temporoparietal junction) have undergone very rapid growth while primary and secondary sensorimotor areas expanded much more slowly (see chapter 9). Likewise, areas related to the DMN and their connections have also undergone a significant increase in human evolution. Some evidence indicates that hippocampal and amygdalar regions associated with memory and emotion became specifically amplified in hominins, a finding that deserves further study. Finally, the human brain also developed structural and functional asymmetries that are only incipient in the ape brain, particularly in the inferior parietal lobe and the inferior frontal areas, providing a substrate for handedness, toolmaking, and the hemispheric lateralization for language, which were the key behavioral features driving our divergence with the rest of animals.[11]

This expansion was concomitant with a more modest growth of the cerebellum and the basal ganglia, which participate in the selection, coordination, and execution of skilled learned movements including toolmaking and speech, but also to other cognitive functions involving planning behavior, social interactions, and even our daily thoughts. As I will explain some paragraphs later, these components (especially the basal ganglia and other deep nuclei) may have been critical elements in promoting social behavior in our ancestors.

Associated with cortical growth, the human brain has gross morphological differences from those of apes and even among Hominins. Paleoanthropologists have been able to identify marks of some brain sulci in the internal cranial surface of fossil hominins, as seen by molds of the cranial vault (called endocasts), to infer how brain shape changed in human evolution. For instance, nonprimates have brain regions homologous to Broca's language area in the human prefrontal cortex that can be identified by a specific cortical folding pattern. *Homo* fossils display a transition from an ape-like configuration of Broca's cap in early specimens to a modern human-like anatomy in younger fossils.[12]

Brain Genes and Cells

Several of the genes involved in determining brain size in vertebrates may have contributed to human brain evolution, either by amplifying their expression (the gene makes more proteins) or by mutations increasing the activity of the proteins they codify. Many of these genes, like ARHGAP11B and others, are regulatory genes related to increasing neuronal progenitor proliferation and cortical folding compared to apes, while other genes contribute to network reorganization with some related to mental disorders. Interestingly, several noncoding DNA regions and RNAs (presumably acting as regulators of gene activity) have been found to undergo an accelerated evolution on the human lineage and have been associated with the expansion of large-scale cognitive and social-related networks in the human brain. More recently, a new evolutionary strategy has been detected for the human brain: noncoding RNA is able to modify into coding messenger RNA, giving rise to new human-specific genes that have increased neuronal production. This mechanism is reminiscent of the RNA-editing processes involved in cephalopod brain evolution (see chapter 6).

There are also speech-related genes, particularly a gene called FOXP2 (see next chapter), which concentrates many of the before-mentioned accelerated regions in its introns (see also chapter 12). Additional genes relate to synaptic function and plasticity like CBLN2 and SRGAP2; the latter underwent two rounds of duplication roughly at the origins of *Homo sapiens*. Likewise, human neurons display increases in dendritic complexity, neurotransmitter signaling, and electrical properties, associated with the expansion of upper cortical layers. Finally, glial cells, particularly astrocytes that participate in brain homeostatic mechanisms, including energy expenditure, show signs of specialization in the human brain, increasing in density in association with brain expansion.[13]

All in all, genetic and cellular modifications in human brain development may have been largely associated with increases in allometric brain expansion and increasing plasticity, with a basically conserved neural architecture. Perhaps one main exception is the development of lateralization and brain asymmetries that, in humans, acquired fundamental relevance in association with toolmaking and language. Some authors have argued that brain lateralization may be partly a consequence of increasing interhemispheric distance in larger brains, which hampers communication between both sides of the brain. However, functional lateralization is not just driven by brain network asymmetries. Handedness, like language dominance, also depends on peripheral asymmetries of the innervation of the motor systems.[14]

Despite all these considerations, the question remains of why we were the only species to evolve such large brains if being so clever is such a good recipe. Homo sapiens coexisted with several other hominin species for about two million years, when population densities allowed for species diversity. But since about 30 KY we have been the only species of *Homo* on the planet. This might be because our lineage took care of extinguishing their close relatives, like we did with Neanderthals and Denisovans. Perhaps there is no space for more than one highly intelligent species on this planet.

Hands, Toolmaking, and Culture

Prodigious Hands

Another striking feature of the *Homo* lineage is the possession of skillful hands that enabled us to use and make sophisticated tools. The human hand is a prodigy of anatomy, allowing a stunning variety of grasping, manipulating, throwing, and other behaviors that undoubtedly have been critical for our evolution. Hand control evolved associated to enhanced cortical control of the forearm musculature (usually the right hand), providing the ability to execute more voluntary and fine-controlled hand movements. Among nonhuman primates, individuals tend to have some hand preferences for manipulation and for social gestures, but there is a high variability and there seems to be no robust tendency for the left or the right hand in different species. In contrast, humans represent an unparalleled extreme in the consistent trend to one-handedness. There may have been unique selective pressures driving this condition, most likely driven by toolmaking behavior, where both hands have to perform complementary tasks.[15]

Even having skillful hands, elaborating stone tools is a painstaking procedure and must have been so to our early ancestors as well, who may have used large amounts of time to make them. Related to these skills, humans have a strong connectivity between parietal and frontal cortices supporting sensorimotor control of the hand, which is related to increasing brain size. However, acquiring familiarity with a tool also requires participation of the object-related ventral visual stream along the inferior temporal lobe. Moreover, like any skilled behavior, tool making and use rely on complex and hierarchically organized movement sequences that are dependent on the basal ganglia. Interestingly, the basal ganglia circuits involved in toolmaking overlap with those related to complex grammatical processing, indicating that these skilled sequences partly share neural substrates. Tool making and use also require a great deal of bimanual coordination, where usually the left hand holds and orients the object while the other usually uses another tool and performs high-frequency rhythmic movements to shape the object, in a hierarchical organization that according to some authors resembles that of language.[16]

Eye and Hand

In addition to handedness and bimanual coordination, a fundamental element in hand use and tool making is visual control. If you want to grasp an object, you not only adjust the arm trajectory to the observed position but your hand extends according to the size and shape of the object to be grasped. At six to eight weeks of age, human babies start watching their hands moving in front of them, in a process of maturation of the hand-eye coordination reflexes.

Perhaps one of the most exciting neurophysiological discoveries of the last thirty years was that of mirror neurons. Studying hand-grasping behavior, researchers found visuomotor neurons in the premotor cortex that fired both when the animal observed the object to be grasped (usually a food pellet) and when it executed the grasping action. To their great surprise, they also found that some of these neurons started to fire when the animal saw a human experimenter picking up the food pellet in the plate. These were called mirror neurons, characterized by firing both when the animal performs an action and when the animal observes the same action made by someone else (see figure 11.3). Mirror neurons were interpreted as recognizing the grasping action made by another individual, enabling us to understand other people's actions or intentions "from within." Mirror neurons have become one of the hottest topics for popular science and nonscience worldwide, and there are many misunderstandings about them. Thus, another group of researchers claims that mirror neurons may not imply a dedicated mechanism of internal representations or models of the others' conducts but emerge from basic associative events that link the sensory context (say, the observation of the experimenter's hand movement) with planned motor goals (the monkey intending to grasp the object). Furthermore, mirror activity may be observed in other brain circuits beside

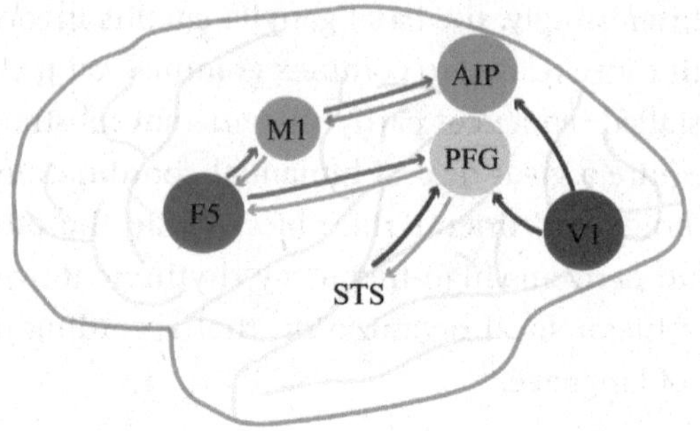

Figure 11.3
The circuits related to hand-grasping mirror neurons in the macaque brain. AIP and PFG represent superior and inferior parietal areas, respectively; F5: hand premotor area; M1: primary motor area; STS: superior temporal sulcus; V1: primary visual area.

hand-grasping, such as the hypothalamus of rodents where neurons code for aggressive behavior performed by the self or by others. Perhaps an important role of the mirror neuron network relates to action observation, linking observed behaviors and one's own motor schemas, which is highly relevant for both social behavior and imitation.[17]

Transmitting Knowledge

Toolmaking behavior spread rapidly in early human groups, being the earliest recorded cultural transmission events in our ancestors. Human culture differs from animal mechanisms of cultural transmission in that it is widespread in the population and especially that it is cumulative, rapidly generating improvements that end up in more efficient results. Still, this may not have been so evident in our early ancestors, as seen from the evolution of lithic technology that improved quite slowly at the beginning. The main process by which culture is transmitted has been proposed to be imitation, which for some authors is an innate mechanism as evidenced by facial and manual imitations of adult gestures made by newborns. However, recent studies have contested this view, again claiming that simple and general associative mechanisms linking one's own motor programs with observed others' behaviors may account for our imitative capacities, which are probably motivated by a strong social reward. As with mirror neurons, perhaps our outstanding skill is to be extremely good in making these associations, rather than having a built-in mechanism that represents the other's behavior.[18]

In this view, our large brains evolved to engage in associative learning and facilitate cultural transmission. Thus, instead of being innate, human cognitive abilities may largely rely on shared cultural knowledge that accumulates across generations. However, other authors argue that it is the other way around: that culture emerged because of our increasing cognitive skills, which were selected in the context of the stringent ecological conditions in which our ancestors had to live. Again, researchers may be artificially making a conceptual divide between two extreme positions, where the most parsimonious explanation may lie somewhere in the middle. The origin of speech and language may be an example of an interplay between both mechanisms, as I will explain in the next two chapters.[19]

Love Thy Neighbor

Social Brains

The cumulative transmission of culture only became possible by virtue of a tight social life where individuals found strong support in the group to care for the young, find food, and protect one another from predators and enemies. Cooperation, and depending on others to survive, became critical aspects for our success, although this did not preclude the existence of competition, rivalry, and aggression between members of the group, and especially between groups. Thus, the right balance between competition and cooperation, both between individuals and between groups, may have been a significant driver of our social behavior.[20]

In mammals, social behavior is regulated by a set of deep brain structures including the amygdala, the basal ganglia, the brainstem, and the hypothalamus. These structures participate in reward, arousal, and behavioral control mechanisms mediated by neurotransmitters like dopamine, norepinephrine, and serotonin, and especially the hypothalamus secreting prosocial hormones like oxytocin and vasopressin that promote social bonding and parental behavior. Another important system contributing to social behavior is the corpus striatum, a large component of the basal ganglia. More specifically, the ventral striatum relates to our desire to affiliate and belong to groups, driven by the neurotransmitter dopamine in concert with hypothalamic hormones. In contrast with other apes, humans have increased the dopaminergic innervation to the striatum, aided by neurotransmitters like serotonin and neuropeptides, all at the expense of decreasing acetylcholine, resulting in decreased aggression and increased social competence. While a balance in favor of acetylcholine in apes drives self-related behaviors and social autonomy, increased dopamine innervation in humans led to social dependency and cooperation, including the formation of pair bonds and social monogamy. The enhanced dopaminergic innervation may have also driven communication behavior and the origin of language, as we will see in the next chapter. A collateral drawback of these innovations may have been a propensity to engage in addictive and overeating behaviors. Yet, some authors claim that humans are not unique in the propensity to conditions like overweight and obesity.[21]

In the human lineage, a complex cortical network has evolved atop of this deep neural scaffolding to regulate social behavior. In fact, many of the rapidly expanding brain regions in the human cerebral cortex are involved in different aspects of social cognition and behavior including reward and motivation mechanisms, cognitive control and inhibition, and systems involved in social and action perception. Mirror neuron networks, language and face perception regions, as well as other multimodal regions encompassing the prefrontal cortex, the orbitofrontal cortex, the temporoparietal junction (TPJ), the superior temporal sulcus (STS), and the DMN make up an extended network of higher-order association areas that regulate social behavior (see chapter 10). This collection of areas has been termed the "social brain," but it is not clear whether they really represent a unified network or is a collection of systems that have been independently recruited for social behavior.[22]

Self-Taming?

Finally, proposing a framework to explain the evolution of human social behavior, some authors claim that besides domesticating other species we also tamed ourselves, decreasing aggressiveness by attenuating the stress response, and by increasing dopaminergic signaling. These features correlate with the maintenance of behavioral and physical juvenile traits, playful behavior, decreased aggression, and several physical traits. Intriguingly, many of these traits are related to the development of the embryonic neural crest, which was critical for vertebrate origins (see chapter 7). In my view, features observed in domestication may be also observed in other social animals where aggression is blunted and social learning plays a relevant role, including birds and primates.[23]

In relation to this hypothesis, some authors documented a paradoxical decrease in brain size from about 1,600 cc in early *sapiens* and Neanderthals to 1,400 cc in extant humans (although a more recent report was unable to confirm this pattern). The most popular interpretation is that this is a consequence of self-domestication, where inhibition of aggressive behavior resulted in smaller brains as a side effect, as it happens in many domestic animals (but not in all). Yet, several authors have disputed this hypothesis. Moreover, domestication features and prosocial behavior like the dependence on social reward were likely evolving in early Hominins during the paleolithic when brain size increased dramatically. It is not straightforward why there would have been a trend inversion if prosocial behaviors and cultural transmission were being selected for a long time before.[24]

Perspective

Homo sapiens emerged from a lineage of tree-descended apes, which evolved bipedalism and increased hand dexterity in association with toolmaking abilities. Many species of *Homo* existed in the last 3 MYA, which may have interbred, but the only extant species is *H. sapiens*. Brain size increased dramatically from early *H. habilis* to late *H. erectus,* probably driven by several genetic modifications. Among the few features deviating from general primate trends is the lateralization of brain functions, notably related to toolmaking and language. Toolmaking was possible by the elaboration of a notable hand dexterity and increasing hand-eye coordination. In this context, mirror neurons are a group of visuomotor neurons involved in hand-grasping that have the property of activating when observing others performing similar behaviors. Mirror neurons have been proposed to participate in behavioral imitation and other kinds of learning, including speech, and to be quite relevant for human social behavior.

The background to all these capacities lies in the intense social life that is characteristic of humans and may have been fundamental for their evolution. Increasing dopaminergic signaling, especially in deep brain centers like the basal ganglia, and other traits like enhanced signaling of prosocial hormones may have provided a strong drive facilitating childcare, social bonds, and collaboration within the group. In the cerebral cortex, a group of dispersed brain regions involved in social behavior have been collectively termed the social brain, but it remains unclear to which extent they make up a cohesive network. All these characters have been associated to the so-called domestication syndrome, featured by decreased aggression and juvenile traits. In the next chapter, I will address the most distinctive characteristic of our species: the emergence of speech.

12 Voices from the Past

Arguably, a large part of our success as a species relies on our great communicative ability that is channeled through language, widely separating us from the rest of the animals. This immense gap has been one of the biggest challenges for human evolutionary theory, as there are no traces of language evolution in the fossil record. Doubtless, language appeared in the context of a toolmaking culture, supported by an increasing sociality, but the details of its emergence are not clear at all. Here I will focus on the evolution of vocal communication in primates and how speech mechanisms might have emerged from it. A popular view asserts that early human communication was predominantly gestural and speech is only a late acquisition, but I will argue that vocal communication has a long history in the human lineage, and a continuity in its neural mechanisms can be traced to nonhuman primates.[1] Rather than a sequence of gestures first, voices later, voice and gestures probably coevolved from very early in *Homo*. In the next chapter, I will address the origin of language, which refers to the emergence of symbolic communication in our species and will discuss the role of gestures in this process.

The Speech Instinct

Before Babel

Darwin highlighted language as a product of biological evolution, proposing that speech had evolved first as a melodic system for social communication, similarly to birdsong, in tandem with the use of communicative hand and body gestures. In fact, speech is different from most other primate vocalizations, as it is acquired from exposure to adults, while the vocalizations of most monkeys and apes are largely innate. Still, Darwin emphasized that unlike other forms of learning, humans are born with an innate tendency to learn it (as opposed to writing, for example, where teachers have to struggle with their pupils).[2]

Thus, a first step in language evolution was a drive to communicate, possibly because it was a highly rewarding behavior. In early hominins, the intense social life required for survival in the open savanna may have put a special pressure on vocal behavior, which was important for communication not only between mother and child but also for group coordination and for strengthening social bonds within the group. In fact, circuits involved in reward mechanisms and prosocial behavior have been related to speech acquisition in humans and to song-learning in birds. Furthermore, this usually takes place during reciprocal interactions, where babbling human infants engage in "conversations" with adults that at least according to parents are highly arousing and rewarding, where they reciprocally exchange gestures (largely facial) and vocalizations that later develop into early speech. In this line, bidirectional vocal behavior, where two individuals take turns to mutually respond to each other, probably evolved as a social glue mechanism in our ancestors, which maintained not only a parental bonding with the child but also group cohesiveness.[3]

From Turn-Taking to Gossip

As adults, most human speech takes place in reciprocal conversations, and most of what we talk about in our conversations is gossip. Robin Dunbar has proposed that gossip (understood not just as commenting what the neighbor did but as a mechanism to maintain a conversation) was among the early functions of language, helping to tighten social bonds and contributing to social learning.[4] While instances of reciprocal vocalizations and duets are commonly observed among nonhuman primates (for example gibbons), recent findings indicate that the new-world marmoset monkeys could best resemble the vocal behavior of our direct ancestors. These tiny monkeys live in highly cohesive groups where they cooperate for breeding and intensely exchange reciprocal vocalizations that can last for several minutes, between adults but also between adults and infants. In these species, youngsters gradually develop their adult calls through listening to elders and are socially reinforced by them. Likewise, marmosets are able to combine different vocalizations, making up short "phrases," and exchange distinct utterances in different social and environmental contexts. However, marmosets are trapped in their ecological settings, while Hominins had the whole African savanna to develop new innovations and behavior.[5]

Making Voices

The Voice Organs

Before going on with the neural mechanisms of Hominin vocal evolution, a brief description of how speech is produced may be needed. Being highly vocal animals, primates display an increased morphological and functional diversification of the larynx compared to other mammals. Yet, the human larynx goes a step further in morphological innovation, attaining a lower position than in other primates, which increases the length of the vocal tract. This condition was initially hypothesized to amplify vowel sound production by increasing the cavity of the upper vocal tract. However, findings indicate that laryngeal descent is used by several species (like deer) to produce the impression of a larger body size with their moo, especially during male-male competition when mating. However, if this is the explanation, it is not clear why human females would have developed this condition too. Whatever the reasons for laryngeal descent, early humans may have taken advantage of this to diversify vocal communication.[6]

Additional studies showed that despite having a shorter vocal tract, monkeys and apes may be mechanically able to produce some speech-like movements and sounds. However, their vocal repertoire is much more limited than predicted by these findings, presumably due to neural limitations. Nonetheless, monkeys, and especially apes, have highly movable lips and tongues and use a variety of consonant sounds like "clicks," "smacks," "kisses," and even "whistles" to communicate. Lip-smacking is a common affiliative behavior used by monkeys and apes, whose rhythmicity and development match those of human lips during speech. For instance, labial consonants like the /p/ sound may be rooted in such behaviors. Some monkeys like the gelada baboons are able to generate speech-like sounds termed "wobbles" by coordinating vocal and lip movements. Likewise, orangutans have impressively movable lips and produce consonant-like and also vowel-like sounds which they can combine, even mimicking the human voice.[7]

Vocal Learning

But more than having a proper vocal tract, the capacity to produce learned vocalizations (like a parrot imitating speech) is the main requirement for speech evolution. Vocal learning is a complex trait found in several species like parrots, songbirds, bats, marine mammals, and a growing list including as said some nonhuman primates. Yet, vocal learning is not an all-or-none trait but is a continuous variable across species, and probably different species may have specialized in different vocal skills. A basic element of vocal learning is the plasticity of vocal development in early life, which presumably depends on a critical period for learning and has been related to the domestication syndrome in social animals. For instance, in marmosets and in bats, vocal behavior develops in pups with an initial babbling-like period that is modified through exposure to adults, similar in a way to the early stages of human speech development.[8]

Vocal Neurons

Vocal learning capacity has been related to the development of specific neuronal circuits. There are two main systems controlling vocalizations in the mammalian brain, one involuntary and the other voluntary that is associated with vocal plasticity in some species. Involuntary vocalizations are stereotyped utterances like human laughter or animal grunts and barks, some of which are highly contagious and may have been used as a strong social binding mechanism. On the other hand, voluntary vocalization circuits depend on the vocal motor cortex, the prefrontal regions and the auditory areas of the cerebral cortex (see figure 12.1). These circuits enable the animal to decide when or not to vocalize, and also permit to modulate some aspects of the vocalization. In a further step, vocal learners (including humans and marmosets) display a direct control from the motor and premotor cortices to the brainstem motor neurons controlling vocal musculature, as opposed to an indirect or polysynaptic control of these motor neurons in vocal nonlearning species.[9] This situation resembles the direct cortical innervation of spinal forelimb motor neurons for hand dexterity (see chapter 11). Like humans, marmoset monkeys deviate from the typical primate pattern, being able to modify their early vocalizations through exposure to adults' voices, and they have enhanced projections from vocal premotor areas in the cerebral cortex to the brainstem, perhaps representing an incipient step in the evolution of vocal plasticity. Humans further

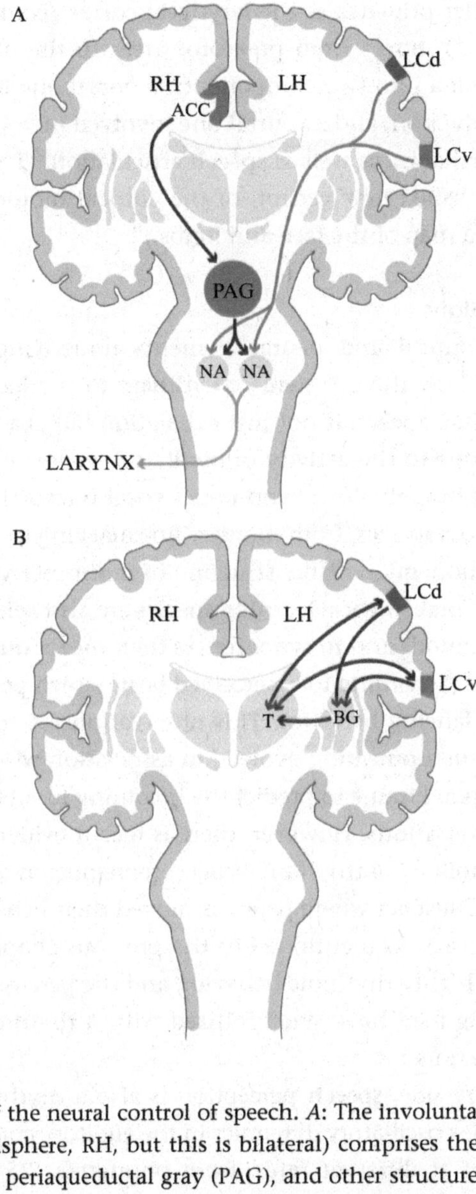

Figure 12.1
Some elements of the neural control of speech. *A*: The involuntary circuit (depicted in the right hemisphere, RH, but this is bilateral) comprises the anterior cingulate cortex (ACC), the periaqueductal gray (PAG), and other structures (not shown). The voluntary vocalization circuit (depicted in the left hemisphere [LH]) includes the dorsal laryngeal cortex (LCd) involved in pitch processing, the ventral laryngeal cortex involved in syllabic processing (LCv), and their direct projections to the nucleus ambiguus controlling the larynx (NA). *B*: Connections between the laryngeal cortices (LCd and LCv) with the thalamus (T) and the basal ganglia (BG) are shown in the LH. There are also contributing cortico-cerebellar circuits involved in speech control that are not shown.

diverge from other primates as the laryngeal cortex (controlling the muscles of the larynx) moves from premotor areas to the motor cortex and becomes subdivided into two components: a dorsal one involved in pitch and melody production, and a ventral one involved in vocal sequencing of phonemes and syllables as well as speech articulation. (This is in line with the recently published new version of the cortical motor "homunculus," with a duplicated map of the face and limbs.)[10]

Voice, Ear, and Sight

Given the anatomical and neural elements underlying voice making, another issue is how these systems coordinate to produce a comprehensive acoustic signal. Speech is not just exhalation but is a highly rhythmic process, largely due to the activity of oscillatory generators located in the brainstem controlling the lower and upper vocal tracts. This is observed at all levels and across species, from monkey lip-smacking to the "wobbles" of geladas to the phonemic-syllabic structure of marmoset vocalizations and human speech. Notably, vocal learning species are also able to "sense" non-vocal rhythmic sounds and to synchronize their motor outputs, even their body movements, according to a perceived beat, which possibly makes the foundations of dancing behavior. This observation led to the hypothesis that rhythmic synchronization evolved in association with vocal learning, providing a temporal frame to predict the incoming inputs and to establish turn-taking conversations. However, there is recent evidence of non-vocal learner animals following rhythmic beats (I remember as a kid a documentary by Jacques Cousteau where pelicans moved their heads synchronizing with a musical beat). As mentioned in the previous chapter, human tool-making is also a highly rhythmic behavior, and the percussion provided by stone hammering may have synchronized with a rhythmic proto-speech stage in our ancestors.[11]

On the sensory side, speech perception is also a rhythmic process that relies on a complex oscillatory dynamics in the auditory cortex that matches the speech input at different rates, from phonemes (25–35 Hz, Gamma activity) to syllables (4–7 Hz; theta frequency) and words and phrases at increasingly lower frequencies, providing a framework for prosodic and grammatical processing. (Syllable-selective neuronal ensembles are also found in the marmoset auditory cortex.) Musical instruments are played at a range of frequencies peaking around 1 or 2 Hz, close to the level of syllable frequencies. Many authors now consider that brain oscillatory patterns are somehow related to the hierarchical structure of grammar, but the details on how this takes place are still debated. In speech perception, this cyclic activity engages with rhythmic visual input conveying face and lip movements, and with the activity of motor cortices that code for the hearer's own speech production. Thus, visual networks connect with auditory areas and vocal systems at different rhythms, making a triangle connecting phonemes, visemes (the visual correlates of phonemes, especially in vocal and labial consonant production), and "articulemes," the hypothetical motor engrams of phonemes (this term was coined by Maëva Michon, then a postdoc at my lab). However, although adults and children may benefit from motor mechanisms and visual input to learn and process speech, they are not strictly necessary for speech development, as seen from early blind people that have normal speech. In the same line, domestic animals like dogs can understand human words, despite being totally unable to produce them. This suggests that the capacity to perceive words may rely on the deployment of general sensory categorization mechanisms partly shared with other species.[12]

Melody

Another unique feature of the human voice is the capacity to sing. Darwin and many recent authors proposed that speech evolved first as primitive melodies, a hypothesis that now is known as the musical protolanguage, including melodic and body rhythmic signals that may have been collectively expressed as dance behavior, and was used for social cohesion and behavioral coordination. The melodic contours of our early vocalizations probably emerged as a form of prosody, modulating pitch and intensity to convey emotional signals or simply to capture the attention of the listener. In modern speech, prosody (which is also cyclical) contributes to parse the speech signal, helping to segment different components in a phrase, and is associated to slow-wave oscillatory activity in the auditory cortex. Notably, although they may share a common origin, music and speech neural circuits seem to have diverged in evolution, as there is a segregation of song-selective and speech-selective neural populations in the human auditory cortex (remember also that the laryngeal motor cortex contains a pitch-related component and an articulatory component). In fact, a premotor region (area 55b) has recently reported to be a hub for musical processing. In most humans, music and prosody are more associated to the right cerebral hemisphere and speech production is biased to the left hemisphere. In a way, the hemispheric separation of pitch and structure may resemble the hand coordination for toolmaking, where both hands specialize in different subtasks to yield a final product. Note however that this lateralization of functions is more a bias rather than an absolute separation. Speech and music perception are largely bilateral processes at least at the lower levels of phoneme and pitch perception. Hemispheric lateralization seems to amplify both at high-level networks that involve working memory and other functions (see the next section), or at the motor level when generating speech structure.[13]

Speech Genes?

Several years ago, a gene called FOXP2 was found to be associated with inherited speech impairments, particularly involving vocal articulation (see chapter 11). FOXP2 and related genes participate in cortico-basal ganglia circuits that contribute to vocal motor learning and perhaps other behaviors. Furthermore, the gene is involved in vocal behavior in other vocal learning species including mice, bats, and songbirds. In humans, beside the noncoding accelerated regions described in chapter 11, FOXP2 has accumulated two point-mutations that are not present in the same gene of monkeys and apes, an event that according to recent evidence may have taken place quite early in Hominin evolution. Thus, increasing hand dexterity and an incipient vocal plasticity could have evolved together in our ancestors, partly driven by selected mutations in genes controlling the development of descending cortical projections.[14] Nonetheless, apart from these punctual findings, the genetic substrate for speech remains elusive. A recent study identified several regulatory genes in mammalian vocal learning species, some of which have been also associated with human speech disability.[15] Hopefully this will provide more insight into the evolution of our most cherished capacity.

Birds Again

Finally in this section, I will make a brief digression to speak about birds again, this time rivalling or excelling mammals in vocal behavior. Vocal learning is particularly noticeable in some species of birds, particularly parrots, songbirds including ravens (oscine birds), and hummingbirds. These three groups are distantly related, and probably evolved this capacity separately. Just like in the evolution of the brain, the vocal organs of birds rely on structures that are not homologous to those in mammals. Instead of a larynx, birds display a syrinx located at the base of the trachea, which is innervated by the hypoglossal nerve and nucleus, instead of the vagal nerve and the ambiguous nucleus as in mammals.[16]

Noteworthy, birdsong and parrot vocal learning develop through stages not unlike infants' speech, where the young first develops a "subsong" (comparable to infant babbling) consisting of a wide repertoire of sounds. This is subsequently refined by narrowing the range vocalizations according to the template provided by the parents, yielding the adult vocal pattern as it happens in marmosets and humans. Together with these behavioral skills, the birdsong brain contains a circuitry intriguingly similar in design (but not homologous) to the human speech networks, consisting of separate but interconnected components involved in song perception and learning, and an output component located in the avian amygdala. As in vocal learning mammals, the latter sends direct projections to the brainstem motor nucleus that controls the syrinx (this projection has not been found in in vocal nonlearning birds). Notably, this circuit is strongly lateralized to one hemisphere, in some species to the left and in others to the right, which points to the possibility that speech lateralization may respond to mechanisms intrinsic to vocal control.[17]

Moreover, birdsong is not only learned but also shows some internal structure that some authors have likened to a rudimentary syntax. Similar to speech, birdsongs are composed of sequences of syllables composed by "motifs," but these are usually arranged in a fixed order, with little signs of combining syllables in different orders. However, some studies have reported the capacity of songbirds to change syllable ordering during song learning and in adulthood, making controversial parallels with early human grammar acquisition (see next chapter).[18] Finally, since birdsong is socially learned, different populations have developed song dialects in different territories (group dialects have been reported in mammals too, particularly in cetaceans). Therefore, Darwin's intuition of convergent mechanisms underlying early human speech and birdsong learning turned out to be largely correct not only at the behavioral but also at the genetic and neuronal levels.

A Brain for Speech

Broca, Wernicke, and Beyond

Human speech substantially diverges from the previously mentioned forms of animal communication by virtue of a much more complex structure that conveys equally complex meanings, enabling us to talk about the world, the future, the past, and ourselves. Besides the auditory and the vocal motor cortices, speech (and language) relies on a large-scale cortical network involving higher-order areas and their long-range connections. The language circuits are typically depicted as a motor-controlling region (Broca's area) in the ventrolateral prefrontal cortex, usually in the left hemisphere, and an auditory component in the left superior temporal lobe (Wernicke's area), both connected via a tract called the arcuate fasciculus (see figure 12.2). However, these regions are not well delimitated anatomically, which has

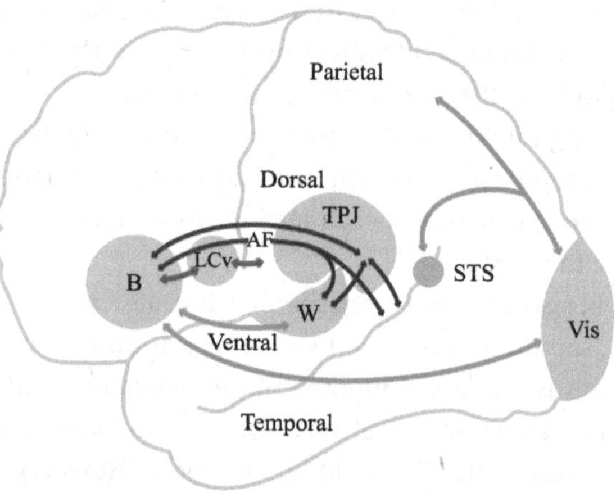

Figure 12.2
Circuits involved in speech production and perception. The so-called "language circuit" encompasses an anterior, predominantly motor component (B: Broca's area, located in the ventrolateral prefrontal cortex) and an auditory-related perceptive region (W: Wernicke's area). Both regions are connected through a ventral stream along the temporal lobe (light gray arrow between B and W), and through a dorsal stream (black arrows), via the parietal lobe. Note that the dorsal stream includes direct projections via the arcuate fasciculus (AF) and indirect projections between W, TPJ, and B. Also shown are the connections between B, the laryngeal cortex (dark gray arrows; for clarity, only LCv is shown), and the somatosensory cortex. The ventral and dorsal visual streams (and the third visual stream to the STS) are shown for comparison.

led to many discussions about their functions and localization practically since they were first discovered. Furthermore, these areas are not strictly speech-dedicated areas but may contribute to several other functions besides speech. Some studies suggest that these brain areas work by coordinating whole-brain distributions of networks for processing language. On the other hand, other authors emphasize the difference between being active during a specific behavior (as seen in brain activation studies, which may display large-scale organizations), and being strictly necessary for that behavior (as evidenced for lesion studies, that show a more restricted set of brain regions). In any case, the "language-related circuit" has served as an extremely useful reference for neuro-linguistic research. Furthermore, some recent analyses have shown an expansion of the areas corresponding to the classical Broca's region in humans compared to other apes, which fits the evidence of Broca's area differentiation from early *Homo* to modern humans (see chapter 10).[19]

More recent findings, largely based on modern brain imaging techniques, have shown that the network connecting both areas is far more complex than was originally envisaged. First of all, early comparative studies had shown that the arcuate fasciculus is not the only tract connecting these areas through the parietal lobe, there being contributions of neighboring tracts as well. Particularly, the temporo-parietal junction (TPJ) is a multimodal region connected to both Wernicke's and Broca's areas, contributing to verbal working memory and other language-relevant functions. Secondly, Broca's region also receives auditory projections (that mix up with visual projections) from the anterior temporal lobe. Thus, similarly to the two-streams organization of visual areas, a double-route has been depicted for language processing, with a dorsal stream via the arcuate fasciculus and related tracts that is related to phonological and auditory-vocal transformations, and a ventral stream that processes words and their meaning. There are also contributions of other neural tracts, particularly from the visual cortex, that will be reviewed in the next chapter.[20]

Before we go on, we have to recall that like all other cortical circuits, these are strongly bidirectional with top-down multimodal contextual influences and bottom-up inputs conveying sensory novelty. This large-scale, bottom-up, and top-down dynamics provides a multimodal context to predict the other's speech during a conversation, which enables not only anticipating the following phonemes or words but also contributing to extract meaning out of the unfolding sentence. Finally, it never hurts to insist that these cortical networks are embedded in large-scale circuits comprising thalamic, basal ganglia, and also cerebellar components besides cross-dialoguing with other cortical areas. Subcortical components like the cerebellum and basal ganglia not only work in learning and coordinating vocal motoric acts but also participate in higher-level processes involving syntactic and conceptual mechanisms. We still do not know the details of how these overall engrams become executed, but these questions promise to become a highly exciting area for future research.

The Speech Loop

The existence of relatively localized brain regions and their connections supporting speech planted a critical but largely ignored question about the evolutionary origin of these areas, and the possibility of corresponding networks in the nonhuman primate brain. In 1997, together with Ricardo García, we proposed an anatomical framework for homology between the human language circuits and auditory-prefrontal circuits of the monkey brain, highlighting a strong similarity of the areas involved and their connections. This circuitry would be characteristic of the primate brain, predating the origin of our species. Furthermore, we postulated that the dorsal auditory stream referred to (see figure 12.2) had critically expanded in human evolution, partly as a result of the allometric expansion of temporoparietal and prefrontal areas but also due to mechanisms that promoted brain lateralization and the anatomical asymmetry of these tracts. This view gained strong support from a series of subsequent comparative studies that evidenced a highly conserved but progressively amplifying pattern of dorsal stream auditory-prefrontal connections from monkey to chimpanzee to human. In addition, these studies have highlighted the strong convergence of ventral visual and auditory streams into the ventrolateral prefrontal cortex (where Broca's area is located) both in monkeys and humans. This may enable animals to associate faces and gestures with vocal sounds, and may also have served as a feedstock for the early emergence of speech and the production of early referential meanings, like associating a word with an object category (i.e., "stone," meaning a piece of rock).[21]

But the striking similarity in connectivity between apes, monkeys, and humans (leaving aside the obvious difference in brain size) does not provide a clue of how these circuits became involved in human speech. In our articles, we proposed that the expanding dorsal stream to Broca's area was relevant for auditory-vocal working memory capacity, just like the dorsal visual stream contributes to visuospatial working memory in monkeys and humans. Particularly, phonological working memory (also referred to as the phonological loop) enables us to maintain phonological information "in mind" for a brief time, as when recalling a sequence of meaningless syllables or a telephone number before dialing it (see chapter 10). This capacity may not be strictly relevant in everyday language when we usually rely on stereotyped speech, but is required for understanding complex sentences and especially for speech learning. Considering this, we hypothesized that increasing auditory-vocal working memory capacity contributed fundamentally to amplify the learned vocal repertoire of our species and was a key acquisition, before the origin of speech. Also supporting this notion, behavioral studies indicate that while monkeys are quite proficient in visual working memory tasks, unlike humans, their auditory short-term memory is quite poor, being unable to recall physical or vocal sounds after a few seconds.[22]

A Left Bias

One of the most striking features of the language network is the strong lateralization of these areas, where most people tend to have left-hemisphere language dominance. Moreover, modern anatomical studies show that the dorsal auditory stream, especially the arcuate fasciculus, involves a larger and more robust tract in the left hemisphere than in the right, a feature that is present since birth. Many authors have related language lateralization to the development of handedness and toolmaking in our ancestors. In this line, there could have been a confluence of hand and vocal brain asymmetries, associated with the development of toolmaking. Yet, the lateralization of song in the songbird brain does not fit this scheme as there is no clear instance of "handedness" ("footed-ness") in these species. This points to mechanisms intrinsic to vocal communication that drive lateralization in songbirds, mechanisms that might have also contributed to lateralization in humans.[23]

Voice and Gestures

Body Language

As said earlier, there is a long-standing hypothesis that proposes that language first arose as manual gestures and that speech is only a recent product of human evolution. This view was reinforced in the 1970s by the finding that apes could be trained (after years of painstaking training) to use hand sign language, being able to utter two-word combinations similar to two-year-old children, but were never able to learn any word order or grammar. Moreover, the discovery of mirror neurons in the monkey, some of which were located in a premotor area called F4, assumed (wrongly in my view) to be homologous to Broca's region, further supported this view. In this perspective, rather than deriving from monkey vocalizations, the neural substrate for human speech would have emerged from mechanisms of hand control that had somehow migrated to the vocal system. Nonetheless, this view contrasts with the extensive recent evidence of homology of vocal control between humans and nonhuman primates that I outlined previously.[24]

In this context, a recent hypothesis proposes that there is a tradeoff between motor dexterity and vocal plasticity in different species (for instance, most nonhuman vocal learners swim or fly), to which humans (and marmosets) are the main exception in which dexterity and vocal learning evolved together.[25] An alternative explanation to this hypothesis is that there are social or ecological drivers for vocal plasticity in fliers and swimmers (like birdsong and echolocation). Besides, despite not having hands, songbirds and parrots can be extremely skilled with their foot and beak when it comes to manipulating objects ("footed-ness" is observed in some species of parrots, but this is not as clear as in humans). In my view, there may not be fundamental neuronal constraints to the evolution of vocal plasticity as the gestural theory or the tradeoff hypothesis imply.

To insist, there is probably a significant cortical and subcortical crossover of the neural circuits involved in manual control and speech. My point is that vocal plasticity in hominins could be a more ancient character than posed by the gestural hypothesis, and that it contributed to reciprocal, socializing behavior both in mother-infant and in adult-to-adult communication before it had any linguistic or proto-linguistic features. Rather than a strict hand-to-mouth sequence as prescribed by the gestural theory, prelinguistic gestural and vocal communication mechanisms may have coevolved from very early in Hominins, yielding increasingly skilled motor control of both mouth and hands, all associated to enhanced sociality and an emerging toolmaking culture. The origin of complex, structured language conveying complex meanings is another story, which I will discuss in the next chapter.

Early Babblers

Supporting the above view, infant development points to a very early onset of vocalizations, starting in the fetus and followed by babbling soon after birth. Babbling represents a first stage in the development of speech and vocal learning, evolving from crying and cooing at birth, simple speech sounds at three months, and the first syllables by five months. Moreover, while speech perception may largely rely on general perceptual mechanisms, infants rapidly develop a robust word recognition system by four to six months of age. Vocal imitation starts at about ten months of age, together with conversational babbling that precedes the first words that appear by one year old. On the other hand, manual gestural development starts at about nine months with simple expressions like give, reach, and so on. Pointing, a critical communicative gesture, appears between seven and fifteen months of age (see the next chapter).[26] This evidence is again consistent with an early development of vocal plasticity in our lineage, as it seems unlikely that the preverbal babbling phase of infants, where they naturally reciprocate prelinguistic vocal utterances with adults, appeared only recently in our history like the gestural hypothesis seems to imply. Finally, in my opinion, the gestural hypothesis fails to account for the evolution of musicality, especially prosody and melody, which may largely derive from the vocal system and, like I have mentioned, may have a long history in Hominins. This pattern of multimodal, prelinguistic communication probably lasted for long in hominin evolution, while the origin of modern symbolic language may be a much more recent event, to which gestural communication may have significantly contributed (to be discussed in the next chapter).

Perspective

Language is probably one of the main innovations that set *Homo sapiens* far apart from the rest of the primates and other animals. There is an instinctive drive to learn to speak in infants, which indicates that together with imitation, social reward is a key factor in communication and was an important driver for the origin of human speech. In early *Homo*, vocal communication was used as a "social glue" between adults and between mother and child that served as a cohesive force within groups and distinguished these groups from others as it happens in other vocalizing primates and mammals. Particularly relevant in this context may be the reciprocal vocal exchanges that presage the development of conversational duets between infants and adult humans, and are also observed in marmoset monkeys. Several authors have suggested that these early vocalizations were a mixture of rhythmic and melodic-prosodic vocal sequences, a sort of "musical protolanguage" from which both language and music emerged.

A prerequisite for speech is vocal learning capacity, a character found not only in humans but in other species like songbirds, bats, marine mammals, and to some degree marmoset monkeys, which is supported by enhanced cortical control of the larynx (and its bird equivalent). In addition, humans have recruited a long-range cortical circuitry encompassing higher-order auditory areas (Wernicke's area) and prefrontal regions (Broca's area), especially in the left hemisphere. These areas are connected through dorsal and ventral streams, emulating the organization of the visual system, a circuit that is also found in other primates. In humans, the dorsal stream has expanded noticeably compared to other primates, contributing to the expansion of the human's vocal repertoire by the generation of a phonological loop.

However, the evolution of human communication has been (and still is) a largely multimodal process where manual, face, and body gestures make an important component. Many authors have argued that human communication emerged first as body and hand signs, with speech appearing as a late acquisition. Among other things, this perspective disregards the evolution of prosody and melody, which are fundamental components of human vocal communication. Alternatively, our view is that gestural and vocal communication have strongly cooperated since very early in hominin evolution, and that they share neuronal mechanisms that have evolved together. In the next chapter, I will go beyond the articulatory and behavioral mechanisms of speech and will try to deepen into the neural mechanisms that enable us to share our minds and knowledge through symbolic language and social life.

13 Sharing the World

In the previous chapter, we discussed the evolution of vocal communication in Hominins, which was largely used for increasing social bonds among members of the group. However, more than a complex vocal system, what makes us different from other animals is that we are able to transmit symbolic contents, or meaning to others. Like Descartes said, only humans can communicate their thoughts. In fact, language enables us to describe the world, to plan complex actions, envision the future, teach to others more efficiently and ultimately, organize in cohesive groups that last through time, constrained by shared oral and written traditions. It even boosts our minds, enabling us to higher levels of consciousness and to construct an autobiographical memory. Thus, language is inextricably bound to our humanity, joining us in a shared reality that emerges through the social transmission of symbols, meanings, and beliefs that are all connected through our brains.[1] How we started to develop a world common to all of us, and how this might take place within and between our brains will be the subject of this chapter.

Look Me in the Eye

Social Life

Many animals share a living. They often cooperate in reproductive tasks, respond to predators together, and hunt and eat in group. Lions, wolves, and chimps hunt in packs, but their cooperative behavior may stem from individual agencies rather than from an agreed plan. When lions ambush zebras that run from another lion, they have not specified what each lion does during the hunt, but given that one is already stalking at the prey, the others simply predict the escape direction of the animals and wait for them there. However, our ancestors gradually evolved to a stage beyond this, in which they became able to consent simple behaviors like toolmaking and planning for the near future. The development of learned cooperation and reciprocity contributed to consolidate shared interests among individuals and helped the emergence of a culture (for example, fire control and toolmaking) that constructed a common niche for living. They implicitly shared a reality, although they were very limited in their ability to communicate about it. The emergence of language provided this, the possibility of making an agreeable universe to all the members of the group. Firstly, language allowed us to give names to things and to other individuals, that is, creating symbols representing specific things or events. Secondly, by generating a grammar, we were able to depict the details of complex actions or events.[2] Thirdly, language spread through social learning mechanisms, where linguistic items (and their meanings) became replicated across individuals and generations, generating powerful cognitive tools that catapulted culture evolution to make us the unique species we now are.

"We" Intentions

Before we became able to manipulate symbols, there is perhaps a basic mechanism that provided a starting point for this vertiginous sequence of events: unlike other animals, our ancestors became able to participate in communal events like dancing and toolmaking, probably aided by vocal and gestural exchanges. We developed what is called a shared intentionality, implying not only that two or more individuals attend to the same thing, like lions hunting a zebra, but also that they know they are all attending to the same thing, so they may cooperate to achieve a common goal. An early stage in the evolution of this capacity was to detect what others were looking at since this provided information about relevant events or things. Actually, apes are able to follow another's gaze for this purpose, indicating that they have a sense of others' perspective. However, similar abilities have been observed in birds and even in fishes (chapters 8 and 10), which suggests that this may not involve high-level neural mechanisms but rather represents a skilled prediction mechanism based on experience. Some reports have proposed that the white sclera in humans' eyes (which is less visible in apes' eyes) evolved to signal an individual's gaze, but other studies have disputed this proposal.

Furthermore, following another's sight does not necessarily mean that both subjects are sharing their perceptions but simply that one may be taking advantage of the other's point of view. Perhaps more than tracking another's gaze, mutual eye contact became important to hominins for "seeing each other." Among primates and in some human cultures, eye-to-eye contact is an intense social signal that may trigger aggressive responses and tends to be avoided. But in highly social primates and especially infants, eye contact works in an opposite way, favoring playful and prosocial behaviors. Our ancestors may have reinforced eye-to-eye contact with each other (perhaps an instance of retaining juvenile traits), associated with a decrease in aggressive behaviors and an increase in mutual engagement between group members. The relevance of eye contact is evident in certain clinical conditions like autism, where mutual gazing tends to be avoided.[3]

There is a third stage following gaze tracking and mutual eye contact that is even more suggestive of the capacity of two individuals to knowingly share their attention to some external object or event. Hand-pointing is a human-specific behavior, probably deriving from a hand-grasping action, where two individuals explicitly agree on what they are looking at (see the previous chapter). Some captive apes have learned to understand and sometimes use pointing when interacting with humans, but they do not do it spontaneously among themselves. In contrast, at seven to fifteen months, human babies actively try to direct others' attention to objects in their surroundings by hand-pointing to things that are of their interest, many times together with babbling. Interestingly, eye contact, babbling, and pointing contribute to the development of the first words in children. As I said in the previous chapter, hand pointing develops at about the same time as babbling, and the coordination of these two features is important for subsequent language and social development. In early Hominins, reciprocal vocalizations, mutual eyesight, and hand pointing may have contributed to establish a common space where two (or more) subjects shared emotional states and the presence of external events. They started to have a world in common.[4]

Minding Others

Many authors have related shared intentionality with the notion of theory of mind (usually termed ToM, also called mentalization), or the capacity to infer mental states in others. It is said that we take for granted that other people have minds like each of us, although we cannot observe their mental states. Thus, we just assume that others have intentions, desires, beliefs, thoughts, and knowledge about things. This capacity is related to a set of components of the "social brain" including the Default Mode Network (DMN) mentioned in chapter 10. The mirror neuron system (chapter 11) has been proposed to be involved in mechanisms of empathy and mind attribution by establishing a link between the observed behaviors (like crying) and our inner states (sorrow). Again, autistic subjects have been claimed to be deficient in their ToM, which would be a consequence of a dysfunctional mirror neuron system, but this is a contentious issue. ToM develops gradually in children; from eighteen months of age (in the two-word stage), toddlers can understand that others may have different taste from theirs, offering others the foods the latter may prefer instead of those they themselves like. At about four years old (when they already understand some grammatical rules) children are able to infer the knowledge of others about more complex situations. In a classic experiment, children are told that two characters, Sally and Ann, put together a toy in a box (box A). When Sally leaves the scene, Ann moves the toy to another box (box B). Then, Sally comes back into the play. The child is asked where will Sally search for the toy, in box A or in box B? Children under four years of age choose box B (where the toy is now), but children over four correctly choose box A, evidencing that they know that Sally has a false belief about the toy's position.

Nonhuman primates are considered to lack a ToM, although some clever experiments have shown that chimpanzees and even crows are able to assess what another subject knows or does not know about certain events, perhaps in a stage comparable to one- or two-year-old children. This indicates that they have subtle observational and predictive capacities, but they may not be actually assuming that others have a mind of their own as we do while using language. As I will argue, while these predictive capacities may have provided the necessary background for the origin of mentalizing, in my view the ToM unfolded together with language and social knowledge, when it became explicit that others may know things one doesn't know, or they may not know things one does.[5]

Words

What's in a Name?

Besides mutual recognition and sharing their attention to external events, at some point our ancestors started to refer to events or things by using vocal or gestural signals that represented these situations, rather than pointing or forcing the other's attention into something. Communicative signals to external events like alarm calls are common among animals, but these behaviors are mostly innate mechanisms produced by natural selection.

There are some examples of learned, referential signals in wild animals, most of these using vocal signals. Vervet monkeys living in the African savanna employ distinct alarm calls for different predators, to which specific escape responses apply. Thus, the best escape from leopards is running to the height of trees, to small branches that can't sustain the predator; while if there is an eagle it is more convenient to hide inside bushes, and when there are snakes the whole group stands on two legs to watch around. Each of these predators is signaled by a different alarm call, whose structure is innate, but young monkeys learn to use them in the correct circumstances. At least in birds, alarm calls have been found to trigger visual search mechanisms for predators, which suggests that the signals may induce visual recognition mechanisms despite being innate in their structure. Additional species with related abilities are marine mammals. Dolphins develop a uniquely personal vocal sequence from early life, which serves as a signature of its identity and contributes to form social alliances and group organization. Other animals like elephants and parrots display vocal signatures as well. Both kinds of signals, vervet calls and individual signatures, are quite different from human words, but they are instances of vocalizations referring to individuals or external events, a capacity that may have been co-opted for the emergence of early proto-words by Hominins.[6] As I said in the previous chapter, our ancestors may have also used hand and body gestures to convey meanings. However, I do not know of instances of gestural referential signals in wild primates that may be comparable to vervet monkeys' or dolphins' calls. Rather, the gestural repertoire of nonhuman primates is largely restricted to behaviors like begging for food or waving to others to follow them.

In our ancestors, a more complex kind of reference signal took place when the vocalization or the gesture resembles the meaning they intend, either by vocal mimicry (the sound of the wind, water running, animal sounds) or by making gestural pantomimes depicting animals, trees, small, big, close, far, and so on. Notably, many kinds of human vocal utterances can be understood across cultures as they bear a sound-object association that may be acquired very early in life, like "r" signaling roughness in many languages. Likewise, individual or group names are likely to have been conveyed by vocal signals with which each subject identified herself or himself, perhaps similarly to what dolphins do. Yet, in humans, gestures can be quite eloquent to transmit simple meanings, and some studies suggest that across cultures, they can be more efficient for this than nonverbal vocalizations. Again, it is most likely that our ancestors used a mixture of gestures and vocal utterances to convey their messages about the world and about different members of the group. In a way, they used their gestures and voices opportunistically, just selecting those signals that better fulfilled the communicative goal in each instance.

A further step was when words (gestural or vocal) acquired meanings whose sound (or gesture) was unrelated, or very distantly related to the meaning they intend, as in "cat" depicting a small felid carnivore. At this point, communication became truly symbolic, where the sign does not resemble the meaning it intends. Notably, hand stencils painted in caves, perhaps made by Neanderthals, sometimes appear missing digits like as if the subject was making a hand-sign by retracting some of her/his fingers, which could represent the earliest form of writing as well.[7]

A fundamental feature of naming objects or events is that we can imagine together things or situations that are not present, sharing only our thoughts about them. In addition, names also implied a social order, where each individual had a role and an identity. In a way, we started what Descartes called the process of communicating thoughts, where objects, individuals, and the relations between them were represented by common signals.

Names in the Brain

One hypothesis for the origin of names is that humans have amplified direct cortical links between different sensory modalities, while nonhuman primates needed to establish these links via connections with the amygdala. This enabled our hominin ancestors to name objects by associating them with sounds or gestures. Furthermore, the superior temporal sulcus (STS, the “third visual stream” referred to in chapter 10) and neighboring areas work as an interface where socially relevant visual and auditory signals converge and could have worked as a link between the auditory cortex and the visual cortex in the translation from sound to visual meaning (see figure 13.1). Still, enculturated animals are able to understand human words, be them spoken or by signs, indicating that they are able to perform cross-modal associations to some degree. However, they are much slower than humans in learning this skill, who do it spontaneously without need of training.

Another region that has been associated with semantic processing is the inferior parietal lobe, especially the temporo-parietal junction (TPJ) that represents another multimodal region where auditory, visual, and somatosensory modalities converge. Yet, while the STS and the TPJ may represent interfaces between the speech or language regions and visual contents (and

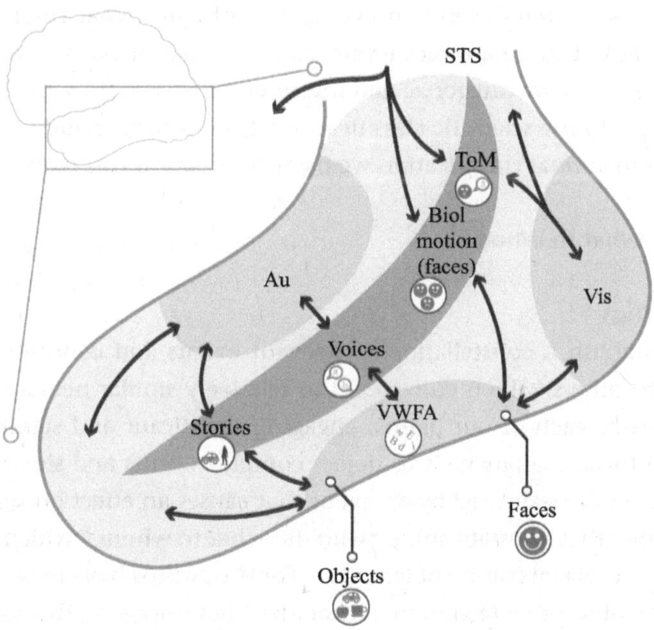

Figure 13.1
Connections of the superior temporal sulcus (STS). The STS (in dark gray) makes a link between auditory (A) and parietal areas with visual regions in the inferior temporal lobe processing faces, objects, and written words (VWFA). The STS is a multimodal region where several socially relevant stimuli like theory of mind (ToM), biological motion of bodies, and faces, voices, and words, including narratives, are being processed.

other modalities too), semantic networks are not restricted to the temporal and parietal lobes but are widespread through the brain including frontal areas, the Default Mode Network (DMN), and also subcortical components involved in emotional and procedural memory.[8]

As the child grows, she learns to associate words with distinct concepts she is building in her mind. Some words may help disentangling some overlapping concepts like "vapor" and "gas," although whether the language system makes a difference in how we internally perceive or categorize the world is a matter of debate. A recent study found that color word learning affects color categorization in children. Other studies indicate that the number of color words in different cultures reflects the use of colors in the culture (like Inuits having about forty words for white), rather than the perceptual ability. Perhaps there are large color categories that remain invariant across cultures, but at a finer level, different idioms may make a difference (sometimes even two people cannot agree what color they are seeing!).[9] But at more abstract levels, language may make a fundamental relevance of how we categorize and interpret the world, from the early creation myths to the scientific classification of the world, including the controversies in animal classification we have discussed in this book.

Who Did What to Whom?

Word Strings

Names establish a constellation of relevant events and concepts that we share with others, which correspond to relatively similar neural representations inside each of our brains, engaging significant and signified. The next step toward language is to depict complex events and situations, for example, actions executed by an agent that causes an effect on something or someone, that is, establishing "who did what to whom," which to some linguists is a central feature of language. For this, words have to be arranged in a hierarchical order (a grammar) that identifies the agent, the object, and the action being performed. This can be viewed as an algorithm that permits the bidirectional translation between a working memory of (usually) visual scenes, on the one hand, and a working memory of word sequences on the other, making up a comprehensible sentence that describes these events. Said another way, word order and the intercalation of connective words specifying roles to each of the elements of a sentence provide the sufficient details to reconstruct a scene. Thus, we can talk and refer to practically everything we see around us.

A fundamental question is whether grammar results from arbitrary convened social rules or is innately embedded in our minds. Supporting the latter view, many linguists assert that children learn grammar with minimal exposure to it. However, children are hearing adults speaking a lot of their time, and it is possible that they make generalizations of some learned rules by using basic cognitive and social learning mechanisms, not requiring an inborn grammar circuit to do so. While there is no discussion as to grammar being the result of human genetics, what are genes precisely doing to achieve our grammatical competence we do not know. Unfortunately, there is no evidence for genes specific for grammar, nor for a discrete language circuitry unique to the human brain (Broca's and Wernicke's areas are highly multimodal regions, with homologues in nonhuman primates). Another approach could be that this capacity depends on general learning and memory capacities that our big brains have acquired through evolution, which have been canalized through cultural development. Again, perhaps a critical factor involved in the acquisition of grammar has been the amplification of associative and working memory systems to maintain active linguistic and cognitive patterns while they are being learned.[10]

Not surprisingly, the processing of distinct grammatical components at different hierarchical levels encompasses the language circuits (especially Broca's area and the auditory-visual ventral streams) and surrounding regions, together with basal ganglia circuits that contribute to sequential programming. Yet, the details of the neural mechanisms for grammar remain a mystery. Following computational models of language production, there may be at least two main operations involved in grammatical processing: first, extracting structural regularities through statistical associations of those sequences that are more commonly presented; and secondly, using top-down contextual information to anticipate and process the incoming words. Notably, recent brain-computer interfaces have been able not only to decode and reproduce both visual images and short phrases from brain activity but also to translate these images or phrases into phrases or images, respectively.[11]

Grammar versus Meaning

Besides claiming that grammar was present only in humans, the great linguist Noam Chomsky separated the syntactical organization of language from the meaning or semantics it conveys. In fact, there can be syntactic organization without explicit semantics. Music has a structure that may be similar in several aspects to that of language, despite not conveying a linguistic meaning but producing a harmonic organization that impinges onto emotion. Similarly, some authors have used so-called artificial grammars, that is, organized sequences of abstract items, to assess syntactic competences in humans and other animals. Using this approach, early experiments reported a limited or simply no capacity to learn grammatical rules in monkeys, which supported Chomsky's view. However, more recent studies found that songbirds, marmosets, whales, and even chimps can use simple rules of vocal sequencing to transmit meaningful messages (like discriminating an alarm call from a gathering call, or chimps using distinct vocal sequences in relation to past events). Other reports indicate that monkeys, apes, and songbirds are able to understand relatively complex sequences of sounds to which they are exposed, and human infants recognize these relations before developing any grammatical competence. Similar findings have been obtained in songbirds and whales, who learn hierarchically organized songs with one or two notes.[12]

In this context, an important property of grammar is combinatoriality. Simply, this refers to the capacity to combine a finite number of syllables into a practically infinite number of meaningful names and sentences. As mentioned, there are reports suggesting combinatoriality in songbirds, and other species including chimpanzees. One possibility is that this function derives from a cognitive process known as "chunking," whereby different elements can be decomposed into basic units and bound again to generate new, organized combinations, as in a Lego game. From a functional perspective, combinatoriality may have emerged as a solution to the constraints of sound (or gesture)-to-meaning translation, where a limited set of motor production mechanisms had to fit increasingly complex semantic representations with minimal ambiguity. Assembling ever longer strings of elements could become highly inefficient for communicating messages, and means to compact and recombine them may have provided great benefit.[13] After all, grammar has evolved in humans by virtue of its explicit communicative properties (it has no other use). In the next section, I will refer to the evolution of a meaningful grammar that permits to depict scenes, actions, and events that are fundamental for complex communication.

Action Observation (and Description)

As children mature, they acquire the concepts and categories not only of objects but also of actions and their consequences, as well as recognizing the agents producing those actions. In other words, they develop a basic understanding of how the world works including cause-and-effect relationships, actions composed of agents and patients, and so on. Some authors have proposed the term "event cognition" for these mechanisms and refer to "mentalese" to refer to the nonlinguistic understanding of world phenomena, which might contain rules similar to those of language but differing in some aspects. Like the conceptual system, this mental organization pattern becomes impregnated with language as the child becomes exposed to speaking adults. As I said earlier, the child learns from adults to translate cognitive categories into distinct names for each class (cups, cars, jumping, walking, and so on) and subsequently learns to order these words in phrases to depict events according to regularities they hear from their parents' talking. The specific grammar or word order used to describe an event may depend on consented to but functionally constrained rules established by the adult speakers and learned through statistical associations (some combinations are more frequent than others), which, as said, can vary enormously across idioms as long as it becomes clear who did what to whom. Thus, the intricacies of grammars in different idioms could result in large part from cultural mechanisms that promote diversification but at the same time are constrained to efficiently transmit coherent meanings.[14]

In my view, provided sufficient learning and memory capacity, cumulative cultural development may have worked to progressively increase the diversity and complexity of communication in our species, establishing binding rules between items (be they syllables, words, or sentences) that emerged from preexisting cognitive mechanisms. In this sense, the linguistic universals referred to by many as reflecting the innateness of grammar might reflect the functional requirements for achieving efficient communication based on conceptual knowledge about things and events. Therefore, learning and mastering complex grammatical rules implies a tight neural organization, but this relies on plastic neural networks that have become more robust and have been coordinated through social exposure to generate coherent behavior.

The Power of Language

We have seen that language evokes mental representations of events and things that we share even if they are not present and may not even exist like in a fable. How do our brains become coupled with each other to do this fantastic feat? As mentioned in the previous chapter, during speech perception, the activity of the auditory-vocal brain circuits matches the acoustic vibrations produced by speech. Furthermore, the fluctuations of long-ranging multimodal networks in the cerebral cortex may contribute not only to process the speech signal but also to generate large-scale semantic representations of the contents of the message, all through a hierarchical organization provided by grammatical structure (as said in the previous chapter, there are intense discussions among scholars on the details of these mechanisms). This creates a symphony of neural activities where distinct networks couple and uncouple their oscillations at different frequencies and in different brain regions. In this way, the language system becomes attached to widespread neuronal networks permeating our perceptions, feelings, intentions, and memories.

This extended networking consolidates our minds into a social narrative, organizing our cognitive mechanisms and enhancing the coherence and temporal continuity of our minds. Through language we can refer not only to past and future events, but we may use it to recollect memories and make up a history of ourselves and of our behaviors, basically narrating our own history. Moreover, we can refer to ourselves and to our own thoughts, contributing to our sense of self provided by a history of recollected events and feelings, all framed in the social environment where we are immersed. In this process, a fundamental component is inner speech, that is, when we talk to ourselves and make up stories of recent or long-past events. This process, which involves the sustained activation not only of the language circuits but also of other systems, particularly the Default Mode Network (DMN; see chapter 10), may be especially relevant for decision-making mechanisms and for consolidating some memories that make our life history.

Synchronous Brains

Furthermore, these complex and nested oscillatory processes inside our brains connect to other brains. The brain activities of two persons (or animals like marmosets and bats) can synchronize at certain frequencies when they participate in a conversation, listen to a story or music, in a psychotherapy session, in political discussions when sharing the same views, and when they engage in some collaborative social interactions. This is reminiscent of the body synchronies during ritualized dances, in which music and rhythm shape collective behaviors. Therefore, large-scale intersubject neuronal synchronization may underlie our shared perceptions, contributing through language to construct our social world. In a way, inter-individual synchrony can be viewed as the extension of our brain activities into others, which together may generate a dynamic beyond our own bodies. In this line, the neuroscientist Christof Koch has toyed with the idea of fusing two minds through a brain–computer–brain interface, just like the two cerebral hemispheres engage to generate a single mental experience. Yet, I am skeptical about the notion of a higher-order mind emerging from our brain couplings. We keep most of our brain's neural dynamics to ourselves, and so to speak, only a surface of these activities becomes coupled with other brains. This may be sufficient to coordinate our behaviors and to share a semantic world of general meanings and beliefs, but it may not give rise to an extended or shared mind.[15]

Homo Loquens

Building Babel's Tower

Modern languages probably appeared shortly before the onset of agriculture. Some authors claim that the mastering of fire and gatherings around it made a perfect setting for the transmission of the early oral traditions, when stories about what happened during the day or more mystical tales about origins and death were told and transmitted from generation to generation. Likewise, melodies and rhythms have served to keep words and verbal strings in cultural memory, but they also have their own function to attract the members of a group.

Furthermore, language quickly diversified into all extant idioms in the world. Like Darwin asserted, the evolution of modern languages has mimicked the divergences and fusions of human groups throughout prehistory. Scientists have established a correlation between the genetic divergences and the separation of languages and folk music across cultures, where both suffer "mutations" in their structure that produce small alterations in the sequences that accumulate over time and can be similarly traced through history.[16]

Writing and Maths

Another crucial cultural invention was writing, whose earliest evidence is a set of clay tokens dating from some 9.5 KYA from Syria (if we exclude the about 40 KYA hand stencils mentioned earlier), with distinct shapes that were apparently used for accounting purposes. A great advance took place when signs came to represent vocal sounds (phonograms), some 5 KYA, and when personal names or authorities such as "Queen" were depicted in the rudimentary texts. Soon after, writing became used to record prayers, legal issues, poetry, and even scholarly texts. Writing provided a strong stability to oral traditions and became an external, solid memory of stories, laws, and meanings that traversed time. However, unlike speech, it was used by a small minority of educated people for a long time until education became institutionalized by the emergence of the modern state. Even today, there are about 780 MM illiterate adults around the world, about 10 percent of the total population. Notably, the advent of writing was taken by skepticism at least by the ancient Greeks, who considered that relying on texts was detrimental for memory capacities and gave the appearance of wisdom without any deep knowledge. Yet, through writing, knowledge has become available to many more people and more in depth than in ancient times, where it was only a privilege of few elites.[17]

The capacity for reading and writing cannot have evolved from natural selection (it was acquired in too short a span for this), and is definitely the result of neural plasticity. In the human inferior temporal cortex, adjacent to the fusiform face area and the object-selective areas lies the visual word form area (VWFA) that is selective for the perception of letter strings. Recent experiments have shown that the VWFA emerges during schooling, as children learn to read and write. In the absence of schooling, the regions that would form the VWFA mature to become face-selective or object-selective patches. In addition, Exner's area is located in the premotor cortex, involved in the manual control for handwriting.[18]

Finally, mathematics is another important product of writing. While several animals have a sense of numerosity and some rudimentary ability to count small numbers (including bees, fish, birds, and monkeys; see chapter 6), humans have uniquely become able to use arithmetical operations and developed a highly complex mathematics. Like in music, the neural populations involved in arithmetic processing are partially segregate to those of language, involving the inferior temporal lobe and importantly parietal areas, possibly associated with visuospatial processing mechanisms.[19]

Language Is Viral

In the 1970s, Richard Dawkins proposed the revolutionary idea that evolution was driven by the replicative properties of genes and extended this view to social behavior, proposing "memes" as units of cultural learning that replicated among societies like genes do for biology. Neither of these views was devoid of controversy, but time has passed, and, like Daniel Dennett said, both ideas remain "alive and well" among many respected scholars. Another approach to describe cultural transmission has been to consider it as a case of Lamarckian, or pseudo-Lamarckian inheritance, where the learned cultural trait is transmitted to other individuals, especially the descendants. However, this differs from Lamarck's theory as the acquired character never becomes heritable by itself, it must be learned generation after generation. In any case, language became a sort of "bug" inside our brains, providing us with what have been called "thinking tools" that enabled us to manipulate the world, organize societies, make artworks, and do science to discover the ultra-microscopic, the ultra-macroscopic, the remote history, and the distant future.[20] In other words, we created a huge and limitless collective world.

Perspective

The common life of our Hominin ancestors included signals like eye gazing, mutual eye-to-eye contact, and hand-pointing that is characteristic of our species, enabling them to knowingly share perceptions and some features of the world. Together with these adaptations, the capacity to predict or anticipate others' conducts and knowledge served as a feedstock for the emergence of a referential communication system where objects or events acquired names. Names for things may have emerged using calls and gestures through multimodal associative mechanisms that evolved in the human brain. The acquisition of meaning or a semantic system possibly evolved through an interface between word or gesture sensorimotor systems and other brain regions, that was executed at least by the STS and the TPJ. An important aspect of naming was that things could be referred to even if they were not present, by producing similar representations in the brains of different individuals.

At some point, communication transitioned from the one- or two-word stage typical of toddlers into coherent strings depicting complex events, for which grammar was required to avoid ambiguity. While a long tradition has considered that grammar is the result of an innate, species-specific module of *H. sapiens*, a more recent perspective considers that it is the result of cultural learning in which our genetic endowment provides the necessary brain plasticity to acquire it. Grammar facilitated behavioral coordination and especially paved the way to sharing stories that made up part of common knowledge in the forms of oral traditions, fables, or myths that were transmitted across generations. Through language, we are able to construct a social reality where we all agree on the basic composition of the world. Another important event is the invention of writing, which permitted us to solidify language, making it an external and enduring memory. Finally, language has spread through our culture, contributing to shaping our minds and giving us conceptual tools that facilitate our cooperation and the generation and further transmission of knowledge across generations. In the next chapter, I will refer to those aspects of our minds that we do not share with others, particularly consciousness and the solitary realm of our perceptions, feelings, and thoughts.

14 Consciousness Unexplained

While language provides a shared reality to which we all can refer to and agree about what we are observing, it also enables us to refer to our inner states, which, although we can name them, we cannot point to them so that the other observes them too. We may say "I feel happy," and others may perceive our facial expression and tone of voice, but they cannot directly perceive our feeling of happiness. Rather, this triggers a happy feeling in the others that they identify with ours. Likewise, when referring to perceptual processes, for example, optical illusions like the well-known "what color is this dress test," we agree that there is a dress, but we cannot observe the other's color perceptions and we have to trust them when they say they see a gold or a brown dress. Theoretically, we could monitor the subject's brain activity to predict which color he or she will report, but we will have no access to his or her subjective perception of the color. In other words, we are positively certain of our internal states including perceptions, feelings, and thoughts, and language has enabled us to make reference to them although these remain in our private realm, unobservable by others. In this chapter, I will first discuss the neural and cognitive aspects of conscious processing, on the understanding that these result from biological evolution. Secondly, the question about whether other animals have consciousness, and whether or when there was a moment in which we acquired consciousness, will be addressed at the end of the chapter.

The Scala Conscientia

What Is Consciousness?

Currently, there are more than twenty theories of consciousness that by themselves show that we are not close to an agreed account for this process.

A common definition is that consciousness is an awareness of internal and external existence. However, this does not allow us to get much deeper as awareness implies conscious knowledge, which puts us in a circular situation. Moreover, there are two perspectives under which we may study consciousness: as the internal experience each of us have, or by attributing it to someone else (remember the ToM). As we will see, different theories address these two perspectives in distinct ways.

According to most neuroscientists, consciousness is intimately bound to the workings of the brain. However, according to some views, consciousness is not just inside the brain but is embodied in the organism's functions and is produced through the enactive interaction between the individual and the world. In fact, consciousness is embodied as long as there is always someone experiencing these events, which are largely sensory (including proprioceptive and interoceptive mechanisms). Moreover, sensorimotor interactions with the environment and with others build up the contents of consciousness along the subject's life (see later in this chapter). Still, practically all conscious states have been experimentally correlated with brain activity, particularly in the cerebral cortex, and disorders of consciousness are mostly attributable to brain conditions. (Some structures like the cerebellum or the basal ganglia may not contribute directly to consciousness, whereas cortical networks [connected with the thalamus and other structures], are required for the maintenance of a unified and continuous mental representation of ourselves and the world around us.) Furthermore, in conditions where the brain is largely disconnected to its sensory-motor systems as in locked-in patients or while dreaming during sleep, people may report vivid conscious experiences.[1] In my view, if we want to understand consciousness, the brain is the first place in which we should focus, which does not undermine the fact that it is grounded in a self-producing body with a long evolutionary history.

There are three issues that are critical for human consciousness, which I will try to keep separate in this chapter. First, consciousness is a subjective phenomenon in which we experience sensations, probably derived from basic perceptual and homeostatic mechanisms. Secondly, is the primary consciousness, which refers to the capacity to form raw representations of oneself and the world. Thirdly, extended consciousness is understood as the set of cognitive and neural mechanisms associated with the capacity to form autobiographical memories, plans for the future, and so on, where we can think about our own thoughts.[2]

The Existential Iceberg

Perhaps Sigmund Freud was the first to make an explicit difference between the conscious and the unconscious, in which the former was only the tip of the iceberg of mental processes. Somehow in line with Freud's insights, Antonio Damasio and others have proposed a hierarchical organization of consciousness (see figure 14.1), starting with the existence of an unconscious "proto-self" involved in regulatory and homeostatic functions. In vertebrates, the proto-self relies on hypothalamic and brainstem circuits that in mammals map into a body scheme in the insular and sensory cortices. According to Damasio, emotions represent complex physiological reactions to external events, triggering reflex homeostatic mechanisms of survival, feeding, and reproduction, which may provide the feedstock for the emergence of sentience. On the other hand, feelings are more elaborate processes consisting of a conscious perception of the changes produced during emotional states. As I mentioned in chapter 6, other authors claim that emotions and feelings are privative of humans, while animals' reactions are in last instance survival-related reflexes.[3]

According to Damasio, feelings are a basic expression of consciousness, that is, the ability to make an image of ourselves and our surroundings. This emerges when signals from the body's interior (visceral or motor-driven) converge with external sensory signals (likely in the cerebral cortex), generating a boundary between self and environment. This way, we can define a core or primary consciousness, which is the capacity to construct integrated, allocentric models of the world including ourselves in it. Primary consciousness refers to our immediate experience of being here and now, and its contents relates to perceptions and predictions of incoming stimuli (internal and external) and activated memories of past situations. According to one view, primary consciousness is largely determined by these predicting, top-down mechanisms that enable us to anticipate forthcoming events (see chapter 10). Similarly, visual imagery (like imagining an elephant or remembering a situation) is an example where images from memory are vividly retrieved through top-down processes, despite us knowing they are not real. On the other hand, hallucinations occur when this boundary becomes blurred and the subject confounds illusion and reality. The point is that illusion and perception are experiential phenomena of the same kind, and we usually distinguish them just by further testing our senses and behavior.

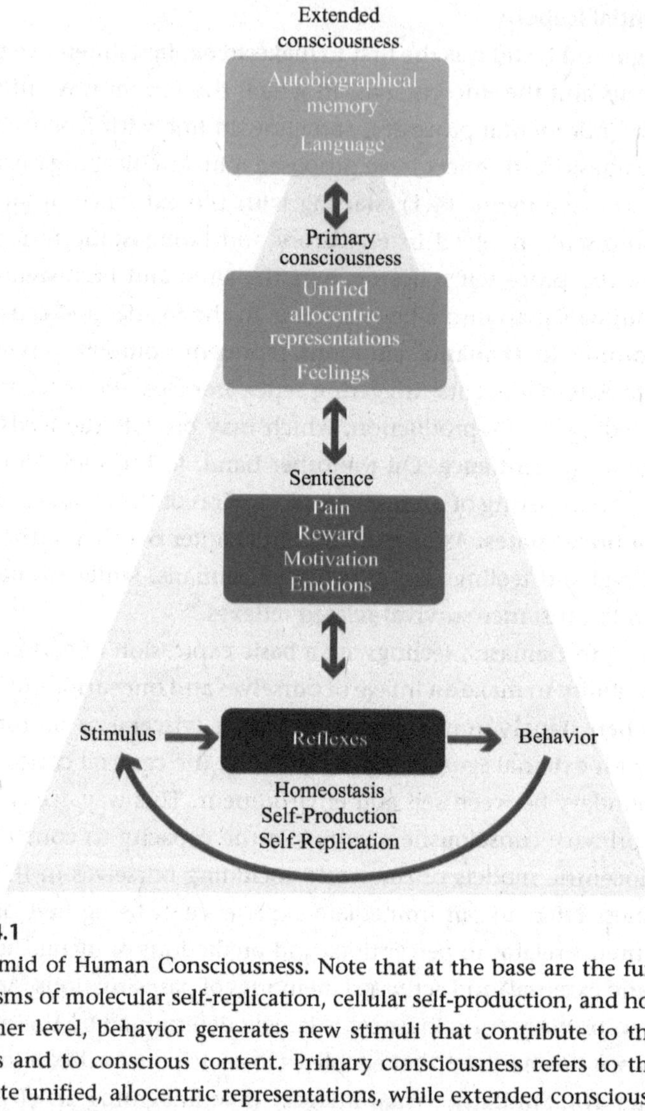

Figure 14.1
The Pyramid of Human Consciousness. Note that at the base are the fundamental mechanisms of molecular self-replication, cellular self-production, and homeostasis. At a higher level, behavior generates new stimuli that contribute to the system's dynamics and to conscious content. Primary consciousness refers to the capacity to generate unified, allocentric representations, while extended consciousness refers to self-reflection and metacognition, especially through language. Sentience, or the subjective experience of perceptions, remains as the most enigmatic aspect of consciousness and permeates our mental experiences at all levels.

Interestingly, we may not always be aware of the events that make up primary consciousness, but they appear in our minds when we refer to them, a phenomenon called metacognition. For many authors, a secondary, higher-level, or extended consciousness is achieved when we reflect about our experiences, becoming acquainted not only with things and ourselves but also with our own thoughts ("I think, therefore I am"). Extended consciousness also implies a sense of continuity of our minds over time, giving rise to autobiographical consciousness, in which our life becomes threaded in a personal history. In my view, extended consciousness is largely a result of us having language, while a limited degree of nonlinguistic metacognition may be present in primary consciousness, as it may happen in infants or some non-human animals. When language is introduced in our lives, it significantly amplifies the primary consciousness by giving names to things, external and internal (including our emotions and thoughts), facilitating the retrieval of memories, giving rise to an expanded notion of our self, feelings, and perceptions, and we ask ourselves why we are conscious. Together with helping articulate our past, language enables us to classify the world beyond basic cognitive mechanisms, to think about the distant future, and to plan projects that provide us a sense of being across time. Without language, it is most likely that consciousness and many other things would not be an issue at all. However, the most mysterious component of consciousness to me is the primary experience of our vivid sensations, from which higher-order mental phenomena are built. Metacognition and language enable us to set outside and think of our perceptions, sensations, and feelings, but this does not address their subjective essence (Why is there a mental experience at all?) nor does it explain why we experience them the way we do, that is, the blueness of the sky and the sense of pain. In other words, metacognition and language allow us to refer to our internal states, but their elusive first-person phenomenology remains indescribable to us. Thus, despite appearing at relatively low levels of consciousness development, the subjective experiences provided by sentience deeply permeate our existence, making us to feel what is being alive [4]

Bricks to Build a Mind

Consciousness is associated with a series of cognitive mechanisms that, although each by itself may not account for the features of our minds, all together may contribute to generate the unified feature that characterizes our subjective experiences. Two basic and closely interrelated neural processes involved in consciousness are attention and working memory. A crucial characteristic of both attention and working memory is that they are limited-capacity processes, being able to handle only a few items online. This is consistent with conscious phenomenology, as we are only aware of a few events at a time. In addition, the insular cortex that monitors internal states providing us with a bodily sensation can be essential for our daily self-awareness. Notably, Cotard syndrome is an extremely rare condition in which patients feel that they are dead or zombie-like beings, which has been attributed to lesions in the insula.

In addition, the Default Mode Network (DMN) participates in daydreaming and introspective attention, which makes up most of our mental lives. We spend long hours every day just thinking, even when we are performing rutinary tasks. Studies have reported that he DMN participates in metacognition and mentalization (thinking of others' thoughts). These functions might derive from the replay function involved in memory consolidation (see chapter 10). I mentioned how the language system contributes to higher-order consciousness, but besides speaking to others, a fundamental aspect of our mental existence is provided by inner speech, which operates in concert with the DMN organizing our thoughts and ideas. Recent studies indicate that there are dynamic interactions between the DMN and the language and executive cortical networks that feed onto each other, generating an alternating flux of introspection and outward-directed attention phenomena that makes up our daily existence. In this line, the temporoparietal component of the DMN (recall figure 10.3) may partly reflect the activation of language-related working memory mechanisms at rest. Beside the cerebral cortex, the hippocampal and limbic components involved in episodic memory and emotional processing provide critical elements to conscious experience and are connected to executive areas of the prefrontal cortex through the temporal pole cortex and orbitofrontal cortex. In addition, some authors have called attention to the reciprocal connectivity between the neocortex and the thalamus as a fundamental aspect for the generation of awareness and consciousness, but in this sense, the neocortex is also reciprocally connected with other structures like the basal ganglia and the cerebellum, which may contribute to the generation of conscious states even if they may not be strictly required for them. Finally, brainstem and basal forebrain nuclei involved in generating an aware state are also fundamental for our neocortex to maintain itself vigil and attentive.[5,6]

Importantly, of all the elements I depicted here, none of them is unique to our species with the exception of language and inner speech. Likewise, we may distinguish different levels or states of consciousness in our daily cycles, neurological patients, anesthesia, under the use of psychedelics, near-death experiences, and mindfulness meditation. Again, notwithstanding these descriptions of conscious and nonconscious states, we find that excepting meditative practices that result from extensive training, the brain states referred to in these cases may also develop in other animals. While their neural activities and their behavioral consequences may be equivalent in animals and in humans, the question remains of whether other animals experience these states in the way we do. Possibly, we share a state of primary consciousness with many animals in which we generate a unified representation of our environment and ourselves, which can be of undoubted adaptive significance. However, we are the only ones that can routinely make reference to them by using language.

Models of the Mind

The Hard Problem and the Hard Question

Now we approach the toughest problem of all: the generation of conscious content. How the materialistic skein of neuronal connections, neurotransmitters, and membrane excitation mechanisms conflate to generate an immaterial subjective experience is in my view still unexplained. David Chalmers has referred to this as the "hard problem" of neuroscience, claiming that qualitative sensations of percepts like the blueness of the sky or the unpleasant sensation of itch (termed qualia) are unexplainable to neuroscience. Colors may produce biological emotions, yet the intimate sensation of colors, and any other percepts, is indescribable to others. There have been attempts to characterize the phenomenology of qualia, but in my view, they are inaccessible to rational analysis—you cannot use terms to describe them. Of course, this view is not without contestants, particularly those who advocate that scientific materialism accounts for all kinds of explanations and deny any kind of Cartesian dualism. Some philosophers like Daniel Dennett have claimed that consciousness is only an illusion deriving from metacognition, where we attribute our perceptions to some kind of immaterial substance in conditions that these are in fact the result of complex mechanisms taking place in the brain, capturing the information provided by the senses. According to Dennett, consciousness has no relation to subjective sensations but rather with cognitive control about our actions and decision-making capacities. Thus, free will is the capacity to select among competing courses of actions in the brain. For Dennett, the real "hard question" is another: "Once some item or content 'enters consciousness,' what does it cause or enable or modify?" This question may be understood from two different perspectives: firstly, it is clear that the cognitive functions that entail consciousness are useful to us. But secondly, it is not totally clear that having a subjective experience will have an impact on our behavior. Could we live as unconscious automata, performing the same behaviors, engaging in language and using metacognition without being sentient? [7]

This problem was addressed by the famous (but flawed) experiment by Benjamin Libet in the 1980s, where an EEG wave called the readiness potential was elicited before the subjects reported to decide making a move. There are several problems with the experiment's design including the delay between the onset of an intention and the execution of a task, the unreliability of the person's report and the inaccuracy of the moment in which he or she actually made the decision, and finally the fact that the readiness potential seems to be a nonspecific indicator of attention rather than on the decision to move. Nonetheless, more recent experiments indicate that brain activity prior to a conscious decision can predict the specific decision the subject will make, ruling out nonspecific attentional effects. Additional studies, mostly performed during neurosurgical procedures, have yielded much insight on the neural mechanisms of voluntary decision-making, which point to good neuronal predictors of conscious decisions. Still, when we come to planned decisions, for example about whether marrying or not, our internal experience seems to be determinant in our choices. However, some might argue that this an illusion, that our choices are taken by unconscious mechanisms and do not strictly depend on our experiential subjectivity. For instance, the replay of past experiences or possible outcomes may be conscious but the final decision-making mechanisms could be driven by unconscious processes.[8]

Even considering the relevance of Dennett's hard question, in my opinion, his arguments are not sufficient to me to ignore Chalmer's hard problem, that is, the existence of subjective experiences and their relevance to our lives, which are ontologically evident for all of us. In this line, one approach to the hard problem has been to consider the identity of mental states with brain mechanisms, just like gravity or energy are properties of matter (remember the vitalistic arguments in chapter 2). In one understanding of the identity hypothesis, neural mechanisms and mental experiences are two sides of the same coin depending on from where you are looking at. Optical illusions, like the transparent Necker's cube where the front and back surfaces alternate successively or the "duck-rabbit" that may alternatively be interpreted as a rabbit or as a duck, have been used as analogies for this mechanism. Thus, you are in the realm of describing neuronal mechanisms, and as if by flipping a coin, you find yourself in a subjective experience of the same mechanisms. However, what kind of phenomena accounts for this coin-flipping, and why it works only for me but will not work if I am observing the brains of others is not explained at all. You might say that this is so because your own mind is the result of a very specific neural mechanism, but what if we could perfectly mimic it? Would this have a mind? Others say that you would have to "be" the other machine—or animal—to see its mind, but the sensation of "being" something is about the same as having a mind embodied in this something, which does not add to the explanation.

More generally, in science, different people have to agree on the phenomena they observe. When dealing with the mind, only the bearer has access to his or her subjective states, while others are blind to them (they can observe their physical correlates). In this way, it may be impossible to verify if some physical mechanism generates consciousness as we will not be able to verify it excepting for personal reports. One theoretical proposal is neurophenomenology, in which subjects keep reporting their experiential states while their neural activity is being recorded. Nonetheless, even if we established a good neuronal correlate of conscious states (called NCC), we may never know if these mechanisms are sufficient to make consciousness (correlation is not causation), nor how the physical mechanism is "flip-flopped" into a subjective sensation. This is particularly relevant when we come to the possibility of nonhuman consciousness, including other animals and machines. Particularly, machines could simulate conscious behavior including metacognition (some of them already do), but there is no way to verify whether they experience anything. Some authors have argued that in other branches of science like particle physics, researchers are used to working with phenomena that cannot be directly observed. However, theories in this field make specific predictions about widely agreed on and precise observations that can be confirmed or rejected, thus validating or discarding alternative hypotheses. At this point, there is no agreement on what consciousness is, and there are only a few (and vague) predictions made by the different theories of consciousness that have been proposed, which up to date have yielded uncertain results. Moreover, I think the epistemological situation is different between physics and consciousness. We are not speaking of unobservable particles or of principles like inertia that serve to account for agreed-on observations made by a community of researchers; we are speaking of a phenomenon that is 100 percent observable for the one who experiences it, while others are blind to it.

Summarizing, while consciousness is most likely linked with brain activity (especially cortical activity), we have no cues about the phenomena involved in its generation, particularly regarding our subjective experiences. In view of this, some decide to consider subjectivity as totally irrelevant, while others propose that our minds are existentially valid but inaccessible to scientific research. Personally, I am biased to the second opinion since, after all, everything starts and ends in our minds, which is to me a sufficiently valid argument for an ontology of sentience. In any case, there is more agreement among scientists in that neural correlates of consciousness (NCC) can theoretically be found, that is, that distinct patterns of brain activity relate to our mental states and that different mental events are associated with specific neural activities. Yet, an NCC first requires an agreed-on definition of consciousness and a way to identify conscious states. Furthermore, an NCC could be a wide variety of things, from the activation of a single cell in the cerebral cortex to a hemodynamic pattern observed with magnetic resonance imaging. Determining which of these correlates is the best and most complete predictor of consciousness remains a daunting challenge.

Calculating Consciousness

There are also some computational and modeling theories about consciousness, which nonetheless remain unfalsifiable in my view, and at best can contribute to establishing NCCs. One of these theories is the Integrated Information Theory (IIT), which points to a sort of primary consciousness (and hence of subjective experience) that appears as an emergent property of complex networks, increasing sharply as reentrant or reciprocal interactions are a robust component of the network. The IIT is based on the assumption that conscious states are highly unified and indissoluble, and features models where the different components are tightly interacting with each other, generating an indivisible but dynamic whole where changes in one component have strong influence on the rest. The IIT claims that the primary conscious experience relates to the activity of sensory-related areas in the temporal, parietal, and occipital areas, which is supported by findings that brain stimulation produces conscious effects in sensory-related areas. However, and this is crucial, according to this theory, sensory or interoceptive stimuli are not required for consciousness to occur; it is just that sensory networks are more tightly organized than other brain networks. Furthermore, the IIT can be used to calculate Φ (phi), a unidimensional measure (that is, a number) that describes the system's interconnectedness, and its value has been proposed as a possible NCC, which would enable us to make comparative studies of consciousness. According to this view, neither the cerebellum nor currently designed computers display this degree of interrelatedness, which gives them a very low Φ value and cannot qualify them as conscious systems. However, presumably lab-cultured neuronal organoids might eventually develop recursive connections, raising their Φ to conscious levels (see the next chapter). Personally, I am not convinced that consciousness may be built devoid of a sensorimotor organization and an interaction with the world. The philosopher John Searle claims that consciousness is a referential phenomenon, in the sense that it always refers to something or is about something, that is, it has a content (be it interoceptive or sensorimotor). However, we may concur with the IIT that the network requires a tight causal organization for consciousness to take place.[9]

Another theory is the Global Workspace Theory (GWS), positing a neural system that gathers partial information from many subsystems including sensory, mnemonic, and interoceptive signals, and binds them into a coherent model of internal or external reality, generating different layers of representations (meta-representations) that develop through life. In this view, multimodal coalitions between sensory systems, memories, emotions, and verbal reports can be maintained in working memory to support the mechanisms of cognitive branching and predictive coding described in previous chapters. This could enable us to plan future events and to construct a continuous thread from the distant past to the present and imagined future. The GWS attempts to explain complex thought and planning, a hierarchy of goals, and the implementation of long-term strategies that characterize our conscious life. Contrary to the IIT, the GWS puts more emphasis on the top-down regulation from the prefrontal cortex to sensory cortices and refers to the executive aspects of consciousness, those that allow us to model the world and ourselves. Still, a problem I see with the GWS, and in general with meta-representation or higher-order theories of consciousness, is that they somehow imply an observer inside our brains that becomes acquainted with all these stimuli and manages this information to organize behavior. Perhaps language, by enabling us to point and name our own thought mechanisms, somehow allows us to reach these higher levels of self-reflection by expressing them in a social, semantic dimension.[10]

Natural Minds

The Birth of Consciousness

When did consciousness emerge in evolution? Which species can be considered to be conscious? When do babies acquire consciousness? Although neuroscience has developed without the need to address these questions, an increasing number of neuroscientists and other scientists are now asking themselves how and when consciousness arises in the natural world.[10]

Perhaps a first insight into these problems can be attained by addressing the development of consciousness in human development, in order to envision a sequence of stages leading to the capacity of conscious report. As said, according to the IIT, consciousness might not require exposure to external stimuli and might be present already in the fetus. Being optimistic, a measure of consciousness like the Φ value of the IIT might yield some clarity as to when the brain network acquires properties that can yield consciousness in human prenatal development (there could even be several thresholds indicating different degrees of integration). Some authors have suggested that human consciousness arises at about 35 weeks of gestation, when the thalamus starts driving cortical activity and the EEG becomes patterned by continuous waves. Another possibility (which I consider more likely) is that while the human fetus might have a basic internal organization as specified by the IIT, consciousness may develop after birth, when the subject engages in interactions with the surroundings and with others. According to some researchers, the postnatal development of consciousness is a process of continuous learning throughout our lifetime, in which, guided by sensorimotor experience, the brain networks feed their activity patterns into other networks as well as into themselves in an expanding spiral, where again interactions with the environment are crucial. As I said, language may provide a mechanism to amplify these phenomena by enabling us to produce increasingly abstract mental events.[11]

Animal Minds?

Another critical and current problem refers to animal sentience or consciousness. Anyone who has had a dog or a cat as a pet can witness their expressions signaling what one could call at least emotions. Furthermore, experiments described in chapter 9 suggest that animals might be able to make "mental maps" of space. This issue has strong implications in the domain of bioethics, where increasing regulations are being imposed about the welfare conditions, not only of animals commonly used as food and in research laboratories but also among pets that by the way have given rise to a profitable industry around them. In its strongest expression, some activists claim for animal (and even nature's) rights, which poses great problems about the limits of these rights in conditions that they cannot be obliged to any responsibilities, and of determining which animals will qualify for which kinds of rights. It may become illegal to boil living lobsters, but will it come a point where we cannot eat them, together with mollusks, fish, chicken, cattle, and even insects? Where do we draw the line?[12]

Again, the key issue is how to detect sentience or consciousness in other animals. The IIT has provided a criterion for basic consciousness (perhaps a form of sentience), but this is experimentally very difficult at this point. Another approach could be to assess behavior, but then the interpretations of the results are usually being contested. The experiments of self-recognition in a mirror were initially used to verify self-consciousness, but it was found that some fishes can pass the mirror test, suggesting that this ability is based on simple brain mechanisms (see chapter 8). Many studies have addressed metacognition in nonhuman animals, like assessing one's chances to solve a task or having a sense of body size, but again there may be simpler explanations for these findings. A more subtle experiment took advantage of a well-described "double dissociation visual awareness" task where people perform in totally opposite ways whether they have consciously processed a stimulus or when they have unconsciously processed it (this is different from not perceiving the event, as there is a stimulus-related response). When applying a version of this task to macaques, these animals developed double-dissociation responses much akin to those in humans, which suggests that there is a difference between conscious and unconscious processing in this species. Yet, the difference between unconscious and conscious perception could rely on different neuronal activities, with consciousness being only an epiphenomenon.[13]

Rather than finding an absolute indicator of nonhuman consciousness, perhaps a more fruitful approach to a comparative study of the mind is to recognize different indicators of consciousness and determine how these parameters are distributed among animals. This implies that consciousness and subjective experiences may not be equivalent for different species, as for example in an echolocating animal versus a highly visual animal like us. Nonetheless, some general mechanisms may be associated with increasing degrees of consciousness. Using a sort of *Scala conscientia*, as indicated in figure 14.1, may work as a good reference guide for these processes.[14]

All in all, research in animal consciousness faces the same challenges as the study of human consciousness. First, Chalmer's hard problem, which recognizes the ontology of mind but claims the incapacity to explain it in materialistic terms, leaves the issue of animal subjectivity as an eternal unknown. Secondly, Dennett's hard question (then what?) could be stated as whether consciousness itself confers any selective advantage or rather if it is the material complexity of the brain and the consequent cognitive capacity that provide evolutionary benefit. Some authors argue that consciousness promotes survival in the face of contingencies and facilitates the evolution of culture and societies, but again this would be very hard to show.

Mind and Matter

In the end, the dualistic dilemma between subjective experience and mechanistic explanations seems to me inescapable. Denying its existence as an illusion appears unsatisfactory to me. Some scholars argue that as with the conception of life, which was once considered to be a spiritual phenomenon and was explained by science, the problem of consciousness will also be resolved with scientific research. However, life is a scientifically observable phenomenon, while our subjective experiences are not accessible to third parties. Furthermore, for conscious subjects, their own life is an experiential phenomenon as well. One alternative to get away from the mind-body conundrum has been pan-psychism, which is an extension of the identity hypothesis, proposing that some degree of consciousness is a feature of the universe like energy or gravity are. As prescribed by the IIT, the subjective experience may be a widespread phenomenon among living beings that amplifies with the increase in complexity, and in particular may have exploded like a "big bang" with the origin of large brains, giving rise to fully subjective consciousness. Some other researchers, particularly physicists, have moved much farther, proposing that the source of consciousness relies on quantum mechanics. In this view, quantal superpositions of states in elementary particles would collapse wave functions into (proto-) moments of experience composed of basic qualia, the fundamental "particles" of consciousness. Many prestigious authors have endorsed this view in one way or another. While quantal panpsychism is highly controversial and not accepted by many scholars, it makes some predictions, one of which is that quantum computers might eventually gain consciousness by this process (but again, how would we know?). Finally, while panpsychism may be an intriguing possibility, this perspective runs the substantial risk of anthropomorphizing nature, projecting human experiences into much simpler beings whose subjectivity, if it exists, might be radically different from ours. We need to keep present that our mind is practically unique (at least on earth), and results from having an extremely complex brain endowed with language. Other kinds of subjective experiences are probably much simpler, or different, or many times simply negligible to be considered as relevant.[15]

Perspective

Consciousness is perhaps the most difficult to define of all biological and cognitive categories, yet it is at the same time essential to our existence. It refers primarily to the qualitative experience of being the protagonists of a world independent from us, which includes the capacity to observe our own bodies and to sense our internal selves. Here I adhere to a sort of *scala conscientia* with different stages of increasing complexity that might give rise to human consciousness, which is vastly amplified with the acquisition of language. However, despite the apparent relation between consciousness and complexity, how material mechanisms become transformed into a first-person qualitative experience poses a major problem for neuroscientific research, which is conceived to deal with second-person observable phenomena. Since the unobservable nature of subjective states by second parties severely limits their scientific study, some have decided to downplay them as an irrelevant illusion generated by our complex brain networks. Whether consciousness is or is not relevant for our decisions and behavior is a debatable issue, but the illusion argument does not really explain to me how subjectivity is in fact produced.

Perhaps acknowledging these difficulties, neuroscientists have started to search for the neural correlates of consciousness (NCCs), which imply patterns of neural activity that fit reported conscious states. Yet, a correlation does not imply causation and is even less an explanation. In this line, two main theories have been proposed to account for the NCCs: the IIT and the GWS. The IIT, on the one hand, prescribes a tightly encapsulated dynamical network that works as an indivisible whole, providing the unified property of conscious states. On the other hand, the GWS refers to a "central executive" system that receives multimodal signals from the body, the exterior, and from memory, and organizes these inputs to generate a simple working model. It is likely that both theories are in fact addressing conscious phenomena at different levels: the IIT refers more to a primary consciousness, while the GWS refers to the executive aspects of extended consciousness.

Finally, another major question is whether consciousness can be found beyond human existence. The claims for animal sentience and consciousness have recently increased, even among scientists. The application of indicators of a *scala conscientia* in the tree of life, considering an important degree of diversity, may provide a useful tool to examine at least the cognitive elements that are associated with conscious states. However, in my opinion, the hard problem of subjectivity remains inescapable. Some authors propose pan-psychism, the notion that consciousness is an everywhere byproduct of matter, which elaborates with material complexity. For some physicists, psychism could be somehow related to quantum uncertainties. In the next and final chapter, I will discuss where evolution may lead our species, our brains, and our minds.

15 Quo Vadis?

Contrary to popular belief, the biological evolution of our species has not halted. However, cultural and technological development are producing much more relevant changes in our life, affecting our epigenetic and plasticity mechanisms, and we may even start altering our genetic makeup. Furthermore, we are living in an increasingly collective realm where we communicate instantly with people all around the world, and many of us already live in a virtual-reality world that is growing faster than we may think. In addition, our species has managed to change the earth more rapidly than any other species has. Our population is rising without signs of slowing down yet (in the last fifty years, it more than doubled from about 3.6 billion to 8 billion), and we are producing a tremendous amount of damage through pollution and environmental destruction. Finally, there is growing fear of possible catastrophes like a nuclear war, a major asteroid impact, and microbial infections that we have seen may spread at rates never heard of. Thus, we are not exempt from the risk of disappearing. The immense majority of species in the history of life have gone extinct, and many presently living ones are likely to become so largely due to us. An average mammal species exists for about one million years (with a wide variability). With some 300 KY of existence, *Homo sapiens* is a young species, and it is difficult to know if we will outlast other species in survival time or if we will become extinct according to other mammals' expectations. If we survive in the long term (which I believe we will), what kind of species will we become in the future? In this last chapter, I will offer some reflections about our future and the explosive rise of technology, and how this is affecting our biology and our evolutionary fate.

Ongoing Evolution

Milk, Vitamins, and Wisdom Tooth

Like any other species, we keep evolving through differences in the reproductive capacities of some individuals over others. In fact, there is a lot of evidence for subtle but relevant genetic changes in our recent history, including lactose tolerance genes in milk-feeding peoples; the evolution of paler skin and increasing production of vitamin D in Britons, a strong selection for the immune response to resist infection; metabolic and respiratory adaptations in populations that live from diving or at high altitudes; and the decreasing frequency of the wisdom tooth, which may increase risk of oral infection. But there are other recent changes like increasing frequency of the median artery in the forearm in the last three centuries, whose selective benefit is not clear and may result from other factors like genetic drift, driven by statistical fluctuations rather than by adaptive reasons.[1]

Nonetheless, natural selection could be very weak in our species compared to others, at least in regard to the adaptation to the environment. We have become largely free of predators and many parasites, we generate our own food supplies, and our technology protects us very well from climatic adversity. Furthermore, in recent years, we have made an astounding progress in the control of emerging infectious diseases (HIV treatments and COVID-19 vaccines are examples of this), and genetic and chronic conditions are being increasingly controlled by modern medicine. It is often said that people with "nonadaptive" characters are now reproducing successfully in conditions in which they would not in the wild, promoting the accumulation of undesirable traits like genetic diseases that will be a burden to our species. Nonetheless, these statements have not been confirmed scientifically, and there is simply no evidence that our genomes are in any way degrading or increasing the species' extinction risk. Instead, what may be happening is an increase of genetic diversity, providing more robustness for variability and adaptive capacity.

Anyhow, none of the genetical changes observed in recent humans seem to forecast a radical transformation of our species. It seems more likely that we have reached a status of evolutionary stasis in our anatomy and physiology as many other species have, like the coelacanth or the platypus, which have survived largely unchanged for several million years. We usually tend to think that our robust cultural and technological endowment may protect us from extinction, but there is a "B" side to our triumph over nature. First, our recently accelerated evolution may not have been without biological side effects and may make us prone to different conditions like cancer and especially brain diseases. Furthermore, technological development is driving environmental destruction and even changing our biology, perhaps transforming our species in completely novel and unexpected ways.

The Price of Success

As we saw in chapter 11, our brains were subject to strong selection in our recent history. Several noncoding genes only found in modern humans are associated with creativity, prosocial behavior, and healthy longevity, suggesting that they reflect a recent selective sweep in our modern lineage. But what has dramatically changed our lives recently is not so much natural selection but the exponential growth of culture and language. Possibly, the recently accelerated cultural development may feed on more plastic rather than on larger brains, while in return cultural improvements might put pressure on increasing brain plasticity. An eloquent example of culture-driven brain plasticity is the recruitment of cortical regions for reading, writing (including musical notations), and mathematics (see chapter 12). In this line, epigenetic mechanisms may have contributed to increase brain plasticity in our recent evolution (note that these epigenetic mechanisms may be genetically driven as well). In fact, there are significant epigenetic differences between modern humans and Neanderthals, many of them associated with vulnerability to brain disease.[2]

This also reveals a dark side of brain evolution and cultural development, where susceptibility to neuropsychiatric conditions including ADHD, dyslexia, autism, psychosis, mood disorders and brain aging may have significantly increased in our recent history due to genetic selection or to increasing epigenetic modifications associated with plasticity and cultural development (see chapter 11). In fact, many of the genes that have shown signs of positive selection in human evolution are also risk factors for neuropsychiatric conditions like micro- and macrocephaly, dyslexia, schizophrenia, autism, and attentional disorders. An example is a variant of the DRD4 gene that codes for a dopaminergic receptor and has been associated with ADHD. According to some authors, this allele favored migratory behavior and enhanced peripheral attention capacities that were useful for hunter-gathering and migratory behavior. Yet, more focused attentional strategies became favored with the advent of civilization, a trend that culminated with the rise of institutional education in the last century, yielding a disadvantage of this gene's bearers when having to stay still and attend to long-lasting classes. A similar situation may happen with dyslexia, which was not a disorder until the educational system massified and required all children to learn to read.[3]

The evolutionary psychiatrist Timothy Crow once asserted that schizophrenia is the price humans pay for acquiring language. Extending on this comment, one could say that mental disorders are the price we pay for having a more complex and plastic brain, being the subject of cultural evolution. According to recent studies, the increasing complexity of human cortical networks implies a tradeoff between circuit efficiency and network stability, which may result in increasing risk of neuropsychiatric conditions. Thus, while our brains have developed a more fine-grained neural circuitry, this makes them more unstable compared to the more redundant networks of the nonhuman primate. Another factor that may be crucial for the development of mental diseases is the stress response at all ages, which affects brain development at multiple levels and is becoming increasingly prevalent, affecting especially the poor populations.[4]

The Rise of Technology

Welcome to the Machines

In early hominins, the brain-culture coevolution resulted in the development of rudimentary technologies that rapidly increased our capacities with the earliest stone tools, sticks, and twigs, including the mastery of fire, which were used as bodily extensions to crack bones, deter predators, hunt prey, dig for edible roots, and cook food. The development of enhancing body protheses is in the essence of our species, extending our phenotypes to acquire skills that our ancestors could not have dreamed of. Many other species have evolved phenotypic extensions, some produced by their own bodies like spiders' webs, mollusk shells, and coral skeletons, but others use objects from their surroundings like hermit crabs' seashells, birds' nests, beavers' dams, and the tools used by some birds and primates. Extended phenotypes are usually coded in the genes (like the spider's net and the bird's nest), but especially humans have done so by amplifying social learning (which is also genetically driven), therefore transmitting knowledge and skills across generations to assemble increasingly powerful tools, and constructing their own niche for living, including the domestication of other species.[5]

The use of technology gave rise to civilizations, and we were able to conquer the earth. More recently, we have started a new era of phenotypic extensions with the development of artificial automata, computers, and digital processing, which have undergone an exponential explosion in the last seventy years. This industry has generated powerful innovations in medicine, science, and probably most human activities, contributing to increase our well-being. Brain–computer interfaces (BCIs) that allow us to write or drive vehicles without even using our voice may become a reality in the next decades. Some people argue that technology is slowing down our mental development, which has been supported by evidence that the Flynn effect, which indicates a continuous rise of IQ across generations in developed countries, has been reversed in the last fifteen to twenty years, with younger generations scoring lower than previous ones. This is not due to "dumber" people having more children but from environmental factors that some attribute to the reliance on external memories and other devices, others to educational policies, and others to environmental pollution.[6] In my view, it is not clear what the reversal of the Flynn effect means, and before that, it is difficult to jump to conclusions. Furthermore, the use of external memories is not so new and dates at least from the origin of writing. Recall that the early Greek philosophers considered that students should entrain their memory and not use written texts or write notes (see chapter 13). Yet, we are here and feel pretty good about our capacities, which admittedly are largely inherited through culture.

Besides these advances, BCIs promise to help physically impaired people live normal lives. Now, if these devices are successful in enhancing the capacities of the disabled, normal people might also want to use them for the same purposes. Suppose there are BCIs that increase our biological memory capacities or our IQ, or just make us physically stronger. Would some people want to use them for their kids to become skilled musicians or great athletes? Will everybody have equal access to these improvements?

In addition, AI can manage a million times more information than humans and much more rapidly, and even learn from its experience. Recent AIs like ChatGPT and others have been able to generate quite good texts, make scientific papers, write music and poetry, engage in conversations, and do fairly well on IQ tests. There is every reason to think that with time, AIs will surpass average human performance in practically all these items. However, AIs still perform poorly in common sense tasks like managing expectations, thinking of others' intentions or desires, understanding social circumstances, peeling oranges, and so on. They lack the capacity to behave in changing contexts and need large amounts of information for doing things that we do effortlessly. But many of these feats will probably be overcome in the following years. AIs that are programmed to learn using basic physics rules like a baby or learn to move emulating cognitive and sensorimotor human development are a matter of intense research these days.[7] Every day we see improvements in this field, and every comment we make becomes soon outdated.

Artificial Minds?

"Robots of the world! The power of man has fallen! A new world has arisen: the Rule of the Robots! March!" Perhaps emulating the *Communist Manifesto* by Marx and Engels, Karel Capek wrote this epigraph to his play "Rossum's Universal Robots" in 1920. In relation to this concern, Isaac Asimov first proposed in the 1940s the three main laws for the regulation of robotics, which are: (i) not hurting or allowing any human to become hurt; (ii) obey human orders except when asked to hurt other humans; (iii) protect themselves except if this contradicts any of the first two laws. Still, if we could make machines that intelligent, it is not clear if they could be effectively programmed to follow these three rules; our own species has bypassed many basic instincts that are "hardwired" in the genes, by the influence of culture.[8] In any case, the dystopic notion that machines may control us is highly unlikely to me, at least in the short- and midterm. For one reason, computers are highly inefficient, they easily overheat, and require an enormous energy supply. For instance, the AI AlphaGo, which defeated the world champion in the Chinese play Go, used about nine thousand times more energy in that game than its human contestant. Moreover, the computer industry requires increasing amounts of commodities, which implies enormous economic and working capacity to be extracted. Machines would have to generate their own resources and self-reproduce in order to dispense with humans.

Another issue is whether machines could ever become conscious. According to the Integrated Information Theory (IIT) (chapter 14), no matter how smart computers become, they will never gain consciousness or sentience, as their design is profoundly different than the brain's. In other words, they lack robust re-entrant connectivity at many different levels that make neural ensembles work as indivisible wholes (which according to the theory is the critical feature of consciousness).[9] The alternative theory, Global Workspace (GWS), is more permissive on this point, granting the possibility that robots might reach consciousness. Likewise, some physicists claim that quantum computers might eventually acquire human-like consciousness. Personally, on this my bet is on the IIT. Computers may manage language, logical operations, and even refer to themselves, but they lack the biological foundations of consciousness, that is, complex homeostatic mechanisms that are at the basis of sentience (figure 14.1). There is no evidence that AIs could develop a will of their own and feel any sensations, even less daydream, feel remorse, or take their own initiative to get benefit from something. In my opinion, while consciousness is deeply bound to physical brain activity so that in principle a conscious mechanism could be designed, we are now simply too far to emulate the intricate self-organizing and recursive complexity of the brain that generates human-like sentience. Moreover, even if we were able to produce such systems, we would never know if we were talking to a "philosophical zombie" or to a sentient being (but on the other hand, would that matter?).

Social Networks

Another dimension of technology is human communication with the recent development of social networks through the internet. Digital media are increasingly impregnating our lives, affecting our privacy, sense of identity, and free will (here I refer to the social interpretation of free will, when there is an external agent that may curb our intentions or forcing us to do something we may not want). People massively follow influencers' ideas and are afraid of having different opinions for fear of being attacked in the same media. Many of us are diving into a collective network where people become anonymous and behave more like a herd than as deliberating and opinion-taking individuals. Furthermore, social media are progressively immersing us into a massive pseudo-reality where the fake news and the unreal beliefs (like the flat earth and anti-vaccine movements) gain support and spread in ways not seen before in our history. The prospect of virtual realities like the Metaverse may only contribute to further detaching some people from the physical world, living in avatars that may be more comfortable than real life.

More worrisome, there is the danger of the misuse of these technologies by big corporations or by the state. If the development of AI is not properly regulated, our thoughts and preferences might be fed into algorithmic processes that direct our lives in order to maximize consumption or some other behavior for the sake of someone else's benefit (or, like Yuval Noah Harari says, perhaps for nobody's benefit in particular, which would be worse). More specifically, recent advances on internal brain protheses and devices that can monitor our inner brain states could put us in increasing danger of losing these basic rights. We have already witnessed the effects of big data manipulation and the vulnerability to filtration of these gigantic databases containing information from all of us. In the Western world, giant technological companies are outpacing traditional institutions in research on robotics and brain–computer interfaces, and social networks already contain just too much information about each of us. In other countries, the state itself is concerned with monitoring people's life and behavior through increasingly sophisticated devices. Regulating AI development by law is a challenging issue, first because law-making is too slow and may become quickly outdated and also because all countries should agree with the legislation. Besides, this would not prevent the emergence of black markets. An initiative promoting the validity of individual rights in the face of rising technologies, including privacy, determinacy, identity, and nondiscrimination, has been issued in many countries, led by the neuroscientist Rafael Yuste. Recently, Florencia Álamos, Leonie Kausel, Yuste, and others have proposed the requirement of a technocratic oath, similar to the Hippocratic oath submitted by doctors, to help limit the misuse of these technologies for non-ethical purposes. At least in France, the necessity for such a vow is under current discussion for young scientists.[10]

Beyond Humanity

Defeating Death

Enthusiasts of AI and robotics, self-called transhumanists, have generated a movement where they prophesize that AI development will reach a Singularity point where it will surpass our human capabilities, starting an unprecedented evolutionary process of bodies together with machines. As crazy as this looks for some people, this group includes many highly skilled computer scientists, physicists, doctors, and biologists that cannot be simply ignored here. According to transhumanists, the fusion with machines would free us from our biological yoke, and we (rather, some of us) would become free of disease and especially of aging.

Furthermore, some futurists have proposed the prospect of "mind uploading," aiming to transfer one's mind into a digital code. However, even if we granted that machines like we know them could gain consciousness (which I strongly doubt), mind uploading poses many unanswered issues. How would the mind be transferred to a digital machine, as if executing a money transference? Or consider if it was possible to make a digital duplicate of ourselves and our sentience. Would you eliminate your biological self? Which would be our needs, feelings, and motivations if there were no biological constraints? Would we be really free of disease or be threatened by the proliferation of informatic viruses?[11]

Editing Life

Another more biologically based revolution is taking place after the development of biotechnologies, including cloning procedures (a great milestone was the famous sheep Dolly in 1996), the sequencing of the whole human genome in the early 2000s, and the more recent development of gene-editing technologies. We have been modifying the genetic makeup of plants and animals since the beginning of domestication by the use of artificial selection. Now, this process can be in a way hacked through direct gene manipulation technology. Genetic engineering is widely used in agriculture, promoting pest-resistant, more nutritive, and cheaper products that, although they have been highly criticized, may heavily contribute to solve the world's famine epidemic (hunger affects about 10 percent of people globally, and this may increase in the following years), saving production costs, generating more nutritious strains, and so on. Furthermore, genetically modified organisms (GMOs) have proven to be safe for human consumption (although care must always be taken to detect any possible damaging effects). Still, there is an ecological concern about GMOs, which could crossbreed with wild varieties or escape and compete with them, an issue that needs to be tightly controlled. Notably, there is significant overproduction of food in developed countries (at least in the United States, about 30–40 percent of the food supply is wasted). Finding more efficient ways to produce and deliver food across the world could be a major advance in our well-being as a species.

Furthermore, biotechnology is being increasingly used to improve human life, developing gene-based therapies to ameliorate a diversity of conditions. Perhaps the most discussed technological innovation in this line concerns the CRISP-R technology (see chapter 3), which has been used to insert or delete genes in different organisms for research and other purposes. This was first applied to human embryos by the Chinese researcher He Jiankui in 2018, who claimed to have created two HIV-resistant babies. In the US, the world's first CRISP-R therapy recently became avaliable to treat sickle cell anemia and beta-thalassemia. Importantly, if gene editing is performed early enough, it may access the germline and the modification may become heritable, which would end up transforming our species' genotypes. If genetic upgrading is accessible to only some of us, this might lead to the dangerous possibility that our species becomes separated in different populations, or even in the reproductive isolation of some groups, leading in an extreme situation to the separation into different species.[12]

Likewise, transhumanists have closely followed research oriented to retard or even revert aging. For instance, a set of proteins called Yamanaka factors are involved in stem cell formation during early embryonic development. These genes can be applied to differentiated adult cells, like skin cells, and induce their re-conversion into pluripotential stem cells. This technique could provide a means to prevent and even revert aging. In addition, future couples may not even need to use their germ cells to reproduce. De-differentiated stem cells (extracted from, say, the skin by manipulating the Yamanaka factors) could be used for human reproduction when transformed into eggs or sperm. This may allow homosexual couples to have children of their own, which could be a very good thing, but also only one person might be able to generate both types of gametes and become a biological single parent. Likewise, parthenogenesis is a mechanism where embryos develop solely from eggs, not requiring sperm-induced fertilization, which naturally takes place in some vertebrate and invertebrate species. In principle, this process might be engineered in humans to generate all-female lineages without the need of males, which would all be clones of their mothers. Needless to say, this would minimize genetic recombination, hampering our evolutionary potential. The idea of how life would be without men has been the subject of several dystopic novels, some envisaging that the world's problems would finally be solved while others claiming that women would rapidly take the lead of the extinguished men. Another issue is that sex determination may become not only manipulated as to preferring which sex your kid will be, but intermediate sex forms could be produced, not fitting the binary male-female notion. Associated to this, the industry of sex robots may find a growing market for increasingly sophisticated products, fulfilling all kinds of customers' fantasies. An extremely delicate issue is the case of virtual pedophilia, in which digital models of children may be used for sexual purposes. In this case, there is nobody being harmed, yet for most of us it is a highly reprehensible behavior.[13]

Other trending topics in biotechnology are synthetic biology, in which cells are designed by the insertion of different genes into a cellular cytoplasmatic device (see chapter 3), and the neuronal organoid cultures, where organs, neural networks, and even rudimentary embryos can develop in *in vitro* assays. But more than this, it might be possible to design not only cells but also complex organisms to serve our needs. Again, according to the IIT, neuronal organoids might eventually gain consciousness if they assemble the proper organization. Provided with bodies and sensory systems and becoming able to reproduce, these organisms might start an evolution of their own.[14]

The Anthropocene

Covering the Land

From a planetary perspective, our extended phenotypes not only concern our own life but we are also making a major impact on the planet we live on. At this point, between a third and a half of earth's land surface has been deeply transformed by humans, and about 97 percent of the land has some signs of human activity, only remaining 3 percent that is still untouched. But we can go even further. Our extended phenotype also includes all the species that are coevolving with us, from cows to the bacteria in our guts. Domesticated animals make about 60 percent of the biomass of all terrestrial vertebrates. Humans sum about 35 percent to this value, which leaves wild vertebrates with only about 5 percent of all the land vertebrate biomass. Terrestrial wild mammals have decreased to one-seventh of what existed in pre-human times, but never reached the numbers raised by domestic animals today.[15]

The Human Age

Besides using space, the environmental alterations of recent human activity are hard to minimize. Since the industrial revolution, we have increased by 50 percent the CO_2 concentrations in the atmosphere by using fossil fuels (a value that keeps on quickly rising). Additional factors are the production of fluorocarbons used for refrigeration, of methane produced by domestic bovids and wetland microbes, and of other compounds that all contribute to the greenhouse effect. Altogether, these events are dramatically promoting global warming. Average temperatures have shown an increase of more than 1°C since preindustrial times that keeps on growing. If the trend continues, world temperatures might have increased by 2°C by 2050 (the average global temperature today is about 14°C and rising). Furthermore, microplastics are spreading all over the planet. Considering these events, Paul Crutzen coined the term Anthropocene to refer to this new geological period that shows unmistakable signs of human activity. Associated to the Anthropocene, we may be witnessing the beginnings of the sixth-largest extinction event of the planet. Previously, there have been five massive life-extinctions in earth's history, including the great Permian extinction and the most recent one being the asteroid impact that wiped out the dinosaurs. Today, the combined effect of climate change, environmental contamination, and habitat destruction has produced a sharp decrease in biodiversity including massive deforestation and the decline of insects, soil organisms, corals, amphibians, and large mammals among other groups. On the other hand, besides domestic and human commensal species like cockroaches, there are many species that have benefited from human intervention, invading different ecosystems and contributing to destroy native species. In today's globalized world, it is easy for fungi, bacteria, plants, and small animals to disperse all around the planet by hitchhiking human transportation.[16]

But let us take some perspective considering earth's history. As said, the largest massive extinction of all times took place at the end of the Permian period, killing about 95 percent of all life. In this event, temperature rose about 5–13 °C (depending on the estimate) in tens of thousands of years, reaching a global average of about 28 °C. (Current global warming is far from those levels but is rising more rapidly and shows no signs of slowing down.) Nonetheless, despite these apparently hostile conditions, biodiversity rapidly increased again throughout the Mesozoic, with proliferating dinosaurs, early mammals, and flowering plants, at average temperatures of about 25° C, and CO_2 levels more than ten times as today. Since the early ice ages (a million years ago) until preindustrial times, temperature and CO_2 levels have been close to the lowest ever in life's history.[17] We are still way below the temperature and CO_2 levels of the Mesozoic and most of the Cenozoic, but this does not mean we should not be concerned. Most living animals and plants are adapted to a rather cool earth and may suffer heavily from relatively small but very quick changes that do not give them a chance to adapt. Yet, in the long term, global warming may generate new forms of warm-adapted species, just like it happened before. We must not forget that modern mammals, and ourselves, are largely the products of the dinosaur-killing mass extinction event, and dinosaurs themselves are the result of the Permian extinction.

All in all, my perspective is that the human impact on earth will not destroy life on earth, and our species may not become extinct by these adversities either. However, it may severely affect the world as we have known it and the way we live, especially those already living at risk, which may generate profound human tragedies. There is a major challenge to produce sustainable economies and new technologies that optimize energy production with minimal environmental impact, contributing to making a more friendly planet for us. In the end, this is the only time in earth's history where a single species may do something about keeping the planet within range (even if it has also caused much damage), ultimately for its own well-being. As I have mentioned, many measures are being proposed to prevent these trends from continuing, but they seem to have only a limited effect yet.

Still, even surviving the Anthropocene, we are by no means eternal and there may be other possibilities of a fatal strike to our species, whether from an asteroid, a nuclear war, or microbial infections. If our species disappears, who would be left? Possibly, microscopic life, together with many plants, may well survive us and will survive many other catastrophes. Among animals, arthropods in the sea and insects on land are likely inheritors of a post-human earth. Considering vertebrates, bony fish in the seas, and small amphibians, reptiles, birds, and mammals (especially rodents) on the ground, are probably those that will rapidly fill the place. How these groups would evolve in the long-distance future is just for speculation.

Perspective

This last chapter of the book concerns the present and future of our species, especially considering how culture and technology are affecting our biology. Cultural and technological evolution have eclipsed biological evolution, putting pressure on neural plasticity as the main mechanism for behavioral innovations. While we develop culture and new technologies, our plastic brain networks must adapt to these innovations. However, this may not come free of charge as we may be also becoming more vulnerable to mental disorders and brain aging.

Very recently, the technological revolution of the last few decades has added a new dimension to cultural evolution, modulating our biology by means of body and brain protheses that have increased our mental and physical capacities. The recent explosion of AI has changed our daily life in ways that few people predicted many years ago. The rise of the internet and digital social media poses concerns about our own individuality, our privacy, and our free will. Associated with this is the recent development of biotechnology driven by gene-editing procedures. While again, this may significantly improve our lives, it may also produce major changes in our genetic endowment, especially if gene manipulation is used to enhance the capabilities of only a few people.

Perhaps the most relevant challenges of the technological revolution for the next decades will have to do with the social domain: reinforcing the access of scientific developments to all the population and preventing the misuse of technology by a few groups to control the rise of inequality. Inequality has increased steadily throughout human evolution if we measure the differences between the very rich and the very poor, which may be perhaps inevitable, but we have already seen that it is a major source of social conflict. Furthermore, this may relegate many of us to live worthless lives. One way to ameliorate these possibilities may be through strict legal controls and strong ethical education, including the possibility of making an ethical oath by professionals and researchers working in the field of technology. Finally, the impact we have made on the planet with our technology has become increasingly evident in the last fifty years or so. There is an urgent need to moderate the consumption of resources from the planet and find new energy sources, as they are not eternal and we are severely limiting the existence of wildlife and biodiversity as we know them.

Final Thoughts

Cosmic Life

In this book, I have attempted to provide a succinct history of life on earth, from its beginnings on this planet to the generation of human consciousness and the possible outcomes of our species. In the present perspective, our minds are inextricably bound to our living condition and have somehow emerged through the history of life in this planet. Critical elements in the beginning of life were heredity and self-production, while common themes through all levels of life are self-maintenance and homeostasis, and the mechanisms to anticipate external perturbations. Thus, the nature of life itself, and how it has evolved and diversified through time, can provide critical insight into our self-knowledge and our vision of the future as a species.

However, we could amplify our perspective to conceive life as a cosmic phenomenon. There are the fundamental questions of whether we could live outside earth, which would definitely be the most radical transition of life's history, and whether there are other worlds with life or inhabited by intelligent beings. In the last twenty years, the discovery of more than five thousand exoplanets just in the Milky Way has fueled people with out-of-earth fantasies. However, extraterrestrial human life will bring many hard challenges to our health, and we will probably need to develop new medical technologies to overcome these difficulties. Perhaps gene-editing technology and AI may make a significant contribution there. In any case, for the time being, space exploration may be largely confined to robots, and likewise, we might expect to get in touch first with alien machines rather than with intelligent aliens themselves.

On the other hand, how likely is it that we will find life outside of earth? Recent research on Mars indicates that it had liquid water and rivers about

3 BYA (other reports suggest about 700 MYA), and the different rovers that have been sent to this planet have been intensely looking for signs of life, still without results. Saturn's and Jupiter's water-containing moons are also candidates for bearing simple life forms. In a cosmological scale, Drake's equation is an optimistic mathematical formula to calculate the probability of life elsewhere in the universe, containing a large set of assumptions that have not been confirmed. Frank Drake founded SETI, an organization devoted to the search of extraterrestrial intelligence that after spending millions of dollars has not yielded any positive findings. These results fit what has been known as Fermi's paradox: If life is a likely physical event and thrives here on earth, how come we have never observed any signs of it in the universe besides ourselves? In principle, highly intelligent life could be detected from much farther away than microscopic life, especially if it is organized in large-scale civilizations like ours, or even at more advanced levels. The Kardashev scale is a hypothetical classification of cosmic civilizations in terms of the amount of energy they consume. On a first level, civilizations use their original planet's resources like we are doing right now. Second-level civilizations profit from the energy provided by their star (we do so but minimally, mostly indirectly from plants or fossil fuels, also derived from ancient photosynthetic activity). A Dyson sphere is a hypothetical megastructure in space that could capture a high fraction of the sun's energy. A large, elongated asteroid termed Oumuamua, which crosses our solar system, was proposed to be a second-level civilization megastructure, but most astronomers disagree. Finally, third-level civilizations are able to capture the energy of many stars, even from a whole galaxy.[1]

More generally, the question of why there is life and even consciousness in the universe at all is simply mind boggling, which has prompted many to argument the need of God. Said the other way, we may ask why the universe is understandable to us. One attempted solution to this is the Anthropic principle, which in simple terms says that the only universe in which we could exist is this, one that allows for the existence of life, consciousness, and the knowledge of it. This notion is compatible with the multiverse theory and with the view of the finely tuned universe, in which there are many possible universes, but only those in which the fundamental physical constraints (like a precise value of Planck's constant) acquire a given value will be able to generate atoms, molecules, life, and consciousness.[2]

Now, given that we live in a life-compatible universe, another intriguing issue is how would life evolve on a different planet. The famous Stephen Jay Gould once asserted that if the tape of life was run again on earth, it would develop different forms but there would be some common themes that would not change. Likewise, there may be many things in common between earthlings and extraterrestrial aliens. Perhaps they would be carbon-based, although silicon-based life has also been hypothesized. A requisite for evolution would be a mechanism or inheritance that could be brought by nucleotide-like molecules, as all the key components of nucleic acids have been found in extraterrestrial meteorites. Likewise, energy might be generated through electron transfer reactions, and photosynthesis or a similar mechanism, possibly deriving from a star's radiation, would be required for ecological relations to develop, promoting the evolution of organismal complexity. Self-producing cells, multicellularity, and animals feeding on others, with sense organs and brains, could evolve, and intelligence and consciousness might emerge but not necessarily in the biological form that we know it. The onset of a civilization would require the capacity to manipulate materials in some way, generate a robust extended phenotype, and amplify communicational skills. Yet, each level higher in complexity would be much more unlikely to achieve, so in the universe there might be perhaps some instances of cellular life but many fewer cases of intelligent life. Intriguingly, we are the only example we know where not only has life become possible but intelligent and conscious life has taken place. Thus, we must accept that in practical terms, we are heavily isolated in a vast universe. The next extraterrestrial civilization could be so far from us in time and space that we may never be able to contact it given what we know about physics today.[3] Still, there is always the hope of finding life elsewhere, which could definitely be the most exciting discovery of all human history.

In any case, all life on earth will probably disappear in some 7–8 BY from now, when the sun consumes itself if there are no other life-threatening cosmical events. If there are still intelligent beings on this planet by that time (it is very unlikely that we, as we know ourselves now, will be present then), they will need to have worked a way out of here. Perhaps life will be so complex then that our difference from these intelligent beings will be even greater than that between us and the earliest cells that appeared on earth some 3 or 4 BYA. Alternatively, all life could have become extinct or exist only at the cellular level.

To end this book, I do hope we can work hard to develop a sustainable planet where we coexist with the wilderness and make this world a good place for many generations to come. This will imply much sacrifice and a strong determination, but in my view, this is not an impossible task. Likewise, we might be able to control the use of technology for the good of all mankind instead of privileging only a few. In the end, it is much better to live trying to make a better world than otherwise.

Notes

Chapter 1

1. Darwin, Charles. 1859. *On the Origins of Species by Means of Natural Selection.* 1st ed. London: John Murray; **Lamarck, Jean Baptiste**. 1984. *Zoological Philosophy*. Chicago: University of Chicago Press.

2. Aboitiz, F. 1992. "Mechanisms of Adaptive Evolution. Darwinism and Lamarckism Restated." *Medical Hypotheses* 38 (3):194–202; **Wang, Shi Qi**, Jie Ye, Jin Meng, Chunxiao Li, Loïc Costeur, Bastien Mennecart, Chi Zhang, et al. 2022. "Sexual Selection Promotes Giraffoid Head-Neck Evolution and Ecological Adaptation." *Science* 376 (6597): 1–10.

3. Standen, Emily M., Trina Y. Du, and Hans C. E. Larsson. 2014. "Developmental Plasticity and the Origin of Tetrapods." *Nature* 513 (7516): 54–58.

4. Dawkins, Richard. 1989. *The Selfish Gene*. New York: Oxford University Press.

5. Misteli, Tom. 2020. "The Self-Organizing Genome: Principles of Genome Architecture and Function." *Cell* 183 (1): 28–45.

6. Loison, Laurent. 2021. "Epigenetic Inheritance and Evolution: A Historian's Perspective." *Philosophical Transactions of the Royal Society of London. Series B, Biological Sciences* 376 (1826): 1–8; **Ashe, Alyson**, Vincent Colot, and Benjamin P. Oldroyd. 2021. "How Does Epigenetics Influence the Course of Evolution?" *Philosophical Transactions of the Royal Society B* 376 (1826): 1–9; **Husby, Arild**. 2022. "Wild Epigenetics: Insights from Epigenetic Studies on Natural Populations." *Proceedings of the Royal Society B* 289 (1968): 1–9; **Yao, N.**, Z. Zhang, L. Yu, R. Hazarika, C. Yu, H. Jang, L. M. Smith, et al. 2023. "An Evolutionary Epigenetic Clock in Plants." *Science (New York, N.Y.)* 381 (6665): 1440–1445.

7. Ragkousi, Katerina, and Matthew C. Gibson. 2014. "Cell Division and the Maintenance of Epithelial Order." *Journal of Cell Biology* 207 (2): 181–188; **Chanda, Sanjay**, and Harihara M. Mehendale. 1996. "Hepatic Cell Division and Tissue Repair: A Key to Survival after Liver Injury." *Molecular Medicine Today* 2 (2): 82–89.

8. Gehring, Walter J., and Kazuho Ikeo. 1999. "Pax 6: Mastering Eye Morphogenesis and Eye Evolution." *Trends in Genetics* 15 (9): 371–377.

9. Gould, Stephen Jay. 1985. *Ontogeny and Phylogeny. Cambridge*: The Belknap Press.

10. Gould, Stephen Jay, and Niles Eldredge. n.d. "Punctuated Equilibria: The Tempo and Mode of Evolution Reconsidered." *Paleobiology* 3 (2): 115–151. Accessed November 17, 2022; **Gould, S. J.**, and R. C. Lewontin. 1979. "The Spandrels of San Marco and the Panglossian Paradigm: A Critique of the Adaptationist Programme." *Proceedings of the Royal Society of London. Series B. Biological Sciences* 205 (1161): 581–598.

11. Xia, Bo, Weimin Zhang, Aleksandra Wudzinska, Emily Huang, Ran Brosh, Maayan Pour, Alexander Miller, et al. 2021. "The Genetic Basis of Tail-Loss Evolution in Humans and Apes." *BioRxiv*, September, 2021.09.14.460388.

12. Simpson, George Gaylord, and L. Simpson. 1949. *The Meaning of Evolution, a Study of the History of Life and of Its Significance for Man.* Vol. 23. New Haven, CT: Yale University Press; **Pagel, Mark**, Ciara O'Donovan, and Andrew Meade. 2022. "General Statistical Model Shows That Macroevolutionary Patterns and Processes Are Consistent with Darwinian Gradualism." *Nature Communications* 13 (1): 1–12.

13. Belan, Mark. 2022. "Visualizing the Evolution of Vision and the Eye." Visual Capitalist, November 4, 2022. https://www.visualcapitalist.com/eye-evolution.

Chapter 2

1. Mariscal, Carlos. 2021. "Life." *Stanford Encyclopedia of Philosophy*. https://plato.stanford.edu/entries/life/.

2. Harrison, Peter. 1992. "Descartes on Animals." *Philosophical Quarterly* 42 (167): 219–227; **Lehewych, Daniel**. 2019. "Aristotle and Descartes on Animals." October 25, 2019.

3. Wikipedia. 2014. "Digesting Duck." January, 3–5. https://en.wikipedia.org/wiki/Digesting_Duck; Wikipedia. 2022. "Mechanical Turk." https://en.wikipedia.org/wiki/Mechanical_Turk.

4. Zimmer, Carl. 2021. *Life's Edge: The Search for What It Means to Be Alive.* Dutton Press. *Vol. 1.* New York: Dutton Press.

5. Lane, Nick. 2022. *TRANSFORMER: The Deep Chemistry of Life and Death. Vol. 1.* London: Profile Books.

6. Schrödinger, Erwin. 1944. *What Is Life? The Physical Aspect of the Living Cell.* Cambridge: Cambridge University Press.

7. Jeffery, Kate, Robert Pollack, and Carlo Rovelli. 2019. "On the Statistical Mechanics of Life: Schrödinger Revisited." *Entropy* 21 (12): 1211.

8. Stafleu, F. A. 1971. "Lamarck: The Birth of Biology." *Taxon* 20: 397–442; **Lamarck, Jean Baptiste**. 1984. *Zoological Philosophy*. Chicago: University of Chicago Press, 260.

9. Luisi, Pier Luigi. 1998. "About Various Definitions of Life." *Origins of Life and Evolution of the Biosphere* 28 (4): 613–622.

10. Maturana R., Humberto, and Francisco Varela G. 1972. *De Máquinas y Seres Vivos: Autopoiesis: La Organización de Lo Vivo.* Santiago, Chile: Editorial Universitaria; **Maturana, Humberto R.**, and Francisco J. Varela. 1980. *Autopoiesis and Cognition: The Realization of the Living. Vol. 42.* Boston Studies in the Philosophy and History of Science. Berlin: Springer Netherlands.

11. Xavier, Joana C., Wim Hordijk, Stuart Kauffman, Mike Steel, and William F. Martin. 2020. "Autocatalytic Chemical Networks at the Origin of Metabolism." *Proceedings. Biological Sciences* 287 (1922): 1–10.

12. Prigogine, Ilya. 1986. "Life and Physics. New Perspectives." *Cell Biophysics* 9 (1–2): 217–224.

13. Jeancolas, C., A. Y. Singh, S. Jain, S. Krishna, and P. Nghe. 2023. "An Interdisciplinary Effort to Understand Chemical Organizations at the Origin of Life." *IScience* 26 (1): 105834; **Goldman, Aaron** D. 2023. "How Did Life Become Cellular?" *Proceedings of the Royal Society B* 290 (1992): 1–3.

14. Friston, Karl. 2010. "The Free-Energy Principle: A Unified Brain Theory?" *Nature Reviews Neuroscience* 11 (2): 127–138.

15. Kanaparthi, Dheeraj, Marko Lampe, Jan-Hagen Krohn, Falk Hildebrand, Thomas Boesen, Andreas Klingl, Prasad Phapale, and Tillmann Lueders. 2022. "On the Reproductive Mechanisms of Protocells." *BioRxiv*, October 2021.

16. Moger-Reischer, R. Z., J. I. Glass, K. S. Wise, L. Sun, D. M. C. Bittencourt, B. K. Lehmkuhl, D. R. Schoolmaster, M. Lynch, and J. T. Lennon. 2023. "Evolution of a Minimal Cell." *Nature*, July, 1–6; **Pelletier, James F.**, Lijie Sun, Kim S. Wise, Nacyra Assad-Garcia, Bogumil J. Karas, Thomas J. Deerinck, Mark H. Ellisman, et al. 2021. "Genetic Requirements for Cell Division in a Genomically Minimal Cell." *Cell* 184 (9): 2430–2440.e16; **Sandberg, Troy E.**, Kim S. Wise, Christopher Dalldorf, Richard Szubin, Adam M. Feist, John I. Glass, and Bernhard O. Palsson. 2023. "Adaptive Evolution of a Minimal Organism with a Synthetic Genome." *IScience* 26 (9): 1–9.

Chapter 3

1. Darwin, Charles. 1871. "Darwin Correspondence Project, 'Letter No. 7471.'" University of Cambridge, February 1, 1871. https://www.darwinproject.ac.uk/letter/?docId=letters/DCP-LETT-7471.xml; **Miller, Stanley L.**, and Harold C. Urey. 1959. "Organic Compound Synthesis on the Primitive Earth." *Science* 130 (3370): 245–251; **Oparin, Ivanovich Aleksandr**. 1938. *Origin of Life*. New York: Dover Publications.

2. Dodd, Matthew S., Dominic Papineau, Tor Grenne, John F. Slack, Martin Rittner, Franco Pirajno, Jonathan O'Neil, and Crispin T. S. Little. 2017. "Evidence for Early Life in Earth's Oldest Hydrothermal Vent Precipitates." *Nature* 543 (7643): 60–64; **Oba, Yasuhiro**, Yoshinori Takano, Yoshihiro Furukawa, Toshiki Koga, Daniel P. Glavin, Jason P. Dworkin, and Hiroshi Naraoka. 2022. "Identifying the Wide Diversity of Extraterrestrial Purine and Pyrimidine Nucleobases in Carbonaceous Meteorites." *Nature Communications* 13 (1): 1–10; **Sasselov, Dimitar D.**, John P. Grotzinger, and John D. Sutherland. 2020. "The Origin of Life as a Planetary Phenomenon." *Science Advances* 6 (6): 1–9.

3. Bailey, Jeremy. 2000. "Chirality and the Origin of Life." *Acta Astronautica* 46 (10–12): 627–631; **Ozturk, S. Furkan**, Ziwei Liu, John D. Sutherland, and Dimitar D. Sasselov. 2023. "Origin of Biological Homochirality by Crystallization of an RNA Precursor on a Magnetic Surface." *Science Advances* 9 (23): 1–10.

4. Babajanyan, Sanasar G., Yuri I. Wolf, Andranik Khachatryan, Armen Allahverdyan, Purificacion Lopez-Garcia, and Eugene V. Koonin. 2023. "Coevolution of Reproducers and Replicators at the Origin of Life and the Conditions for the Origin of Genomes." *Proceedings of the National Academy of Sciences* 120 (14): e2301522120; **Epstein, Irving R.**, and Manfred Eigen. 1979. "Selection and Self-Organization of Self-Reproducing Macromolecules under the Constraint of Constant Flux." *Biophysical Chemistry* 10 (2): 153–160; **Huson, Daniel**, Joana Xavier, Allen Rodrigo, and Mike Steel. 2023. "Self-Generating Autocatalytic Networks: Structural Results, Algorithms, and Their Relevance to Evolutionary Processes." *BioRxiv*, September, 2023.09.01.556005: 1–24.

5. Lane, Nick. 2022. *Transformer: The Deep Chemistry of Life and Death. Vol. 1.* London: Profile Books.

6. Jordan, Sean F., Hanadi Rammu, Ivan N. Zheludev, Andrew M. Hartley, Amandine Maréchal, and Nick Lane. 2019. "Promotion of Protocell Self-Assembly from Mixed Amphiphiles at the Origin of Life." *Nature Ecology and Evolution* 3 (12): 1705–1714.

7. McSwiggen, David T., Mustafa Mir, Xavier Darzacq, and Robert Tjian. 2019. "Evaluating Phase Separation in Live Cells: Diagnosis, Caveats, and Functional Consequences." *Genes & Development* 33 (23–24): 1619–1634.

8. Woese, Carl R. 2002. "On the Evolution of Cells." *Proceedings of the National Academy of Sciences of the United States of America* 99 (13): 8742–8747.

9. Weiss, Madeline C., Martina Preiner, Joana C. Xavier, Verena Zimorski, and William F. Martin. 2018. "The Last Universal Common Ancestor between Ancient Earth Chemistry and the Onset of Genetics." *PLoS Genetics* 14 (8): 1–19.

10. Kanaparthi, Dheeraj, Marko Lampe, Baoli Zhu, Thomas Boesen, Andreas Klingl, Petra Schwille, and Tillmann Lueders. 2021. "On the Nature of the Earliest Known Lifeforms." *BioRxiv*, November, 2021.08.16.456462.

11. Service, Robert F. 2021. "Life Could Use Oxygen Long before It Was Abundant." *Science* 371 (6533): 974.

12. Schirrmeister, Bettina E., Jurriaan M. De Vos, Alexandre Antonelli, and Homayoun C. Bagheri. 2013. "Evolution of Multicellularity Coincided with Increased Diversification of Cyanobacteria and the Great Oxidation Event." *Proceedings of the National Academy of Sciences of the United States of America* 110 (5): 1791–1796.

13. Cohen, Phoebe A., and Robin B. Kodner. 2022. "The Earliest History of Eukaryotic Life: Uncovering an Evolutionary Story through the Integration of Biological and Geological Data." *Trends in Ecology & Evolution* 37 (3): 246–256.

14. Roughgarden, Joan. 2022. "Holobiont Evolution: Population Theory for the Hologenome." *BioRxiv*, May, 2020.04.10.036350.

15. Kazmierczak, Jozef, Stephan Kempe, and Barbara Kremer. 2013. "Calcium in the Early Evolution of Living Systems: A Biohistorical Approach." *Current Organic Chemistry* 17 (16): 1738–1750.

16. Richards, Thomas A., and Suely L. Gomes. 2015. "How to Build a Microbial Eye." *Nature* 523 (7559): 166–167; **Zabelskii, Dmitrii**, Natalia Dmitrieva, Oleksandr Volkov, Vitaly Shevchenko, Kirill Kovalev, Taras Balandin, Dmytro Soloviov, et al. 2021. "Structure-Based Insights into Evolution of Rhodopsins." *Communications Biology* 4 (1): 1–12.

17. Schaap, Pauline. 2021. "From Environmental Sensing to Developmental Control: Cognitive Evolution in Dictyostelid Social Amoebas." *Philosophical Transactions of the Royal Society B* 376 (1820): 1–8; **Brette, Romain**. 2021. "Integrative Neuroscience of Paramecium, a 'Swimming Neuron.'" *ENeuro* 8 (3): 1–30; **Kramar, Mirna**, and Karen Alim. 2021. "Encoding Memory in Tube Diameter Hierarchy of Living Flow Network." *Proceedings of the National Academy of Sciences of the United States of America* 118 (10): 1–6; **Baluška, František**, William B. Miller, and Arthur S. Reber. 2021. "Biomolecular Basis of Cellular Consciousness via Subcellular Nanobrains." *International Journal of Molecular Sciences* 22 (5): 1–15.

Chapter 4

1. Dinet, Céline, Alphée Michelot, Julien Herrou, and Tâm Mignot. 2021. "Linking Single-Cell Decisions to Collective Behaviours in Social Bacteria." *Philosophical Transactions of the Royal Society B* 376 (1820): 1–8; **Tagkopoulos, Ilias**, Yir Chung Liu, and Saeed Tavazoie. 2008. "Predictive Behavior within Microbial Genetic Networks." *Science* 320 (5881): 1313–1317; **Schaap, Pauline**. 2021. "From Environmental Sensing to Developmental Control: Cognitive Evolution in Dictyostelid Social Amoebas." *Philosophical Transactions of the Royal Society of London. Series B, Biological Sciences* 376 (1820): 1–8.

2. Howe, Jack, Jochen C. Rink, Bo Wang, and Ashleigh S. Griffin. 2022. "Multicellularity in Animals: The Potential for within-Organism Conflict." *Proceedings of the National Academy of Sciences of the United States of America* 119 (32): 1–10.

3. Solana, Jordi. 2013. "Closing the Circle of Germline and Stem Cells: The Primordial Stem Cell Hypothesis." *EvoDevo* 4 (1): 1–16.

4. Baluška, František, and Stefano Mancuso. 2021. "Individuality, Self and Sociality of Vascular Plants." *Philosophical Transactions of the Royal Society of London. Series B, Biological Sciences* 376 (1821): 1–10; **Klein, Tamir**, Rolf T. W. Siegwolf, and Christian Körner. 2016. "Belowground Carbon Trade among Tall Trees in a Temperate Forest." *Science* 352 (6283): 342–344.

5. Segundo-Ortin, Miguel, and Paco Calvo. 2022. "Consciousness and Cognition in Plants." *Wiley Interdisciplinary Reviews. Cognitive Science* 13 (2): 1–23; **Khait, Itzhak**, Ohad Lewin-Epstein, Raz Sharon, Kfir Saban, Revital Goldstein, Yehuda Anikster, Yarden Zeron, et al. 2023. "Sounds Emitted by Plants under Stress Are Airborne and Informative." *Cell* 186 (7): 1328–1336.e10.

6. Sheldrake, Merlin. 2020. *Entangled Life: How Fungi Make Our Worlds, Change Our Minds & Shape Our Futures*. Random House Publishing Group, London UK.

7. Larson, Ben T., Teresa Ruiz-Herrero, Stacey Lee, Sanjay Kumar, L. Mahadevan, and Nicole King. 2020. "Biophysical Principles of Choanoflagellate Self-Organization." *Proceedings of the National Academy of Sciences of the United States of America* 117 (3): 1303–1311; **Göhde, Ronja**, Benjamin Naumann, Davis Laundon, Cordelia Imig, Kent McDonald, Benjamin H. Cooper, Frédérique Varoqueaux, Dirk Fasshauer, and Pawel Burkhardt. 2021. "Choanoflagellates and the Ancestry of Neurosecretory Vesicles." *Philosophical Transactions of the Royal Society B* 376 (1821): 1–12; **Ocaña-Pallarès, Eduard**, Tom A. Williams, David López-Escardó, Alicia S. Arroyo, Jananan S. Pathmanathan, Eric Bapteste, Denis V. Tikhonenkov, Patrick J. Keeling, Gergely J. Szöllősi, and Iñaki Ruiz-Trillo. 2022. "Divergent Genomic Trajectories Predate the Origin of Animals and Fungi." *Nature* 609 (7928): 747–753.

8. Grazhdankin, Dmitriy. 2011. "Ediacaran Biota." *Encyclopedia of Geobiology*, no. 9781402092114: 342–348; **Gibson, Brandt M.**, Imran A. Rahman, Katie M. Maloney, Rachel A. Racicot, Helke Mocke, Marc Laflamme, and Simon A.F. Darroch. 2019. "Gregarious Suspension Feeding in a Modular Ediacaran Organism." *Science Advances* 5 (6): 1–9.

9. Mah, Jasmine L., and Sally P. Leys. 2017. "Think like a Sponge: The Genetic Signal of Sensory Cells in Sponges." *Developmental Biology* 431 (1): 93–100; **Brokhman, Irina**, Jie Xu, Brenda L.K. Coles, Rozita Razavi, Silvia Engert, Heiko Lickert, Robert Babona-Pilipos, et al. 2019. "Dual Embryonic Origin of the Mammalian Enteric Nervous System." *Developmental Biology* 445 (2): 256–270.

10. Fritz-Laylin, Lillian K. 2020. "The Evolution of Animal Cell Motility." *Current Biology* 30 (10): R477–R482; **Nakanishi, Nagayasu**, Shunsuke Sogabe, and Bernard M. Degnan. 2014. "Evolutionary Origin of Gastrulation: Insights from Sponge Development." *BMC Biology* 12 (March): 1–9; **Brunet, Thibaut**. 2023. "Cell Contractility in Early Animal Evolution." *Current Biology* 33 (18): R966–85.; **Hall, Thomas E.**, Nicholas Ariotti, Harriet P. Lo, James Rae, Charles Ferguson, Nick Martel, Ye-Wheen Lim, Jean Giacomotto, and Robert G. Parton. 2023. "Cell Surface Plasticity in Response to Shape Change in the Whole Organism." *Current Biology* 33 (19): 4276–4284.e4.

11. Sperling, Erik A., and Jakob Vinther. 2010. "A Placozoan Affinity for Dickinsonia and the Evolution of Late Proterozoic Metazoan Feeding Modes." *Evolution & Development* 12 (2): 201–209; **Maegele, Ira**, Sebastian Rupp, Suat Özbek, Annika Guse, Elizabeth A. Hambleton, and Thomas W. Holstein. 2023. "A Predatory Gastrula Leads to Symbiosis-Independent Settlement in Aiptasia." *BioRxiv*, May, 2023.05.26.542442; **Najle S. R.**, X. Grau-Bové, A. Elek, C. Navarrete, D. Cianferoni, C. Chiva, D. Cañas-Armenteros, A. Mallabiabarrena, K. Kamm, E. Sabidó, H. Gruber-Vodicka, B. Schierwater, L. Serrano, and A. Sebé-Pedrós. 2023. "Stepwise Emergence of the Neuronal Gene Expression Program in Early Animal Evolution." *Cell* 186(21): 4676–4693.e29.

12. Willmer, Pat. 1990. *Invertebrate Relationships: Patterns in Animal Evolution*. Cambridge: Cambridge University Press.

13. Whelan, Nathan V., Kevin M. Kocot, Tatiana P. Moroz, Krishanu Mukherjee, Peter Williams, Gustav Paulay, Leonid L. Moroz, and Kenneth M. Halanych. 2017. "Ctenophore Relationships and Their Placement as the Sister Group to All Other Animals." *Nature Ecology & Evolution* 1 (11): 1737–1746; **Whelan, Nathan V.**, and Kenneth M. Halanych. 2023. "Available Data Do Not Rule out Ctenophora as the Sister Group to All Other Metazoa." *Nature Communications*; **Schultz, Darrin T.**, Steven H. D. Haddock, Jessen V. Bredeson, Richard E. Green, Oleg Simakov, and Daniel S. Rokhsar. 2023. "Ancient Gene Linkages Support Ctenophores as Sister to Other Animals." *Nature* 618 (7963): 110.

14. Sachkova, Maria Y. 2022. "Neuropeptides at the Origin of Neurons." *Nature Ecology & Evolution* 6 (10): 1410–1411; **Arendt, Detlev**. 2021. "Elementary Nervous Systems." *Philosophical Transactions of the Royal Society B* 376 (1821): 1–13; **Burkhardt, Pawel**, Jeffrey Colgren, Astrid Medhus, Leonid Digel, Benjamin Naumann, Joan J. Soto-Angel, Eva-Lena Nordmann, Maria Y. Sachkova, and Maike Kittelmann. 2023. "Syncytial Nerve Net in a Ctenophore Adds Insights on the Evolution of Nervous Systems." *Science* 380 (6642): 293–297; **Feuda, Roberto**, and Isabelle S. Peter. 2022. "Homologous Gene Regulatory Networks Control Development of Apical Organs and Brains in Bilateria." *Science Advances* 8 (44): 1–13; **Thoma, Vladimiros**, Shuhei Sakai, Koki Nagata, Yuu Ishii, Shinichiro Maruyama, Ayako Abe, Shu Kondo, et al. 2023. "On the Origin of Appetite: GLWamide in Jellyfish Represents an Ancestral

Satiety Neuropeptide." *Proceedings of the National Academy of Sciences of the United States of America* 120 (15): 1–11.

Chapter 5

1. Arendt, Detlev. 2020. "The Evolutionary Assembly of Neuronal Machinery." *Current Biology* 30 (10): R603–R616.

2. Cornejo, Victor Hugo, Netanel Ofer, and Rafael Yuste. 2022. "Voltage Compartmentalization in Dendritic Spines in Vivo." *Science* 375 (6576): 82–86.

3. Kristan, William B. 2016. "Early Evolution of Neurons." *Current Biology* 26 (20): R949–R954.

4. Ryan, Tomás J., and Seth G. N. Grant. 2009. "The Origin and Evolution of Synapses." *Nature Reviews. Neuroscience* 10 (10): 701–712; **Arendt, Detlev**, Paola Yanina Bertucci, Kaia Achim, and Jacob M. Musser. 2019. "Evolution of Neuronal Types and Families." *Current Opinion in Neurobiology* 56: 144–152; **Goulty, Matthew**, Gaelle Botton-Amiot, Ezio Rosato, Simon G. Sprecher, and Roberto Feuda. 2023. "The Monoaminergic System Is a Bilaterian Innovation." *Nature Communications* 14 (1): 1–15.

5. Guthrie, G. D., and C. S. Nicholson-Guthrie. 1989. "Gamma-Aminobutyric Acid Uptake by a Bacterial System with Neurotransmitter Binding Characteristics." *Proceedings of the National Academy of Sciences of the United States of America* 86 (19): 7378–7381; **Ramos-Vicente**, David, Seth G. N. Grant, and Àlex Bayés. 2021. "Metazoan Evolution and Diversity of Glutamate Receptors and Their Auxiliary Subunits." *Neuropharmacology* 195: 1–15.

6. Jékely, Gáspár, Peter Godfrey-Smith, and Fred Keijzer. 2021. "Reafference and the Origin of the Self in Early Nervous System Evolution." *Philosophical Transactions of the Royal Society B* 376 (1821): 1–13.

7. Bosman, Conrado A., and Francisco Aboitiz. 2015. "Functional Constraints in the Evolution of Brain Circuits." *Frontiers in Neuroscience* 9: 1–13; **Hanson, Alison**. 2021. "Spontaneous Electrical Low-Frequency Oscillations: A Possible Role in Hydra and All Living Systems." *Philosophical Transactions of the Royal Society of London. Series B, Biological Sciences* 376 (1820): 1–15; **Lovas, Jonathan R.**, and Rafael Yuste. 2021. "Ensemble Synchronization in the Reassembly of Hydra's Nervous System." *Current Biology* 31 (17): 3784–3796.e3.

8. Nave, Klaus Armin, Iva D. Tzvetanova, and Stefanie Schirmeier. 2017. "Glial Cell Evolution: The Origins of a Lipid Store." *Cell Metabolism* 26 (5): 701–702; **Helm, Conrad**, Anett Karl, Patrick Beckers, Sabrina Kaul-Strehlow, Elke Ulbricht, Ioannis Kourtesis, Heidrun Kuhrt, et al. 2017. "Early Evolution of Radial Glial Cells in Bilateria." *Proceedings. Biological Sciences* 284 (1859): 1–10; **Ghosh, Tanay**, Rafael G. Almeida, Chao Zhao, Abdelkrim Mannioui, Elodie Martin, Alex Fleet, Ginez

Gonzalez M, et al. 2022. "A Retroviral Origin of Vertebrate Myelin." *BioRxiv*, October, 2022.01.24.477350.

9. Kelava, Iva, Fabian Rentzsch, and Ulrich Technau. 2015. "Evolution of Eumetazoan Nervous Systems: Insights from Cnidarians." *Philosophical Transactions of the Royal Society of London. Series B, Biological Sciences* 370 (1684): 1–9; **Hartenstein, Volker**, and Angelika Stollewerk. 2015. "The Evolution of Early Neurogenesis." *Developmental Cell* 32 (4): 390–407.

10. Roeder, Sebastian S., Petra Burkardt, Fabian Rost, Julian Rode, Lutz Brusch, Roland Coras, Elisabet Englund, et al. 2022. "Evidence for Postnatal Neurogenesis in the Human Amygdala." *Communications Biology* 5 (1): 1–8; **Franjic, Daniel**, Mario Skarica, Shaojie Ma, Jon I. Arellano, Andrew T. N. Tebbenkamp, Jinmyung Choi, Chuan Xu, et al. 2022. "Transcriptomic Taxonomy and Neurogenic Trajectories of Adult Human, Macaque, and Pig Hippocampal and Entorhinal Cells." *Neuron* 110 (3): 452–469.e14.

11. Rao, Meenakshi, and Michael D. Gershon. 2018. "Enteric Nervous System Development: What Could Possibly Go Wrong?" *Nature Reviews. Neuroscience* 19 (9): 552–565.

12. Roig-Puiggros, Sergi, Robin J. Vigouroux, Danielle Beckman, Nadia I. Bocai, Brian Chiou, Joshua Davimes, Gimena Gomez, et al. 2020. "Construction and Reconstruction of Brain Circuits: Normal and Pathological Axon Guidance." *Journal of Neurochemistry* 153 (1): 10–32.

13. Edelman, Gerald. 1987. *Neural Darwinism: The Theory of Neuronal Group Selection*. First Edition. New York: Basic Books.

14. Levi-Montalcini, Rita. 2004. "The Nerve Growth Factor and the Neuroscience Chess Board." *Progress in Brain Research* 146: 523–527.

15. Espinosa, J. Sebastian, and Michael P. Stryker. 2012. "Development and Plasticity of the Primary Visual Cortex." *Neuron* 75 (2): 230–249.

16. Collingridge, G. L., and W. C. Abraham. 2022. "Glutamate Receptors and Synaptic Plasticity: The Impact of Evans and Watkins." *Neuropharmacology* 206: 1–12.

17. Maturana R., Humberto, and Francisco Varela G. 1972. *De Máquinas y Seres Vivos: Autopoiesis: La Organización de Lo Vivo*. Santiago, Chile: Editorial Universitaria; **Maturana, Humberto R.**, and Francisco J. Varela. 1980. *Autopoiesis and Cognition: The Realization of the Living. Vol. 42.* Boston Studies in the Philosophy and History of Science. Berlin: Springer Netherlands.

18. Byrne, John H., Evangelos G. Antzoulatos, and Diasinou Fioravante. 2004. "Aplysia: Neural and Molecular Mechanisms of Simple Forms of Learning: Sensitization, Classical Conditioning, and Operant Conditioning." In *Encyclopedia of Neuroscience—Article 275*, 3rd ed. Edited by G. Adelman and B. H. Smith. Amsterdam: Elsevier Science: 1–5.

19. Lind, Johan, and Vera Vinken. 2021. "Can Associative Learning Be the General Process for Intelligent Behavior in Non-Human Animals?" *BioRxiv*, December, 2021.12.15.472737; **Antonov, Igor**, Irina Antonova, Eric R. Kandel, and Robert D. Hawkins. 2003. "Activity-Dependent Presynaptic Facilitation and Hebbian LTP Are Both Required and Interact during Classical Conditioning in Aplysia." *Neuron* 37 (1): 135–147; **Bielecki, Jan**, Sofie Katrine Dam Nielsen, Gösta Nachman, and Anders Garm. 2023. "Associative Learning in the Box Jellyfish Tripedalia Cystophora." *Current Biology* 33 (19): 4150–4159.e5; **Botton-Amiot**, Gaelle, Pedro Martinez, and Simon G. Sprecher. 2023. "Associative Learning in the Cnidarian Nematostella Vectensis." *Proceedings*.

20. Scaplen, Kristin M., and Karla R. Kaun. 2016. "Reward from Bugs to Bipeds: A Comparative Approach to Understanding How Reward Circuits Function." *Journal of Neurogenetics* 30 (2): 133–148; **Friston, Karl**, Philipp Schwartenbeck, Thomas Fitzgerald, Michael Moutoussis, Timothy Behrens, and Raymond J. Dolan. 2014. "The Anatomy of Choice: Dopamine and Decision-Making." *Philosophical Transactions of the Royal Society of London. Series B, Biological Sciences* 369 (1655): 1–12; **Solié, Clément**, Benoit Girard, Beatrice Righetti, Malika Tapparel, and Camilla Bellone. 2022. "VTA Dopamine Neuron Activity Encodes Social Interaction and Promotes Reinforcement Learning through Social Prediction Error." *Nature Neuroscience* 25 (1): 86–97; **Jeong, Huijeong**, Annie Taylor, Joseph R Floeder, Martin Lohmann, Stefan Mihalas, Brenda Wu, Mingkang Zhou, Dennis A Burke, and Vijay Mohan K Namboodiri. 2022. "Mesolimbic Dopamine Release Conveys Causal Associations." *Science* 378 (6626): 1–10.

Chapter 6

1. Erwin, Douglas H. 2020. "The Origin of Animal Body Plans: A View from Fossil Evidence and the Regulatory Genome." *Development* 147 (4): 1–14.

2. Robertis, E. M. De, and Yoshiki Sasai. 1996. "A Common Plan for Dorsoventral Patterning in Bilateria." *Nature* 380 (6569): 37–40.

3. Levit, Georgy S., Uwe Hoßfeld, Benjamin Naumann, Paul Lukas, and Lennart Olsson. 2022. "The Biogenetic Law and the Gastraea Theory: From Ernst Haeckel's Discoveries to Contemporary Views." *Journal of Experimental Zoology. Part B, Molecular and Developmental Evolution* 338 (1–2): 13–27; **Nielsen, Claus**, Thibaut Brunet, and Detlev Arendt. 2018. "Evolution of the Bilaterian Mouth and Anus." *Nature Ecology & Evolution* 2 (9): 1358–1376.

4. Cannon, Johanna Taylor, Bruno Cossermelli Vellutini, Julian Smith, Fredrik Ronquist, Ulf Jondelius, and Andreas Hejnol. 2016. "Xenacoelomorpha Is the Sister Group to Nephrozoa." *Nature* 530 (7588): 89–93; **Kapli, Paschalia**, and Maximilian J. Telford. 2020. "Topology-Dependent Asymmetry in Systematic Errors Affects

Phylogenetic Placement of Ctenophora and Xenacoelomorpha." *Science Advances* 6 (50): 1–11; **Perea-Atienza**, Elena, Brenda Gavilán, Marta Chiodin, Josep F. Abril, Katharina J. Hoff, Albert J. Poustka, and Pedro Martinez. 2015. "The Nervous System of Xenacoelomorpha: A Genomic Perspective." *Journal of Experimental Biology* 218 (Pt 4): 618–628; **Redmond, Anthony K.** 2023. "Acoelomorph Flatworm Monophyly Is a Severe Long Branch-Attraction Artefact Obscuring a Clade of Acoela and Xenoturbellida." *BioRxiv*, August, 2023.08.27.555014: 1–32.

5. Martindale, Mark Q., and Andreas Hejnol. 2009. "A Developmental Perspective: Changes in the Position of the Blastopore during Bilaterian Evolution." *Developmental Cell* 17 (2): 162–174.

6. Gaunt, Stephen J. 2018. "Hox Cluster Genes and Collinearities throughout the Tree of Animal Life." *The International Journal of Developmental Biology* 62 (11–12): 673–683.

7. Drost, Hajk Georg, Philipp Janitza, Ivo Grosse, and Marcel Quint. 2017. "Cross-Kingdom Comparison of the Developmental Hourglass." *Current Opinion in Genetics & Development* 45 (August): 69–75.

8. Martín-Durán, José M., Kevin Pang, Aina Børve, Henrike Semmler Lê, Anlaug Furu, Johanna Taylor Cannon, Ulf Jondelius, and Andreas Hejnol. 2018. "Convergent Evolution of Bilaterian Nerve Cords." *Nature* 553 (7686): 45–50; **Arendt, Detlev**. 2018. "Animal Evolution: Convergent Nerve Cords?" *Current Biology* 28 (5): R225–R227; **Schuster, Helen C.**, and Frank Hirth. 2023. "Phylogenetic Tracing of Midbrain-Specific Regulatory Sequences Suggests Single Origin of Eubilaterian Brains." *Science Advances* 9 (21): eade8259.

9. Zhao, Di, Siyu Chen, and Xiao Liu. 2019. "Lateral Neural Borders as Precursors of Peripheral Nervous Systems: A Comparative View across Bilaterians." *Development, Growth & Differentiation* 61 (1): 58–72.

10. Tarazona, Oscar A., Davys H. Lopez, Leslie A. Slota, and Martin J. Cohn. 2019. "Evolution of Limb Development in Cephalopod Mollusks." *ELife* 8: 1–19.

11. Koenig, Kristen M., and Jeffrey M. Gross. 2020. "Evolution and Development of Complex Eyes: A Celebration of Diversity." *Development* 147 (19): 1–19; **Belan, Mark.** 2022. "Visualizing the Evolution of Vision and the Eye." Visual Capitalist, November 4, 2022. https://www.visualcapitalist.com/eye-evolution/.

12. Kooi, Casper J. Van Der, Doekele G. Stavenga, Kentaro Arikawa, Gregor Belušič, and Almut Kelber. 2021. "Evolution of Insect Color Vision: From Spectral Sensitivity to Visual Ecology." *Annual Review of Entomology* 66: 435–461; **Yoshimatsu, Takeshi**, Philipp Bartel, Cornelius Schröder, Filip K. Janiak, François St-Pierre, Philipp Berens, and Tom Baden. 2021. "Ancestral Circuits for Vertebrate Color Vision Emerge at the First Retinal Synapse." *Science Advances* 7 (42): 1–18.

13. Giesen, Lena van, Peter B. Kilian, Corey A. H. Allard, and Nicholas W. Bellono. 2020. "Molecular Basis of Chemotactile Sensation in Octopus." *Cell* 183 (3): 594–604.e14.

14. Maza, Francisco Javier, Julieta Sztarker, Avishag Shkedy, Valeria Natacha Peszano, Fernando Federico Locatelli, and Alejandro Delorenzi. 2016. "Context-Dependent Memory Traces in the Crab's Mushroom Bodies: Functional Support for a Common Origin of High-Order Memory Centers." *Proceedings of the National Academy of Sciences of the United States of America* 113 (49): E7957–E7965; **Wolff, Gabriella H.**, and Nicholas J. Strausfeld. 2016. "Genealogical Correspondence of a Forebrain Centre Implies an Executive Brain in the Protostome–Deuterostome Bilaterian Ancestor." *Philosophical Transactions of the Royal Society B: Biological Sciences* 371 (1685): 1–15; **Hartse, Kristyna M.** 2009. "Sleep in Insects." In *Evolution of Sleep: Phylogenetic and Functional Perspectives*, 34–56. Cambridge: Cambridge University Press; **Roßler, Daniela C.**, Kris Kim, Massimo De Agro, Alex Jordan, C. Giovanni Galizia, and Paul S. Shamble. 2022. "Regularly Occurring Bouts of Retinal Movements Suggest an REM Sleep-like State in Jumping Spiders." *Proceedings of the National Academy of Sciences of the United States of America* 119 (33): 1–3.

15. Chittka, Lars, and Natacha Rossi. 2022. "Social Cognition in Insects." *Trends in Cognitive Sciences* 26 (7): 578–592; **Dong, Shihao**, Tao Lin, James C. Nieh, and Ken Tan. 2023. "Social Signal Learning of the Waggle Dance in Honey Bees." *Science* 379 (6636): 1015–1018; **Bridges, Alice D.**, Ha Di MaBouDi, Olga Procenko, Charlotte Lockwood, Yaseen Mohammed, Amelia Kowalewska, José Eric Romero González, Joseph L. Woodgate, and Lars Chittka. 2023. "Bumblebees Acquire Alternative Puzzle-Box Solutions via Social Learning." *PLOS Biology* 21 (3): e3002019.

16. Perry, Clint J., Andrew B. Barron, and Lars Chittka. 2017. "The Frontiers of Insect Cognition." *Current Opinion in Behavioral Sciences* 16: 111–118.

17. Alem, Sylvain, Clint J. Perry, Xingfu Zhu, Olli J. Loukola, Thomas Ingraham, Eirik Søvik, and Lars Chittka. 2016. "Associative Mechanisms Allow for Social Learning and Cultural Transmission of String Pulling in an Insect." *PLOS Biology* 14 (10): e1002564; **Tibbetts, Elizabeth A.**, Juanita Pardo-Sanchez, Julliana Ramirez-Matias, and Aurore Avarguès-Weber. 2021. "Individual Recognition Is Associated with Holistic Face Processing in Polistes Paper Wasps in a Species-Specific Way." *Proceedings. Biological Sciences* 288 (1943): 1–6; **Nieder, Andreas**. 2021. "Neuroethology of Number Sense across the Animal Kingdom." *Journal of Experimental Biology* 224 (Pt 6): 1–15; **Galpayage Dona**, Hiruni Samadi, Cwyn Solvi, Amelia Kowalewska, Kaarle Mäkelä, Ha Di MaBouDi, and Lars Chittka. 2022. "Do Bumble Bees Play?" *Animal Behaviour* 194: 239–251.

18. Anderson, David J. 2016. "Circuit Modules Linking Internal States and Social Behaviour in Flies and Mice." *Nature Reviews. Neuroscience* 17 (11): 692–704; **Chittka, Lars.** 2023. "Do Insects Feel Joy and Pain?" *Scientific American* 329 (1): 26–33;

Oikawa, Izumi, Shu Kondo, Kao Hashimoto, Akiho Yoshida, Megumi Hamajima, Hiromu Tanimoto, Katsuo Furukubo-Tokunaga, and Ken Honjo. 2023. "A Descending Inhibitory Mechanism of Nociception Mediated by an Evolutionarily Conserved Neuropeptide System in Drosophila." *ELife* 12: 1–12.

19. Chung, Wen-Sung, Alejandra L. Galan, Nyoman D. Kurniawan, and N. Justin Marshall. 2022. "Comparative Brain Structure and the Neural Network Features of Cuttlefish and Squid." *BioRxiv*, May, 2022.05.08.491098; **Kuuspalu, Adam**, Samantha Cody, and Melina E. Hale. 2022. "Multiple Nerve Cords Connect the Arms of Octopuses, Providing Alternative Paths for Inter-Arm Signaling." *Current Biology* 32 (24): 5415–5421.e3; **Flash, Tamar**, and Letizia Zullo. 2023. "Biomechanics, Motor Control and Dynamic Models of the Soft Limbs of the Octopus and Other Cephalopods." *Journal of Experimental Biology* 226: 1–13; **Nilsson, Dan E.**, Sönke Johnsen, and Eric Warrant. 2023. "Cephalopod versus Vertebrate Eyes." *Current Biology* 33 (20): 1100–1105.; **Pungor, Judit R.**, and Cristopher M. Niell. 2023. "The Neural Basis of Visual Processing and Behavior in Cephalopods." *Current Biology* 33 (20): 1106–1118.

20. Schnell, Alexandra K., and Nicola S. Clayton. 2019. "Cephalopod Cognition." *Current Biology* 29 (15): R726–R732; **Medeiros, Sylvia Lima de Souza**, Mizziara Marlen Matias de Paiva, Paulo Henrique Lopes, Wilfredo Blanco, Françoise Dantas de Lima, Jaime Bruno Cirne de Oliveira, Inácio Gomes Medeiros, et al. 2021. "Cyclic Alternation of Quiet and Active Sleep States in the Octopus." *IScience* 24 (4): 1–19; **Poncet, Lisa**, Coraline Desnous, Cécile Bellanger, and Christelle Jozet-Alves. 2022. "Unruly Octopuses Are the Rule: Octopus Vulgaris Use Multiple and Individually Variable Strategies in an Episodic-like Memory Task." *Journal of Experimental Biology* 225 (19): 1–9; **Jozet-Alves, Christelle**, Alexandra K. Schnell, and Nicola S. Clayton. 2023. "Cephalopod Learning and Memory." *Current Biology* 33 (20): 1091–1095.

21. Schmidbaur, Hannah, Akane Kawaguchi, Tereza Clarence, Xiao Fu, Oi Pui Hoang, Bob Zimmermann, Elena A. Ritschard, et al. 2022. "Emergence of Novel Cephalopod Gene Regulation and Expression through Large-Scale Genome Reorganization." *Nature Communications* 13 (1): 1–11; **Zolotarov, Grygoriy**, Bastian Fromm, Ivano Legnini, Salah Ayoub, Gianluca Polese, Valeria Maselli, Peter J. Chabot, et al. 2022. "MicroRNAs Are Deeply Linked to the Emergence of the Complex Octopus Brain." *Science Advances* 8 (47): 1–12; **Martínez-Redondo, Gemma I.**, Carolina Simón Guerrero, Leandro Aristide, Pau Balart-García, Vanina Tonzo, and Rosa Fernández. 2022. "Parallel Duplication and Loss of Aquaporin-Coding Genes during the 'out of the Sea' Transition as Potential Key Drivers of Animal Terrestrialization." *BioRxiv*, October, 2022.07.25.501387; **Zolotarov, Grygoriy**, Bastian Fromm, Ivano Legnini, Salah Ayoub, Gianluca Polese, Valeria Maselli, Peter J. Chabot, et al. 2022. "MicroRNAs Are Deeply Linked to the Emergence of the Complex Octopus Brain." *Science Advances* 8 (47): 1–12; **Albertin, Caroline B.**, and Paul S. Katz. 2023. "Evolution of Cephalopod Nervous Systems." *Current Biology* 33 (20): 1087–1091.

Chapter 7

1. Lowe, Christopher J. 2021. "Molecular Insights into Deuterostome Evolution from Hemichordate Developmental Biology." *Current Topics in Developmental Biology* 141 (January): 75–117; **Li, Yujing**, Frances S. Dunn, Duncan J. E. Murdock, Jin Guo, Imran A. Rahman, and Peiyun Cong. 2023. "Cambrian Stem-Group Ambulacrarians and the Nature of the Ancestral Deuterostome." *Current Biology* 33 (12): 2359–2366. e2; **Andrade Ló Pez Id**, José M, Ariel M Pani, Mike Wu, John Gerhart, and Christopher J Loweid. 2023. "Molecular Characterization of Nervous System Organization in the Hemichordate Acorn Worm Saccoglossus Kowalevskii." *PLOS Biology* 21 (9): e3002242: 1–37; **Formery, L.**, P. Peluso, I. Kohnle, J. Malnick, J. R. Thompson, M. Pitel, K. R. Uhlinger, D. S. Rokhsar, D. R. Rank, and C. J. Lowe. 2023. "Molecular Evidence of Anteroposterior Patterning in Adult Echinoderms." *Nature 2023*, November: 1–7 (online ahead of print; doi: 10.1038/s41586-023-06669-2).

2. Shu, D. G., S. Conway Morris, Z. F. Zhang, and J. Han. 2010. "The Earliest History of the Deuterostomes: The Importance of the Chengjiang Fossil-Lagerstatte." *Proceedings. Biological Sciences* 277 (1679): 165–174.

3. Aboitiz, Francisco, and Juan F. Montiel. 2021. "The Enigmatic Reissner's Fiber and the Origin of Chordates." *Frontiers in Neuroanatomy* 15: 1–8; **Adameyko, Igor.** 2023. "Evolutionary Origin of the Neural Tube in Basal Deuterostomes." *Current Biology* 33 (8): R319–R331.

4. Lacalli, Thurston. 2018. "Amphioxus, Motion Detection, and the Evolutionary Origin of the Vertebrate Retinotectal Map." *EvoDevo* 9 (1): 1–5; **Lacalli, Thurston.** 2022. "An Evolutionary Perspective on Chordate Brain Organization and Function: Insights from Amphioxus, and the Problem of Sentience." *Philosophical Transactions of the Royal Society B* 377 (1844): 20200520.

5. Geoffroy Saint-Hilaire, Étienne. 1822. "Considérations Générales Sur La Vertèbre." *Mem. Mus. Hist. Nat. Paris* 9: 89–119; **Satoh, Nori.** 2008. "An Aboral-Dorsalization Hypothesis for Chordate Origin." *Genesis* 46 (11): 614–622.

6. Martik, Megan L., and Marianne E. Bronner. 2021. "Riding the Crest to Get a Head: Neural Crest Evolution in Vertebrates." *Nature Reviews Neuroscience* 22 (10): 616–626; **Gillis, J. Andrew.** 2023. "Scales, Scutes, and Embryonic Origins of the Vertebrate Dermal Skeleton." *Proceedings of the National Academy of Sciences of the United States of America* 120 (33): e2310552120: 1–2.

7. Simakov, Oleg, Ferdinand Marlétaz, Jia Xing Yue, Brendan O'Connell, Jerry Jenkins, Alexander Brandt, Robert Calef, et al. 2020. "Deeply Conserved Synteny Resolves Early Events in Vertebrate Evolution." *Nature Ecology & Evolution* 4 (6): 820–830.

8. Jacobs, Lucia F. 2012. "From Chemotaxis to the Cognitive Map: The Function of Olfaction." *Proceedings of the National Academy of Sciences of the United States of America* 109 Suppl 1: 10693–10700; **Töle, Jonas C.**, Maik Behrens, and Wolfgang

Meyerhof. 2019. "Taste Receptor Function." *Handbook of Clinical Neurology* 164 (January): 173–185.

9. Collin, Shaun P. 2010. "Evolution and Ecology of Retinal Photoreception in Early Vertebrates." *Brain, Behavior and Evolution* 75 (3): 174–185; **Sanes, Joshua R.**, and S. Lawrence Zipursky. 2010. "Design Principles of Insect and Vertebrate Visual Systems." *Neuron* 66 (1): 15–36; **Ellis, Erika M.**, Rikard Frederiksen, Ala Morshedian, Gordon L. Fain, and Alapakkam P. Sampath. 2020. "Separate ON and OFF Pathways in Vertebrate Vision First Arose during the Cambrian." *Current Biology* 30 (11): R633–R634; **Clark, Damon A.**, and Jonathan B. Demb. 2016. "Parallel Computations in Insect and Mammalian Visual Motion Processing." *Current Biology* 26 (20): R1062–R1072.

10. Maturana, Humberto R., J. Y. Lettvin, W. S. McCulloch, and W. H. Pitts. 1960. "Anatomy and Physiology of Vision in the Frog (Rana Pipiens)." *Journal of General Physiology* 43(6): 129–175.

11. Streit, Andrea. 2001. "Origin of the Vertebrate Inner Ear: Evolution and Induction of the Otic Placode." *Journal of Anatomy* 199 (1–2): 99–103.

12. Donoghue, Philip C. J., Peter L. Forey, and Richard J. Aldridge. 2000. "Conodont Affinity and Chordate Phylogeny." *Biological Reviews of the Cambridge Philosophical Society* 75 (2): 191–251; **Tian, Qingyi**, Fangchen Zhao, Han Zeng, Maoyan Zhu, and Baoyu Jiang. 2022. "Ultrastructure Reveals Ancestral Vertebrate Pharyngeal Skeleton in Yunnanozoans." *Science* 377 (6602): 218–222.

13. Carroll, Robert. L. 1990. *Vertebrate Paleontology and Evolution. Vertebrate Paleontology and Evolution.* 1st ed. W. H. Freeman and Company, New York.

14. Pieretti, Joyce, Andrew R. Gehrke, Igor Schneider, Noritaka Adachi, Tetsuya Nakamura, and Neil H. Shubin. 2015. "Organogenesis in Deep Time: A Problem in Genomics, Development, and Paleontology." *Proceedings of the National Academy of Sciences of the United States of America* 112 (16): 4871–4876; **Gai, Zhikun**, Qiang Li, Humberto G. Ferrón, Joseph N. Keating, Junqing Wang, Philip C. J. Donoghue, and Min Zhu. 2022. "Galeaspid Anatomy and the Origin of Vertebrate Paired Appendages." *Nature* 609 (7929): 959–963; **Tzung, Keh Weei**, Robert L. Lalonde, Karin D. Prummel, Harsha Mahabaleshwar, Hannah R. Moran, Jan Stundl, Amanda N. Cass, et al. 2023. "A Median Fin Derived from the Lateral Plate Mesoderm and the Origin of Paired Fins." *Nature* 618 (7965): 543–549; **Miyamoto, Kazuhide**, Koichi Kawakami, Koji Tamura, and Gembu Abe. 2022. "Developmental Independence of Median Fins from the Larval Fin Fold Revises Their Evolutionary Origin." *Scientific Reports* 12 (1): 1–13; **Brazeau, Martin D.**, Marco Castiello, Amin El Fassi El Fehri, Louis Hamilton, Alexander O. Ivanov, Zerina Johanson, and Matt Friedman. 2023. "Fossil Evidence for a Pharyngeal Origin of the Vertebrate Pectoral Girdle." *Nature*. 1–5.

15. Long, John A. 2011. *The Rise of Fishes: 500 Million Years of Evolution.* Baltimore, MD: Johns Hopkins University Press; **Zhu, You-an**, Qiang Li, Jing Lu, Yang Chen,

Jianhua Wang, Zhikun Gai, Wenjin Zhao, et al. 2022. "The Oldest Complete Jawed Vertebrates from the Early Silurian of China." *Nature* 609 (7929): 954–958.

16. Hirasawa, Tatsuya, Yuzhi Hu, Kentaro Uesugi, Masato Hoshino, Makoto Manabe, and Shigeru Kuratani. 2022. "Morphology of Palaeospondylus Shows Affinity to Tetrapod Ancestors." *Nature* 606 (7912): 109–112; **Stewart, Thomas A.**, Justin B. Lemberg, Ailis Daly, Edward B. Daeschler, and Neil H. Shubin. 2022. "A New Elpistostegalian from the Late Devonian of the Canadian Arctic." *Nature* 608 (7923): 563–568; **Jorgewich-Cohen**, Gabriel, Simon William Townsend, Linilson Rodrigues Padovese, Nicole Klein, Peter Praschag, Camila R. Ferrara, Stephan Ettmar, et al. 2022. "Common Evolutionary Origin of Acoustic Communication in Choanate Vertebrates." *Nature Communications* 13 (1): 1–7.

17. Stewart, Thomas A., Ramray Bhat, and Stuart A. Newman. 2017. "The Evolutionary Origin of Digit Patterning." *EvoDevo* 8 (1): 1–7; **Stewart, Thomas A.**, Justin B. Lemberg, Natalia K. Taft, Ihna Yoo, Edward B. Daeschler, and Neil H. Shubin. 2020. "Fin Ray Patterns at the Fin-to-Limb Transition." *Proceedings of the National Academy of Sciences of the United States of America* 117 (3): 1612–1620; **Vargas, Alexander O.**, and Francisco Aboitiz. 2005. "How Ancient Is the Adult Swimming Capacity in the Lineage Leading to Euchordates?" *Evolution & Development* 7 (3): 171–174; **Pennisi, Elizabeth.** 2023. "Tales of the Tongue." *Science* 380 (6647): 786–791; **Young, John J.** 2023. "In Preprints: Of Genitalia and Six-Legged Mice." *Development (Cambridge)* 150 (16): 1–2.

18. Clack, Jennifer A. 2002. *Gaining Ground: The Origin and Evolution of Tetrapods. Gaining Ground Second Edition: The Origin and Evolution of Tetrapods.* Bloomington: Indiana University Press; **Carroll, Robert. L.** 1990. *Vertebrate Paleontology and Evolution. Vertebrate Paleontology and Evolution.* 1st ed. W. H. Freeman and Company, New York.

Chapter 8

1. Pani, Ariel M., Erin E. Mullarkey, Jochanan Aronowicz, Stavroula Assimacopoulos, Elizabeth A. Grove, and Christopher J. Lowe. 2012. "Ancient Deuterostome Origins of Vertebrate Brain Signalling Centres." *Nature* 483 (7389): 289–294; **Puelles, Luis.** 2021. "Recollections on the Origins and Development of the Prosomeric Model." *Frontiers in Neuroanatomy* 15: 94: 1–25.

2. Aboitiz, F., and J. Montiel. 2007. "Origin and Evolution of the Vertebrate Telencephalon, with Special Reference to the Mammalian Neocortex." *Advances in Anatomy, Embryology, and Cell Biology* 193: 1–112; **Lacalli, Thurston**. 2017. "Interpreting Amphioxus, and Thoughts on Ancestral Chordate Mouths and Brains." *International Journal of Developmental Biology* 61 (10-11–12): 649–654; **Butler, Ann B.**, and William. Hodos. 1996. *Comparative Vertebrate Neuroanatomy: Evolution and Adaptation.* New York: Wiley-Liss.

3. Essen, David C. Van, Chad J. Donahue, and Matthew F. Glasser. 2018. "Development and Evolution of Cerebral and Cerebellar Cortex." *Brain, Behavior and Evolution* 91 (3): 158–169.

4. Striedter, Georg F., and R. Glenn Northcutt. 2020. *Brains through Time: A Natural History of Vertebrates*. Oxford: Oxford University Press; **Suryanarayana, Shreyas M.**, Brita Robertson, Peter Wallén, and Sten Grillner. 2017. "The Lamprey Pallium Provides a Blueprint of the Mammalian Layered Cortex." *Current Biology* 27 (21): 3264–3277.e5.

5. Aboitiz, Francisco, and Juan F. Montiel. 2019. "Morphological Evolution of the Vertebrate Forebrain: From Mechanical to Cellular Processes." *Evolution & Development* 21 (6): 330–341; **Briscoe, Steven D.**, and Clifton W. Ragsdale. 2019. "Evolution of the Chordate Telencephalon." *Current Biology* 29 (13): R647–R662.

6. Marín, Oscar, and John L.R. Rubenstein. 2003. "Cell Migration in the Forebrain." *Annual Review of Neuroscience* 26: 441–483.

7. Striedter, Georg F. 2016. "Evolution of the Hippocampus in Reptiles and Birds." *Journal of Comparative Neurology* 524 (3): 496–517.

8. Striedter, Georg F., and R. Glenn Northcutt. 2020. *Brains through Time: A Natural History of Vertebrates*. Oxford University Press; **Smeets, Wilhelmus J. A. J.**, Oscar Marín, and Agustín González. 2000. "Evolution of the Basal Ganglia: New Perspectives through a Comparative Approach." *Journal of Anatomy* 196 (4): 501–517.

9. Striedter, Georg F., 2022. "The Independent Evolution of Dorsal Pallia in Multiple Vertebrate Lineages." *Brain, Behavior and Evolution* 96 (4–6): 200–211.

10. Bshary, Redouan, and Culum Brown. 2014. "Fish Cognition." *Current Biology* 24 (19): R947–R950; **Kohda, Masanori**, Redouan Bshary, Naoki Kubo, Satoshi Awata, Will Sowersby, Kento Kawasaka, Taiga Kobayashi, and Shumpei Sogawa. 2023. "Cleaner Fish Recognize Self in a Mirror via Self-Face Recognition like Humans." *Proceedings of the National Academy of Sciences of the United States of America* 120 (7): e2208420120.

11. Karten, Harvey J. 2015. "Vertebrate Brains and Evolutionary Connectomics: On the Origins of the Mammalian 'Neocortex.'" *Philosophical Transactions of the Royal Society B: Biological Sciences* 370 (1684): 1–12.

12. Aboitiz, F. 1992. "The Evolutionary Origin of the Mammalian Cerebral Cortex." *Biological Research* 25 (1): 41–49; **Aboitiz, Francisco**, Daniver Morales, and Juan Montiel. 2003. "The Evolutionary Origin of the Mammalian Isocortex: Towards an Integrated Developmental and Functional Approach." *Behavioral and Brain Sciences* 26 (5): 535–552; **Puelles, Luis**, Ellen Kuwana, Eduardo Puelles, Alessandro Bulfone, Kenji Shimamura, Jerry Keleher, Susan Smiga, and John L. R. Rubenstein. 2000. "Pallial and Subpallial Derivatives in the Embryonic Chick and Mouse Telencephalon, Traced by the Expression of the Genes Dlx-2, Emx-1, Nkx-2.1, Pax-6, and Tbr-1." *Journal of Comparative Neurology* 424: 409–438; **Striedter, Georg F.** 1997. "The Telencephalon of Tetrapods in Evolution." *Brain, Behavior and Evolution* 49 (4): 179–213; **Striedter, Georg F.**, T. Alejandro Marchant, and Sarah Beydler. 1998. "The 'Neostriatum' Develops as Part of the Lateral Pallium in Birds." *Journal of Neuroscience: The Official Journal of the Society for Neuroscience* 18 (15): 5839–5849.

13. Montiel, Juan F., and Francisco Aboitiz. 2015. "Pallial Patterning and the Origin of the Isocortex." *Frontiers in Neuroscience* 9, 377: 1–10; **Montiel, Juan F.**, and Francisco Aboitiz 2018. "Homology in Amniote Brain Evolution: The Rise of Molecular Evidence." *Brain, Behavior and Evolution* 91 (2): 59–64.

14. Puelles, L., A. Ayad, A. Alonso, J. E. Sandoval, M. Martínez-de-la-Torre, L. Medina, and J. L. Ferran. 2016. "Selective Early Expression of the Orphan Nuclear Receptor Nr4a2 Identifies the Claustrum Homolog in the Avian Mesopallium: Impact on Sauropsidian/Mammalian Pallium Comparisons." *Journal of Comparative Neurology* 524 (3): 665–703; **Wulliman, Mario.** 2017. "Should we redefine the classic lateral pallium?" *Journal of Comparative Neurology* 525(6): 1509-1513.

15. Jerison, Harry J. 1973. *Evolution of the Brain and Intelligence.* New York: Academic Press.

16. Heldstab, Sandra A., Karin Isler, Sereina M. Graber, Caroline Schuppli, and Carel P. van Schaik. 2022. "The Economics of Brain Size Evolution in Vertebrates." *Current Biology* 32 (12): R697–R708; **Namba, Takashi**, Jeannette Nardelli, Pierre Gressens, and Wieland B. Huttner. 2021. "Metabolic Regulation of Neocortical Expansion in Development and Evolution." *Neuron* 109 (3): 408–419; **Martínez-Cerdeño, Verónica**, Fernando García-Moreno, Maria Antonietta Tosches, András Csillag, Paul R. Manger, and Zoltán Molnár. 2018. "Update on Forebrain Evolution: From Neurogenesis to Thermogenesis." *Seminars in Cell & Developmental Biology* 76: 15–22.

17. Herculano-Houzel, Suzana, Débora J. Messeder, Karina Fonseca-Azevedo, and Nilma A. Pantoja. 2015. "When Larger Brains Do Not Have More Neurons: Increased Numbers of Cells Are Compensated by Decreased Average Cell Size across Mouse Individuals." *Frontiers in Neuroanatomy* 9: 1–12; **Olkowicz, Seweryn**, Martin Kocourek, Radek K. Luèan, Michal Porteš, W. Tecumseh Fitch, Suzana Herculano-Houzel, and Pavel Nemec. 2016. "Birds Have Primate-like Numbers of Neurons in the Forebrain." *Proceedings of the National Academy of Sciences of the United States of America* 113 (26): 7255–7256.

18. Aboitiz, Francisco. 1996. "Does Bigger Mean Better? Evolutionary Determinants of Brain Size and Structure." *Brain, Behavior and Evolution* 47 (5): 225–235.

19. Finlay, B. L., R. B. Darlington, and N. Nicastro. 2001. "Developmental Structure in Brain Evolution." *Behavioral and Brain Sciences* 24 (2): 263–278; **Montgomery, Stephen H.**, Nicholas I. Mundy, and Robert A. Barton. 2016. "Brain Evolution and Development: Adaptation, Allometry and Constraint." *Proceedings of the Royal Society B: Biological Sciences* 283 (1838): 1–9.

Chapter 9

1. Kielan-Jaworowska, Zofia., Richard. Cifelli, and Zhe-Xi. Luo. 2004. *Mammals from the Age of Dinosaurs: Origins, Evolution, and Structure.* New York: Columbia University

Press; **Panciroli, Elsa**. 2021. *Beasts Before Us: The Untold Story of Mammal Origins and Evolution.* 1st ed. Bloomsbury Sigma; **Brusatte, Stephen**. 2022. *The Rise and Reign of the Mammals: A New History, from the Shadow of the Dinosaurs to Us.* New York: Harper Collins.

2. Benoit, J., P. R. Manger, and B. S. Rubidge. 2016. "Palaeoneurological Clues to the Evolution of Defining Mammalian Soft Tissue Traits." *Scientific Reports* 6 (1): 1–10; **Perry, Steven F.**, Thomas Similowski, Wilfried Klein, and Jonathan R. Codd. 2010. "The Evolutionary Origin of the Mammalian Diaphragm." *Respiratory Physiology & Neurobiology* 171 (1): 1–16; **Seifert, Marvin**, Paul A. Roberts, George Kafetzis, Daniel Osorio, and Tom Baden. 2023. "Birds Multiplex Spectral and Temporal Visual Information via Retinal On- and Off-Channels." *Nature Communications* 14 (1):1–19.

3. Schultz, Julia A. 2020. "Eat and Listen—How Chewing and Hearing Evolved." *Science* 367 (6475): 244–246; **Lautenschlager, Stephan**, Michael J. Fagan, Zhe Xi Luo, Charlotte M. Bird, Pamela Gill, and Emily J. Rayfield. 2023. "Functional Reorganisation of the Cranial Skeleton during the Cynodont–Mammaliaform Transition." *Communications Biology* 6 (1): 1–13.

4. Poo, Cindy, Gautam Agarwal, Niccolò Bonacchi, and Zachary F. Mainen. 2021. "Spatial Maps in Piriform Cortex during Olfactory Navigation." *Nature* 601 (7894): 595–599; **Rowe, Timothy B.**, and Gordon M. Shepherd. 2016. "Role of Ortho-Retronasal Olfaction in Mammalian Cortical Evolution." *Journal of Comparative Neurology* 524 (3): 471–495.

5. Higashiyama, Hiroki, Daisuke Koyabu, Tatsuya Hirasawa, Ingmar Werneburg, Shigeru Kuratani, and Hiroki Kurihara. 2021. "Mammalian Face as an Evolutionary Novelty." *Proceedings of the National Academy of Sciences* 118 (44): e2111876118; **An, Xu,** Katherine Matho, Yi Li, Hemanth Mohan, X. Hermione Xu, Ian Q. Whishaw, Adam Kepecs, and Z. Josh Huang. 2022. "A Cortical Circuit for Orchestrating Oromanual Food Manipulation." *BioRxiv*, December, 2022.12.03.518964.

6. Moser, May Britt, David C. Rowland, and Edvard I. Moser. 2015. "Place Cells, Grid Cells, and Memory." *Cold Spring Harbor Perspectives in Biology* 7 (2): 1–15; **Shahbaba, Babak**, Lingge Li, Forest Agostinelli, Mansi Saraf, Keiland W. Cooper, Derenik Haghverdian, Gabriel A. Elias, Pierre Baldi, and Norbert J. Fortin. 2022. "Hippocampal Ensembles Represent Sequential Relationships among an Extended Sequence of Nonspatial Events." *Nature Communications* 13 (1): 1–17; **Cothi, William de**, Nils Nyberg, Eva Maria Griesbauer, Carole Ghanamé, Fiona Zisch, Julie M. Lefort, Lydia Fletcher, et al. 2022. "Predictive Maps in Rats and Humans for Spatial Navigation." *Current Biology* 32 (17): 3676–3689.e5; **Lai, Chongxi**, Shinsuke Tanaka, Timothy D. Harris, and Albert K. Lee. 2023. "Volitional Activation of Remote Place Representations with a Hippocampal Brain–Machine Interface." *Science* 382 (6670): 566–73; Nadel, Lynn. 2021. "The Hippocampal Formation and Action at a Distance." *Proceedings of the National Academy of Sciences* 118 (51): e2119670118: 1–5.

7. Aboitiz, Francisco, Daniver Morales, and Juan Montiel. 2003. "The Evolutionary Origin of the Mammalian Isocortex: Towards an Integrated Developmental and Functional Approach." *Behavioral and Brain Sciences* 26 (5): 535–552; **Aboitiz, Francisco**, and Juan F. Montiel. 2015. "Olfaction, Navigation, and the Origin of Isocortex." *Frontiers in Neuroscience* 9: 402.

8. Nadel, Lynn. 2021. "The Hippocampal Formation and Action at a Distance." *Proceedings of the National Academy of Sciences* 118 (51): e2119670118; **Whittington, James C. R.**, David McCaffary, Jacob J. W. Bakermans, and Timothy E. J. Behrens. 2022. "How to Build a Cognitive Map." *Nature Neuroscience* 25 (10): 1257–1272.

9. Rakic, Pasko. 2009. "Evolution of the Neocortex: A Perspective from Developmental Biology." *Nature Reviews. Neuroscience* 10 (10): 724–735; **Cárdenas, Adrián**, and Víctor Borrell. 2020. "Molecular and Cellular Evolution of Corticogenesis in Amniotes." *Cellular and Molecular Life Sciences* 77 (8): 1435–1460.

10. Aboitiz, Francisco, Daniver Morales, and Juan Montiel. 2003. "The Evolutionary Origin of the Mammalian Isocortex: Towards an Integrated Developmental and Functional Approach." *Behavioral and Brain Sciences* 26 (5): 535–552; **Aboitiz, Francisco**. 1999. "Evolution of Isocortical Organization. A Tentative Scenario Including Roles of Reelin, P35/Cdk5 and the Subplate Zone." *Cerebral Cortex* 9 (7): 655–661; **Briscoe, Steven D.**, and Clifton W. Ragsdale. 2018. "Homology, Neocortex, and the Evolution of Developmental Mechanisms." *Science* 362 (6411): 190–193.

11. Aboitiz, Francisco, Juan Montiel, and Ricardo R. García. 2005. "Ancestry of the Mammalian Preplate and Its Derivatives: Evolutionary Relicts or Embryonic Adaptations?" *Reviews in the Neurosciences* 16 (4): 359–376.

12. Mountcastle, Vernon B. 1997. "The Columnar Organization of the Neocortex." *Brain* 120 (4): 701–722; **Briscoe, Steven D.**, and Clifton W. Ragsdale. 2018. "Homology, Neocortex, and the Evolution of Developmental Mechanisms." *Science* 362 (6411): 190–193; **Aboitiz, Francisco**, 2019. "Morphological Evolution of the Vertebrate Forebrain: From Mechanical to Cellular Processes." *Evolution & Development* 21 (6): 330–341.

13. Kaas, Jon H. 2019. "The Origin and Evolution of Neocortex: From Early Mammals to Modern Humans." *Progress in Brain Research* 250: 61–81.

14. Ebbesson, Sven O.E. 1980. "The Parcellation Theory and Its Relation to Interspecific Variability in Brain Organization, Evolutionary and Ontogenetic Development, and Neuronal Plasticity." *Cell and Tissue Research* 213 (2): 179–212; **Krubitzer, Leah A.** 2000. "How Does Evolution Build a Complex Brain?" *Novartis Foundation Symposium* 228: 206–226.

15. Wagner, Nikolaus R., Ashis Sinha, Verl Siththanandan, Angelica N. Kowalchuk, Jessica Macdonald, and Suzanne Tharin. 2021. "MiR-409-3p Represses Cited2 at the

Evolutionary Emergence of the Callosal and Corticospinal Projections." *BioRxiv*, November, 2021.11.03.467191.

16. Gu, Zirong, John Kalamboglas, Shin Yoshioka, Wenqi Han, Zhuo Li, Yuka Imamura Kawasawa, Sirisha Pochareddy, et al. 2017. "Control of Species-Dependent Cortico-Motoneuronal Connections Underlying Manual Dexterity." *Science* 357 (6349): 400–404; **Welniarz, Quentin**, Isabelle Dusart, and Emmanuel Roze. 2017. "The Corticospinal Tract: Evolution, Development, and Human Disorders." *Developmental Neurobiology* 77 (7): 810–829; **Gu, Zirong**, Masaki Ueno, Kelsey Klinefelter, Madhulika Mamidi, Takeshi Yagi, and Yutaka Yoshida. 2019. "Skilled Movements in Mice Require Inhibition of Corticospinal Axon Collateral Formation in the Spinal Cord by Semaphorin Signaling." *Journal of Neuroscience* 39 (45): 8885–8899.

17. Suárez, Rodrigo, Han Gobius, and Linda J. Richards. 2014. "Evolution and Development of Interhemispheric Connections in the Vertebrate Forebrain." *Frontiers in Human Neuroscience* 8: 497.

18. Aboitiz, Francisco, Javier López, and Juan Montiel. 2003. "Long Distance Communication in the Human Brain: Timing Constraints for Inter-Hemispheric Synchrony and the Origin of Brain Lateralization." *Biological Research* 36 (1): 89–99; **Gebhardt, Christoph**, Thomas O. Auer, Pedro M. Henriques, Gokul Rajan, Karine Duroure, Isaac H. Bianco, and Filippo Del Bene. 2019. "An Interhemispheric Neural Circuit Allowing Binocular Integration in the Optic Tectum." *Nature Communications* 10 (1): 1–12; **Wang, Bor Shuen**, Maria Sol Bernardez Sarria, Xu An, Miao He, Nazia M. Alam, Glen T. Prusky, Michael C. Crair, and Z. Josh Huang. 2021. "Retinal and Callosal Activity-Dependent Chandelier Cell Elimination Shapes Binocularity in Primary Visual Cortex." *Neuron* 109 (3): 502–515.e7; **Peiker, Christiane**, Thomas Wunderle, David Eriksson, Anne Schmidt, and Kerstin E. Schmidt. 2013. "An Updated Midline Rule: Visual Callosal Connections Anticipate Shape and Motion in Ongoing Activity across the Hemispheres." *Journal of Neuroscience* 33 (46): 18036–18046.

Chapter 10

1. Essen, David C. Van, Chad J. Donahue, Timothy S. Coalson, Henry Kennedy, Takuya Hayashi, and Matthew F. Glasser. 2019. "Cerebral Cortical Folding, Parcellation, and Connectivity in Humans, Nonhuman Primates, and Mice." *Proceedings of the National Academy of Sciences of the United States of America* 116 (52): 26173–26180; **Liao, Xuhong**, Athanasios V. Vasilakos, and Yong He. 2017. "Small-World Human Brain Networks: Perspectives and Challenges." *Neuroscience & Biobehavioral Reviews* 77 (June): 286–300; **Grosu, George F.**, Alexander V. Hopp, Vasile V. Moca, Harald Bârzan, Andrei Ciuparu, Maria Ercsey-Ravasz, Mathias Winkel, Helmut Linde, and Raul C. Mureşan. 2023. "The Fractal Brain: Scale-Invariance in Structure and Dynamics." *Cerebral Cortex* 33 (8): 4574–4605.

2. Hubel, D. H., and T. N. Wiesel. 1977. "Ferrier Lecture—Functional Architecture of Macaque Monkey Visual Cortex." *Proceedings of the Royal Society of London. Series B. Biological Sciences* 190 (1130): 1–59; **Quiroga, R. Quian**, G. Kreiman, C. Koch, and I. Fried. 2008. "Sparse but Not 'Grandmother-Cell' Coding in the Medial Temporal Lobe." *Trends in Cognitive Sciences* 12 (3): 87–91.

3. Ungerleider, Leslie G., and James V. Haxby. 1994. "'What' and 'Where' in the Human Brain." *Current Opinion in Neurobiology* 4 (2): 157–165; **Kaas, Jon H.**, Hui Xin Qi, and Iwona Stepniewska. 2022. "Escaping the Nocturnal Bottleneck, and the Evolution of the Dorsal and Ventral Streams of Visual Processing in Primates." *Philosophical Transactions of the Royal Society B* 377 (1844): 20210293; **Zeki, Semir.** 1993. *A Vision of the Brain.* Oxford: Blackwell Scientific Publications.

4. Kosakowski, Heather L., Michael A. Cohen, Lyneé Herrara, Isabel Nichoson, Nancy Kanwisher, and Rebecca Saxe. 2021. "Face-Selective Responses Present in Multiple Regions of the Human Infant Brain." *BioRxiv*, December, 2021.12.04.471085; **Vinken, Kasper**, Jacob S. Prince, Talia Konkle, and Margaret S. Livingstone. 2023. "The Neural Code for 'Face Cells' Is Not Face-Specific." *Science Advances* 9 (35): eadg1736: 1–14.; **Pitcher, David**, Geena R Ianni, Kelsey Holiday, and Leslie G Ungerleider. 2023. "Identifying the Cortical Face Network with Dynamic Face Stimuli: A Large Group FMRI Study." *BioRxiv*, September, 2023.09.26.559583: 1–18.

5. Pitcher, David, and Leslie G. Ungerleider. 2021. "Evidence for a Third Visual Pathway Specialized for Social Perception." *Trends in Cognitive Sciences* 25 (2): 100–110; **Wurm, Moritz F.**, and Alfonso Caramazza. 2022. "Two 'What' Pathways for Action and Object Recognition." *Trends in Cognitive Sciences* 26 (2): 103–116.

6. Bosman, Conrado A., and Francisco Aboitiz. 2015. "Functional Constraints in the Evolution of Brain Circuits." *Frontiers in Neuroscience* 9: 303; **Seth, Anil K.**, and Tim Bayne. 2022. "Theories of Consciousness." *Nature Reviews. Neuroscience* 23 (7): 439–452.

7. Namboodiri, Vijay Mohan K., and Garret D. Stuber. 2020. "Interoceptive Inception in Insula." *Neuron* 105 (6): 959–960.

8. Barbas, Helen, and Basilis Zikopoulos. 2016. "The Prefrontal Cortex and Flexible Behavior." *Neuroscientist* 13 (5): 532–545; **Fine, Justin M.**, and Benjamin Y. Hayden. 2022. "The Whole Prefrontal Cortex Is Premotor Cortex." *Philosophical Transactions of the Royal Society B* 377 (1844).

9. Patel, Gaurav H., Danica Yang, Emery C. Jamerson, Lawrence H. Snyder, Maurizio Corbetta, and Vincent P. Ferrera. 2015. "Functional Evolution of New and Expanded Attention Networks in Humans." *Proceedings of the National Academy of Sciences of the United States of America* 112 (30): 9454–9459.

10. Baddeley, Alan. 2011. "Working Memory: Theories, Models, and Controversies." *Annual Review of Psychology* 63: 1–29; **Miller, Earl K.**, Mikael Lundqvist, and André M.

Bastos. 2018. "Working Memory 2.0." *Neuron* 100 (2): 463–475; **Curtis, Clayton E.**, and Thomas C. Sprague. 2021. "Persistent Activity during Working Memory from Front to Back." *Frontiers in Neural Circuits* 15: 72.

11. Damasio, Antonio. 1995. *Descartes' Error: Emotion, Reason, and the Human Brain.* Penguin Books, New York; **Joyce, Mary Kate P.**, Miguel Ángel García-Cabezas, Yohan J. John, and Helen Barbas. 2020. "Serial Prefrontal Pathways Are Positioned to Balance Cognition and Emotion in Primates." *Journal of Neuroscience* 40 (43): 8306–8328; **Basu, Raunak**, Robert Gebauer, Tim Herfurth, Simon Kolb, Zahra Golipour, Tatjana Tchumatchenko, and Hiroshi T. Ito. 2021. "The Orbitofrontal Cortex Maps Future Navigational Goals." *Nature* 599 (7885): 449–452.

12. Hogeveen, Jeremy, Maria Medalla, Matthew Ainsworth, Juan M. Galeazzi, Colleen A. Hanlon, Farshad Alizadeh Mansouri, and Vincent D. Costa. 2022. "What Does the Frontopolar Cortex Contribute to Goal-Directed Cognition and Action?" *Journal of Neuroscience* 42 (45): 8508–8513.

13. Raichle, Marcus E, and Abraham Z Snyder. 2007. "A Default Mode of Brain Function: A Brief History of an Evolving Idea." *NeuroImage* 37 (4): 1083–1090; discussion 1097–1099; **Smallwood, Jonathan**, Boris C. Bernhardt, Robert Leech, Danilo Bzdok, Elizabeth Jefferies, and Daniel S. Margulies. 2021. "The Default Mode Network in Cognition: A Topographical Perspective." *Nature Reviews Neuroscience* 22 (8): 503–513; **Kaefer, Karola**, Federico Stella, Bruce L. McNaughton, and Francesco P. Battaglia. 2022. "Replay, the Default Mode Network and the Cascaded Memory Systems Model." *Nature Reviews Neuroscience* 23 (10): 628–640; **Mantini, D.**, M. G. Perrucci, C. Del Gratta, G. L. Romani, and M. Corbetta. 2007. "Electrophysiological Signatures of Resting State Networks in the Human Brain." *Proceedings of the National Academy of Sciences of the United States of America* 104 (32): 13170–13175; **Menon, Vinod**. 2023. "20 Years of the Default Mode Network: A Review and Synthesis." *Neuron* 111 (16): 2469–2487.; **Correa, Susana**, Emily S. Nichols, Megan E. Mueller, Barbra de Vrijer, Roy Eagleson, Charles A. McKenzie, Sandrine de Ribaupierre, and Emma G. Duerden. 2023. "Default Mode Network Functional Connectivity Strength in Utero and the Association with Fetal Subcortical Development." *Cerebral Cortex* 33 (14): 9144–9153.

14. Varela, Francisco, Jean Philippe Lachaux, Eugenio Rodriguez, and Jacques Martinerie. 2001. "The Brainweb: Phase Synchronization and Large-Scale Integration." *Nature Reviews Neuroscience* 2 (4): 229–239; **Buzsáki, György**, Nikos Logothetis, and Wolf Singer. 2013. "Scaling Brain Size, Keeping Timing: Evolutionary Preservation of Brain Rhythms." *Neuron* 80 (3): 751–764.

15. Raven, Frank, Eddy A. Van der Zee, Peter Meerlo, and Robbert Havekes. 2018. "The Role of Sleep in Regulating Structural Plasticity and Synaptic Strength: Implications for Memory and Cognitive Function." *Sleep Medicine Reviews* 39 (June): 3–11; **Wienke, Christian**, Mandy V. Bartsch, Lena Vogelgesang, Christoph Reichert,

Hermann Hinrichs, Hans-Jochen Heinze, and Stefan Dürschmid. 2021. "Mind-Wandering Is Accompanied by Both Local Sleep and Enhanced Processes of Spatial Attention Allocation." *Cerebral Cortex Communications* 2 (1): 1–13; **Sämann, Philipp G.**, Renate Wehrle, David Hoehn, Victor I. Spoormaker, Henning Peters, Carolin Tully, Florian Holsboer, and Michael Czisch. 2011. "Development of the Brain's Default Mode Network from Wakefulness to Slow Wave Sleep." *Cerebral Cortex* 21 (9): 2082–2093; **Hong, Jiso**, David E. Lozano, Kevin T. Beier, Shinjae Chung, and Franz Weber. 2023. "Prefrontal Cortical Regulation of REM Sleep." *Nature Neuroscience* 26 (10): 1820–1832.

16. Baciadonna, Luigi, Francesca M. Cornero, Nathan J. Emery, and Nicola S. Clayton. 2021. "Convergent Evolution of Complex Cognition: Insights from the Field of Avian Cognition into the Study of Self-Awareness." *Learning and Behavior* 49 (1): 9–22; **Rook, Noemi**, John Michael Tuff, Julian Packheiser, Onur Güntürkün, and Christian Beste. 2021. "A Hierarchical Processing Unit for Multi-Component Behavior in the Avian Brain." *IScience* 24 (10): 103195; **Hahn, Lukas Alexander**, Dmitry Balakhonov, Erica Fongaro, Andreas Nieder, and Jonas Rose. 2021. "Working Memory Capacity of Crows and Monkeys Arises from Similar Neuronal Computations." *ELife* 10: 10–21; **Payne, H. L.**, G. F. Lynch, and D. Aronov. 2021. "Neural Representations of Space in the Hippocampus of a Food-Caching Bird." *Science* 373 (6552): 343–348.

Chapter 11

1. Fleagle, John G. 2013. *Primate Adaptation and Evolution: Third Edition. Primate Adaptation and Evolution: Third Edition.* 3rd ed. Academic Press; **Wilson Mantilla, Gregory P.**, Stephen G. B. Chester, William A. Clemens, Jason R. Moore, Courtney J. Sprain, Brody T. Hovatter, William S. Mitchell, Wade W. Mans, Roland Mundil, and Paul R. Renne. 2021. "Earliest Palaeocene Purgatoriids and the Initial Radiation of Stem Primates." *Royal Society Open Science* 8 (2): 1–10; **Lazar, Leslee**, Prem Chand, Radhika Rajan, Hisham Mohammed, and Neeraj Jain. 2022. "Somatosensory Cortex of Macaque Monkeys Is Designed for Opposable Thumb." *Cerebral Cortex* 33 (1): 195–206.

2. Böhme, Madelaine, Nikolai Spassov, Jochen Fuss, Adrian Tröscher, Andrew S. Deane, Jérôme Prieto, Uwe Kirscher, Thomas Lechner, and David R. Begun. 2019. "A New Miocene Ape and Locomotion in the Ancestor of Great Apes and Humans." *Nature* 575 (7783): 489–493; **MacLatchy, Laura M.**, Susanne M. Cote, Alan L. Deino, Robert M. Kityo, Amon A.T. Mugume, James B. Rossie, William J. Sanders, et al. 2023. "The Evolution of Hominoid Locomotor Versatility: Evidence from Moroto, a 21 Ma Site in Uganda." *Science* 380 (6641): eabq2835; Fannin, Luke D., Mary S. Joy, Nathaniel J. Dominy, W. Scott McGraw, and Jeremy M. DeSilva. 2023. "Downclimbing and the Evolution of Ape Forelimb Morphologies." *Royal Society Open Science* 10 (9): 1–9.

3. Suntsova, Maria V., and Anton A. Buzdin. 2020. "Differences between Human and Chimpanzee Genomes and Their Implications in Gene Expression, Protein

Functions and Biochemical Properties of the Two Species." *BMC Genomics* 21 (7): 1–12; **Suwa, Gen**, Tomohiko Sasaki, Sileshi Semaw, Michael J. Rogers, Scott W. Simpson, Yutaka Kunimatsu, Masato Nakatsukasa, et al. 2021. "Canine Sexual Dimorphism in Ardipithecus Ramidus Was Nearly Human-Like." *Proceedings of the National Academy of Sciences of the United States of America* 118 (49): e2116630118; **Almécija, Sergio**, Ashley S. Hammond, Nathan E. Thompson, Kelsey D. Pugh, Salvador Moyà-Solà, and David M. Alba. 2021. "Fossil Apes and Human Evolution." *Science* 372 (6542): 1–12.

4. Prang, Thomas C., Kristen Ramirez, Mark Grabowski, and Scott A. Williams. 2021. "Ardipithecus Hand Provides Evidence That Humans and Chimpanzees Evolved from an Ancestor with Suspensory Adaptations." *Science Advances* 7 (9): eabf2474.

5. Parkinson, Jennifer A., Thomas W. Plummer, James S. Oliver, and Laura C. Bishop. 2022. "Meat on the Menu: GIS Spatial Distribution Analysis of Bone Surface Damage Indicates That Oldowan Hominins at Kanjera South, Kenya Had Early Access to Carcasses." *Quaternary Science Reviews* 277: 107314; **Stepka, Zane**, Ido Azuri, Liora Kolska Horwitz, Michael Chazan, and Filipe Natalio. 2022. "Hidden Signatures of Early Fire at Evron Quarry (1.0 to 0.8 Mya)." *Proceedings of the National Academy of Sciences of the United States of America* 119 (25): e2123439119; **Zohar, Irit,** Nira Alperson-Afil, Naama Goren-Inbar, Marion Prévost, Thomas Tütken, Guy Sisma-Ventura, Israel Hershkovitz, and Jens Najorka. 2022. "Evidence for the Cooking of Fish 780,000 Years Ago at Gesher Benot Ya'aqov, Israel." *Nature Ecology & Evolution* 6 (12): 2016–2028; **Plummer, Thomas W.**, James S. Oliver, Emma M. Finestone, Peter W. Ditchfield, Laura C. Bishop, Scott A. Blumenthal, Cristina Lemorini, et al. 2023. "Expanded Geographic Distribution and Dietary Strategies of the Earliest Oldowan Hominins and Paranthropus." *Science* 379 (6632): 561–566; **Fuentes, Agustin**, Marc Kissel, Penny Spikins, Keneiloe Molopyane, John Hawks, and Lee R. Berger. 2023. "Burials and Engravings in a Small-Brained Hominin, Homo Naledi, from the Late Pleistocene: Contexts and Evolutionary Implications." *BioRxiv* 17 (June): 2023.06.01.543135.

6. Leder, Dirk, Raphael Hermann, Matthias Hüls, Gabriele Russo, Philipp Hoelzmann, Ralf Nielbock, Utz Böhner, et al. 2021. "A 51,000-Year-Old Engraved Bone Reveals Neanderthals' Capacity for Symbolic Behaviour." *Nature Ecology & Evolution* 5 (9); **Groucutt, Huw S.**, Tom S. White, Eleanor M.L. Scerri, Eric Andrieux, Richard Clark-Wilson, Paul S. Breeze, Simon J. Armitage, et al. 2021. "Multiple Hominin Dispersals into Southwest Asia over the Past 400,000 Years." *Nature* 597 (7876): 376–380.

7. Vespasiani, Davide M., Guy S. Jacobs, Laura E. Cook, Nicolas Brucato, Matthew Leavesley, Christopher Kinipi, Francois-Xavier Ricaut, Murray P. Cox, and Irene Gallego Romero. 2021. "Denisovan Introgression Has Shaped the Immune System of Present-Day Papuans." *BioRxiv*, September, 2020.07.09.196444; **Reilly, Patrick F.**, Audrey Tjahjadi, Samantha L. Miller, Joshua M. Akey, and Serena Tucci. 2022. "The Contribution of Neanderthal Introgression to Modern Human Traits." *Current Biology* 32 (18): R970–R983.

8. Larson, Greger, and Dorian Q. Fuller. 2014. "The Evolution of Animal Domestication." *Annual Review of Ecology, Evolution, and Systematics* 45: 115–136; **Spinney, Laura**. 2021. "The Rise and Fall of the Mysterious Culture That Invented Civilisation." *New Scientist*, February; **Langley, Michelle C.**, and Thomas Suddendorf. 2022. "Archaeological Evidence for Thinking about Possibilities in Hominin Evolution." *Philosophical Transactions of the Royal Society of London. Series B, Biological Sciences* 377 (1866): 1–11.

9. Aiello, Leslie C., and Peter Wheeler. 1995. "The Expensive-Tissue Hypothesis: The Brain and the Digestive System in Human and Primate Evolution." *Current Anthropology* 36 (2): 199–221; **Cornélio, Alianda M.**, Ruben E. de Bittencourt-Navarrete, Ricardo B. de Bittencourt Brum, Claudio M. Queiroz, and Marcos R. Costa. 2016. "Human Brain Expansion during Evolution Is Independent of Fire Control and Cooking." *Frontiers in Neuroscience* 10: 167.

10. Stansfield, Ekaterina, Krishna Kumar, Philipp Mitteroecker, and Nicole D.S. Grunstra. 2021. "Biomechanical Trade-Offs in the Pelvic Floor Constrain the Evolution of the Human Birth Canal." *Proceedings of the National Academy of Sciences of the United States of America* 118 (16): 1–8; **Falk, Dean**. 2019. "How Australopithecus Provided Insight into Human Evolution." *Nature* 575 (7781): 41–42.

11. Pang, James C., James K. Rilling, James A. Roberts, Martijn P. Van Den Heuvel, and Luca Cocchi. 2022. "Evolutionary Shaping of Human Brain Dynamics." *ELife* 11: 1–62; **Caspar, Kai R.**, Fabian Pallasdies, Larissa Mader, Heitor Sartorelli, and Sabine Begall. 2022. "The Evolution and Biological Correlates of Hand Preferences in Anthropoid Primates." *ELife* 11: 77875; **Heuvel, Martijn P. van den**, Dirk Jan Ardesch, Lianne H. Scholtens, Siemon C. de Lange, Neeltje E. M. van Haren, Iris E. C. Sommer, Udo Dannlowski, et al. 2023. "Human and Chimpanzee Shared and Divergent Neurobiological Systems for General and Specific Cognitive Brain Functions." *Proceedings of the National Academy of Sciences of the United States of America* 120 (22): 1–8.

12. Zollikofer, Christoph P.E., Thibault Bienvenu, Yonas Beyene, Gen Suwa, Berhane Asfaw, Tim D. White, and Marcia S. Ponce de Leon. 2022. "Endocranial Ontogeny and Evolution in Early Homo Sapiens: The Evidence from Herto, Ethiopia." *Proceedings of the National Academy of Sciences of the United States of America* 119 (32): 1–7; **Beaudet, Amélie**, and Edwin de Jager. 2023. "Broca's Area, Variation and Taxic Diversity in Early Homo from Koobi Fora (Kenya)." *ELife* 12 (September): 1–7.

13. Franchini, Lucía Florencia. 2021. "Genetic Mechanisms Underlying Cortical Evolution in Mammals." *Frontiers in Cell and Developmental Biology* 9 (February): 73; **Berg, Jim**, Staci A. Sorensen, Jonathan T. Ting, Jeremy A. Miller, Thomas Chartrand, Anatoly Buchin, Trygve E. Bakken, et al. 2021. "Human Neocortical Expansion Involves Glutamatergic Neuron Diversification." *Nature* 598 (7879): 151–158; **Vanderhaeghen, Pierre**, and Franck Polleux. 2023. "Developmental Mechanisms Underlying

the Evolution of Human Cortical Circuits." *Nature Reviews Neuroscience* 24 (4): 213–232; **Keough, Kathleen C.**, Sean Whalen, Fumitaka Inoue, Pawel F. Przytycki, Tyler Fair, Chengyu Deng, Marilyn Steyert, et al. 2023. "Three-Dimensional Genome Rewiring in Loci with Human Accelerated Regions." *Science* 380 (6643): eabm1696; **Morales-Vicente, David A.**, Ana C. Tahira, Daisy Woellner-Santos, Murilo S. Amaral, Maria G. Berzoti-Coelho, and Sergio Verjovski-Almeida. 2023. "The Human Developing Cerebral Cortex Is Characterized by an Increased de Novo Expression of LncRNAs in Excitatory Neurons." *BioRxiv*, October, 2023.10.26.564246: 1–40; **Wang, Li,** Kaifang Pang, Li Zhou, Arantxa Cebrián-Silla, Susana González-Granero, Shaohui Wang, Qiuli Bi, et al. 2023. "A Cross-Species Proteomic Map Reveals Neoteny of Human Synapse Development." *Nature* 622 (7981): 1–8.; **Zhu, Xiaoquan**, Chao Chen, Dong Wei, Yong Xu, Siying Liang, Wenlong Jia, Jian Li, et al. 2023. "FOXP2 Confers Oncogenic Effects in Prostate Cancer." *ELife* 12 (September): 1–23.

14. Aboitiz, Francisco. 1988. "Epigenesis and the Evolution of the Human Brain." *Medical Hypotheses* 25 (1): 55–59; **Changeux, Jean Pierre**, Alexandros Goulas, and Claus C. Hilgetag. 2021. "A Connectomic Hypothesis for the Hominization of the Brain." *Cerebral Cortex* 31 (5): 2425–2449.

15. Sobinov, Anton R., and Sliman J. Bensmaia. 2021. "The Neural Mechanisms of Manual Dexterity." *Nature Reviews Neuroscience* 22 (12): 741–757; **Caspar, Kai R.**, Fabian Pallasdies, Larissa Mader, Heitor Sartorelli, and Sabine Begall. 2022. "The Evolution and Biological Correlates of Hand Preferences in Anthropoid Primates." *ELife* 11: 77875.

16. Stout, Dietrich, and Erin E. Hecht. 2017. "Evolutionary Neuroscience of Cumulative Culture." *Proceedings of the National Academy of Sciences of the United States of America* 114 (30): 7861–7868; **Thibault, Simon**, Raphaël Py, Angelo Mattia Gervasi, Romeo Salemme, Eric Koun, Martin Lövden, Véronique Boulenger, Alice C. Roy, and Claudio Brozzoli. 2021. "Tool Use and Language Share Syntactic Processes and Neural Patterns in the Basal Ganglia." *Science* 374 (6569): 1–14; **Uomini, Natalie Thaïs**, and Georg Friedrich Meyer. 2013. "Shared Brain Lateralization Patterns in Language and Acheulean Stone Tool Production: A Functional Transcranial Doppler Ultrasound Study." *PLoS One* 8 (8): e72693; **Seidel, Gundula**, Michel Rijntjes, Daniel Göllmar, Cornelius Weiller, and Farsin Hamzei. 2023. "Understanding the Concept of a Novel Tool Requires Interaction of the Dorsal and Ventral Streams." *Cerebral Cortex* 33 (16): 9652–9663.

17. Rizzolatti, Giacomo, and Leonardo Fogassi. 2014. "The Mirror Mechanism: Recent Findings and Perspectives." *Philosophical Transactions of the Royal Society B: Biological Sciences* 369 (1644): 1–12; **Cook, Richard**, Geoffrey Bird, Caroline Catmur, Clare Press, and Cecilia Heyes. 2014. "Mirror Neurons: From Origin to Function." *Behavioral and Brain Sciences* 37 (2): 177–192; **Hickok, Gregory.** 2014. *The Myth of Mirror Neurons: The Real Neuroscience of Communication and Cognition.* 1st ed. New York: W. W. Norton & Company; **Yang, Taehong**, Daniel W. Bayless, Yichao Wei,

Dan Landayan, Ivo M. Marcelo, Yangpeng Wang, Laura A. DeNardo, Liqun Luo, Shaul Druckmann, and Nirao M. Shah. 2023. "Hypothalamic Neurons That Mirror Aggression." *Cell* 186 (6): 1195–1211.

18. Heyes, Cecilia 2021. "Imitation." *Current Biology* 31 (5): R228–32; **Heyes, Cecilia.** 2020. "Culture." *Current Biology* 30 (20): R1246–R1250.

19. Scott-Phillips, Thom. 2022. "Biological Adaptations for Cultural Transmission?" *Biology Letters* 18 (11): 20220439.

20. Gopnik, Alison, Willem E. Frankenhuis, and Michael Tomasello. 2020. "Introduction to Special Issue: 'Life History and Learning: How Childhood, Caregiving and Old Age Shape Cognition and Culture in Humans and Other Animals.'" *Philosophical Transactions of the Royal Society B* 375 (1803); **Dunbar, R. I. M.** 2009. "The Social Brain Hypothesis and Its Implications for Social Evolution." *Annals of Human Biology* 36 (5): 562–572.

21. Raghanti, Mary Ann, Melissa K. Edler, Alexa R. Stephenson, Emily L. Munger, Bob Jacobs, Patrick R. Hof, Chet C. Sherwood, Ralph L. Holloway, and C. Owen Lovejoy. 2018. "A Neurochemical Hypothesis for the Origin of Hominids." *Proceedings of the National Academy of Sciences of the United States of America* 115 (6): E1108–E1116; **Raghanti, Mary Ann**, Melissa K. Edler, Alexa R. Stephenson, Lakaléa J. Wilson, William D. Hopkins, John J. Ely, Joseph M. Erwin, Bob Jacobs, Patrick R. Hof, and Chet C. Sherwood. 2016. "Human-Specific Increase of Dopaminergic Innervation in a Striatal Region Associated with Speech and Language: A Comparative Analysis of the Primate Basal Ganglia." *Journal of Comparative Neurology* 524 (10): 2117–2129; Raghanti, Mary Ann, Elaine N. Miller, Danielle N. Jones, Heather N. Smith, Emily L. Munger, Melissa K. Edler, Kimberley A. Phillips, et al. 2023. "Hedonic Eating, Obesity, and Addiction Result from Increased Neuropeptide Y in the Nucleus Accumbens during Human Brain Evolution." *Proceedings of the National Academy of Sciences of the United States of America* 120 (38): e2311118120: 1–9.; Pontzer, Herman. 2023. "The Provisioned Primate: Patterns of Obesity across Lemurs, Monkeys, Apes and Humans." *Philosophical Transactions of the Royal Society B* 378 (1888): 20220218.

22. Dunbar, R. I. M. 2009. "The Social Brain Hypothesis and Its Implications for Social Evolution." *Annals of Human Biology* 36 (5): 562–572; **Atzil, Shir**, Wei Gao, Isaac Fradkin, and Lisa Feldman Barrett. 2018. "Growing a Social Brain." *Nature Human Behaviour* 2 (9): 624–636.

23. Wrangham, Richard W. 2019. "Hypotheses for the Evolution of Reduced Reactive Aggression in the Context of Human Self-Domestication." *Frontiers in Psychology* 10: 1914; **Wilkins, Adam S.**, Richard W. Wrangham, and W. Tecumseh Fitch. 2014. "The 'Domestication Syndrome' in Mammals: A Unified Explanation Based on Neural Crest Cell Behavior and Genetics." *Genetics* 197 (3): 795–808; **Raghanti, Mary Ann.** 2019. "Domesticated Species: It Takes One to Know One." *Proceedings of the National Academy of Sciences of the United States of America* 116 (29): 14401–14403.

24. Henneberg, Maciej. 1988. "Decrease of Human Skull Size in the Holocene." *Human Biology* 60 (3): 395–405; **Villmoare, Brian**, and Mark Grabowski. 2022. "Did the Transition to Complex Societies in the Holocene Drive a Reduction in Brain Size? A Reassessment of the DeSilva et al. (2021) Hypothesis." *Frontiers in Ecology and Evolution* 10 (July): 737: 1–5; **Hecht, Erin E.**, Sophie A. Barton, Christina N. Rogers Flattery, and Araceli Meza Meza. 2023. "The Evolutionary Neuroscience of Domestication." *Trends in Cognitive Sciences* 27 (6): 553–567; **Racicot, Kelsey J.**, Jackson R. Ham, Jacqueline Augustine, Rie Henriksen, Dominic Wright, and Andrew N. Iwaniuk. 2023. "A Comparison of Telencephalon Composition among Chickens, Junglefowl, and Wild Galliforms." *BioRxiv*, September, 2023.09.10.557096:1–35.

Chapter 12

1. Aboitiz, Francisco. 2017. *A Brain for Speech: A View from Evolutionary Neuroanatomy*. 1st ed. London: Palgrave Macmillan.

2. Darwin, Charles. 1981. *The Descent of Man, and Selection in Relation to Sex*. Vol. 1–2. Princeton, NJ: Princeton University Press. First published 1871; **Fitch W.**, Tecumseh. 2010. *The Evolution of Language. The Evolution of Language*. Cambridge: Cambridge University Press.

3. Falk, Dean. 2004. "Prelinguistic Evolution in Early Hominins: Whence Motherese?" *Behavioral and Brain Sciences* 27 (4): 491–503; **Theofanopoulou, Constantina**, Cedric Boeckx, and Erich D. Jarvis. 2017. "A Hypothesis on a Role of Oxytocin in the Social Mechanisms of Speech and Vocal Learning." *Proceedings of the Royal Society B: Biological Sciences* 284 (1861): 1–8; **Simonyan, Kristina**, and Uwe Jürgens. 2003. "Efferent Subcortical Projections of the Laryngeal Motorcortex in the Rhesus Monkey." *Brain Research* 974 (1–2): 43–59; **Nguyen, Trinh**, Lucie Zimmer, and Stefanie Hoehl. 2023. "Your Turn, My Turn. Neural Synchrony in Mother–Infant Proto-Conversation." *Philosophical Transactions of the Royal Society B* 378 (1875): 20210488.

4. Dunbar, Robin. 1998. *Grooming, Gossip, and the Evolution of Language*. Cambridge, MA: Harvard University Press; **Chereskin, Emma**, Richard C. Connor, Whitney R. Friedman, Frants H. Jensen, Simon J. Allen, Pernille M. Sørensen, Michael Krützen, and Stephanie L. King. 2022. "Allied Male Dolphins Use Vocal Exchanges to 'Bond at a Distance.'" *Current Biology* 32 (7): 1657–1663.e4.

5. Takahashi, Daniel Y., Darshana Z. Narayanan, and Asif A. Ghazanfar. 2013. "Coupled Oscillator Dynamics of Vocal Turn-Taking in Monkeys." *Current Biology* 23 (21): 2162–2168; **Varella, Thiago T.**, and Asif A. Ghazanfar. 2021. "Cooperative Care and the Evolution of the Prelinguistic Vocal Learning." *Developmental Psychobiology* 63 (5): 1583–1588; **Gultekin, Yasemin B.**, David G. C. Hildebrand, Kurt Hammerschmidt, and Steffen R. Hage. 2021. "High Plasticity in Marmoset Monkey Vocal Development from Infancy to Adulthood." *Science Advances* 7 (27): 1–10; **Brügger, R. K., E. P.** Willems, and J. M. Burkart. 2021. "Do Marmosets Understand Others'

Conversations? A Thermography Approach." *Science Advances* 7 (6): 1–11; **Dureux, Audrey**, Alessandro Zanini, and Stefan Everling. 2023. "Mapping of Facial and Vocal Processing in Common Marmosets with Ultra-High Field FMRI." *BioRxiv*, August, 2023.08.09.552659:1–41; **Phaniraj, Nikhil**, Kaja Wierucka, and Judith M. Burkart. 2023. "Dynamic Vocal Learning in Adult Marmoset Monkeys." *BioRxiv*, September, 2023.09.22.559020: 1–17.

6. Bowlingi, Daniel L., Jacob C. Dunn, Jeroen B. Smaers, Maxime Garcia, Asha Sato, Georg Hantke, Stephan Handschuh, et al. 2020. "Rapid Evolution of the Primate Larynx?" *PLoS Biology* 18 (8): 1–21; **Pisanski, Katarzyna**, Andrey Anikin, and David Reby. 2022. "Vocal Size Exaggeration May Have Contributed to the Origins of Vocalic Complexity." *Philosophical Transactions of the Royal Society of London. Series B, Biological Sciences* 377 (1841): 1–11.

7. Lameira, Adriano R., Ian Maddieson, and Klaus Zuberbühler. 2014. "Primate Feedstock for the Evolution of Consonants." *Trends in Cognitive Sciences* 18 (2): 60–62; **Pereira, André S.**, Eithne Kavanagh, Catherine Hobaiter, Katie E. Slocombe, and Adriano R. Lameira. 2020. "Chimpanzee Lip-Smacks Confirm Primate Continuity for Speech-Rhythm Evolution." *Biology Letters* 16 (5): 1-6; **Fitch, W. Tecumseh**, Bart De Boer, Neil Mathur, and Asif A. Ghazanfar. 2016. "Monkey Vocal Tracts Are Speech-Ready." *Science Advances* 2 (12): 1–7; **Gustison, Morgan L.**, and Thore J. Bergman. 2017. "Divergent Acoustic Properties of Gelada and Baboon Vocalizations and Their Implications for the Evolution of Human Speech." *Journal of Language Evolution* 2 (1): 20–36; **Lameira, Adriano R.**, Madeleine E Hardus, and Stephen Fleming. 2023. "Wild Orangutans Can Simultaneously Use Two Independent Vocal Sound Sources Similarly to Songbirds and Human Beatboxers." *PNAS Nexus* 2 (6): 1–4; **Lameira, Adriano R.**, and Steven Moran. 2023. "Life of p: A Consonant Older than Speech." *BioEssays* 45 (4): 2200246: 1–5.

8. Martins, Pedro Tiago, and Cedric Boeckx. 2020. "Vocal Learning: Beyond the Continuum." *PLoS Biology* 18 (3): 1–18; **Thomas, James**, and Simon Kirby. 2018. "Self-Domestication and the Evolution of Language." *Biology & Philosophy* 33 (1): 1–30; **Ghazanfar, Asif A.**, Lauren M. Kelly, Daniel Y. Takahashi, Sandra Winters, Rebecca Terrett, and James P. Higham. 2020. "Domestication Phenotype Linked to Vocal Behavior in Marmoset Monkeys." *Current Biology* 30 (24): 5026–5032.e3.

9. Simonyan, Kristina, and Uwe Jürgens. 2003. "Efferent Subcortical Projections of the Laryngeal Motorcortex in the Rhesus Monkey." *Brain Research* 974 (1–2): 43–59; **Hickok, Gregory.** 2017. "A Cortical Circuit for Voluntary Laryngeal Control: Implications for the Evolution Language." *Psychonomic Bulletin and Review* 24 (1): 56–63; **Cerkevich, Christina M.**, Jean Alban Rathelot, and Peter L. Strick. 2022. "Cortical Basis for Skilled Vocalization." *Proceedings of the National Academy of Sciences of the United States of America* 119 (19).

10. Belyk, Michel, Nicole Eichert, and Carolyn McGettigan. 2021. "A Dual Larynx Motor Networks Hypothesis." *Philosophical Transactions of the Royal Society B*

376 (1840): 1–11; **Hickok, Gregory**, Jonathan Venezia, and Alex Teghipco. 2023. "Beyond Broca: Neural Architecture and Evolution of a Dual Motor Speech Coordination System." *Brain* 146 (5): 1775–1790; **Gordon, Evan M.**, Roselyne J. Chauvin, Andrew N. Van, Aishwarya Rajesh, Ashley Nielsen, Dillan J. Newbold, Charles J. Lynch, et al. 2023. "A Somato-Cognitive Action Network Alternates with Effector Regions in Motor Cortex." *Nature* 617 (7960): 351–359.

11. Patel, Aniruddh D. 2021. "Vocal Learning as a Preadaptation for the Evolution of Human Beat Perception and Synchronization." *Philosophical Transactions of the Royal Society B* 376 (1835): 1–15; **Smith, Heather N.**, Alyssa Perrone, Michael Wilson, Mary Ann Raghanti, C. Owen Lovejoy, Merri J. Rosen, Sharad J. Shanbhag, David S. DeForrest, R. Lee Lyman, and Metin I. Eren. 2021. "Rock Music: An Auditory Assessment of Knapping." *Lithic Technology*, 46 (4): 320–335.

12. Poeppel, David. 2014. "The Neuroanatomic and Neurophysiological Infrastructure for Speech and Language." *Current Opinion in Neurobiology* 28: 142–149; **Aboitiz, Francisco**, Sergio Osorio, and Rodrigo A Henríquez-Ch. 2020. "A Neural Code for Multimodal Language Processing and Its Origins." *Paradigmi* XXXVIII (2/2020): 187–206; **Michon, Maëva**, Vladimir López, and Francisco Aboitiz. 2019. "Origin and Evolution of Human Speech: Emergence from a Trimodal Auditory, Visual and Vocal Network." *Progress in Brain Research* 250: 345–371; **Michon, Maëva**, José Zamorano-Abramson, and Francisco Aboitiz. 2022. "Faces and Voices Processing in Human and Primate Brains: Rhythmic and Multimodal Mechanisms Underlying the Evolution and Development of Speech." *Frontiers in Psychology* 13: 1–15; **Reeve, Catherine**, and Sophie Jacques. 2022. "Responses to Spoken Words by Domestic Dogs: A New Instrument for Use with Dog Owners." *Applied Animal Behaviour Science* 246: 105513; **Kazanina, Nina**, and Alessandro Tavano. 2022. "What Neural Oscillations Can and Cannot Do for Syntactic Structure Building." *Nature Reviews Neuroscience* 24 (2): 113–128.

13. Albouy, Philippe, Lucas Benjamin, Benjamin Morillon, and Robert J. Zatorre. 2020. "Distinct Sensitivity to Spectrotemporal Modulation Supports Brain Asymmetry for Speech and Melody." *Science* 367 (6481): 1043–1047; **Hausen, Maija**, Ritva Torppa, Viljami R. Salmela, Martti Vainio, and Teppo Särkämö. 2013. "Music and Speech Prosody: A Common Rhythm." *Frontiers in Psychology* 4: 1–16; **Norman-Haignere, Sam V.**, Jenelle Feather, Dana Boebinger, Peter Brunner, Anthony Ritaccio, Josh H. McDermott, Gerwin Schalk, and Nancy Kanwisher. 2022. "A Neural Population Selective for Song in Human Auditory Cortex." *Current Biology* 32 (7): 1470–1484. e12; **Siman-Tov, Tali**, Carlos R. Gordon, Netanell Avisdris, Ofir Shany, Avigail Lerner, Omer Shuster, Roni Y. Granot, and Talma Hendler. 2022. "The Rediscovered Motor-Related Area 55b Emerges as a Core Hub of Music Perception." *Communications Biology* 5 (1): 1–13; **Labache, Loïc**, Tian Ge, B. T. Thomas Yeo, and Avram J. Holmes. 2023. "Language Network Lateralization Is Reflected throughout the Macroscale Functional Organization of Cortex." *Nature Communications* 14 (1): 1–13.

14. Enard, Wolfgang, Molly Przeworski, Simon E. Fisher, Cecilia S. L. Lai, Victor Wiebe, Takashi Kitano, Anthony P. Monaco, and Svante Pääbo. 2002. "Molecular

Evolution of FOXP2, a Gene Involved in Speech and Language." *Nature* 418 (6900): 869–872; **Atkinson, Elizabeth Grace**, Amanda Jane Audesse, Julia Adela Palacios, Dean Michael Bobo, Ashley Elizabeth Webb, Sohini Ramachandran, and Brenna Mariah Henn. 2018. "No Evidence for Recent Selection at FOXP2 among Diverse Human Populations." *Cell* 174 (6): 1424–1435.e15; **Martins, Pedro Tiago**, Maties Marí, and Cedric Boeckx. 2018. "SRGAP2 and the Gradual Evolution of the Modern Human Language Faculty." *Journal of Language Evolution* 3 (1): 67–78.

15. Unger, Nina, Stefan Heim, Dominique I. Hilger, Sebastian Bludau, Peter Pieperhoff, Sven Cichon, Katrin Amunts, and Thomas W. Mühleisen. 2021. "Identification of Phonology-Related Genes and Functional Characterization of Broca's and Wernicke's Regions in Language and Learning Disorders." *Frontiers in Neuroscience* 15: 1–17.

16. Nieder, Andreas, and Richard Mooney. 2020. "The Neurobiology of Innate, Volitional and Learned Vocalizations in Mammals and Birds." *Philosophical Transactions of the Royal Society of London. Series B, Biological Sciences* 375 (1789): 1–21.

17. Petkov, Christopher I., and Erich D. Jarvis. 2012. "Birds, Primates, and Spoken Language Origins: Behavioral Phenotypes and Neurobiological Substrates." *Frontiers in Evolutionary Neuroscience* 4 (August): 12; **Petkov, Christopher**. 2014. "The Basal Ganglia within a Cognitive System in Birds and Mammals." *The Behavioral and Brain Sciences* 37 (6): 568–569.

18. Searcy, William A., Jill Soha, Susan Peters, and Stephen Nowicki. 2022. "Long-Distance Dependencies in Birdsong Syntax." *Proceedings of the Royal Society B* 289 (1967): 1–8.

19. Aliko, Sarah, Bangjie Wang, Steven L Small, and Jeremy I Skipper. 2023. "The Entire Brain, More or Less, Is at Work: 'Language Regions' Are Artefacts of Averaging." *BioRxiv*, October, 2023.09.01.555886: 1–53.; **Amiez, Céline**, Charles Verstraete, Jérôme Sallet, Fadila Hadj-Bouziane, Suliann Ben Hamed, Adrien Meguerditchian, Emmanuel Procyk, et al. 2023. "The Relevance of the Unique Anatomy of the Human Prefrontal Operculum to the Emergence of Speech." *Communications Biology* 6 (1):1–12.; **Gallardo, Guillermo**, Cornelius Eichner, Chet C. Sherwood, William D. Hopkins, Alfred Anwander, and Angela D. Friederici. 2023. "Morphological Evolution of Language-Relevant Brain Areas." *PLOS Biology* 21 (9): e3002266: 1–13; **Rogalsky, Corianne**, Alexandra Basilakos, Chris Rorden, Sara Pillay, Arianna N. Lacroix, Lynsey Keator, Soren Mickelsen, et al. 2022. "The Neuroanatomy of Speech Processing: A Large-Scale Lesion Study." *Journal of Cognitive Neuroscience* 34 (8): 1355–1375.

20. Aboitiz, Francisco, and Ricardo García V. 1997. "The Evolutionary Origin of the Language Areas in the Human Brain. A Neuroanatomical Perspective." *Brain Research Reviews* 25 (3): 381–396; **Petrides, Michael.** 2014. *Neuroanatomy of Language Regions of the Human Brain. Neuroanatomy of Language Regions of the Human Brain.* New York:

Academic Press; **Catani, Marco**, and Dominic H. Ffytche. 2005. "The Rises and Falls of Disconnection Syndromes." *Brain: A Journal of Neurology* 128 (Pt 10): 2224–2239.

21. Aboitiz, Francisco, 2018. "A Brain for Speech. Evolutionary Continuity in Primate and Human Auditory-Vocal Processing." *Frontiers in Neuroscience* 12: 1–14; **Rilling, James K.**, Matthew F. Glasser, Todd M. Preuss, Xiangyang Ma, Tiejun Zhao, Xiaoping Hu, and Timothy E. J. Behrens. 2008. "The Evolution of the Arcuate Fasciculus Revealed with Comparative DTI." *Nature Neuroscience* 11 (4): 426–428; **Sierpowska, Joanna**, Katherine L. Bryant, Nikki Janssen, Guilherme Blazquez Freches, Manon Romkens, Margot Mangnus, Rogier B. Mars, and Vitoria Piai. 2022. "Comparing Human and Chimpanzee Temporal Lobe Neuroanatomy Reveals Modifications to Human Language Hubs beyond the Frontotemporal Arcuate Fasciculus." *Proceedings of the National Academy of Sciences of the United States of America* 119 (28): 1–9; **Romanski, Lizabeth M.**, and Keshov K. Sharma. 2023. "Multisensory Interactions of Face and Vocal Information during Perception and Memory in Ventrolateral Prefrontal Cortex." *Philosophical Transactions of the Royal Society B* 378 (1886). 20220343.

22. Aboitiz, F. 1995. "Working Memory Networks and the Origin of Language Areas in the Human Brain." *Medical Hypotheses* 44 (6): 504–506; **Aboitiz, Francisco**, 2018. "A Brain for Speech. Evolutionary Continuity in Primate and Human Auditory-Vocal Processing." *Frontiers in Neuroscience* 12: 1–14; **Barbeau, Elise B.**, Shanna Kousaie, Kanontienentha Brass, Maxime Descoteaux, Michael Petrides, and Denise Klein. 2023. "The Importance of the Dorsal Branch of the Arcuate Fasciculus in Phonological Working Memory." *Cerebral Cortex* 33 (16): 9554–9565.

23. Schmitz, Judith, Stephanie Lor, Rena Klose, Onur Güntürkün, and Sebastian Ocklenburg. 2017. "The Functional Genetics of Handedness and Language Lateralization: Insights from Gene Ontology, Pathway and Disease Association Analyses." *Frontiers in Psychology* 8: 1–12.

24. Rizzolatti, Giacomo, and Michael A. Arbib. 1998. "Language within Our Grasp." *Trends in Neurosciences* 21 (5): 188–194; **Hobaiter, Catherine**, Kirsty E. Graham, and Richard W. Byrne. 2022. "Are Ape Gestures Like Words? Outstanding Issues in Detecting Similarities and Differences between Human Language and Ape Gesture." *Philosophical Transactions of the Royal Society of London. Series B, Biological Sciences* 377 (1860): 1–9.

25. Martins, Pedro Tiago 2023. "The Trade-off between Vocal Learning and Dexterity: A Balancing Act," May 2023, *EcoEvoRxiv*, 1–15. https://doi.org/10.32942/X2F88F.

26. Narayanan, Darshana Z., Daniel Y. Takahashi, Lauren M. Kelly, Sabina I. Hlavaty, Junzhou Huang, and Asif A. Ghazanfar. 2022. "Prenatal Development of Neonatal Vocalizations." *ELife* 11; **Burkhardt-Reed, Megan M.**, Helen L. Long, Dale D. Bowman, Edina R. Bene, and D. Kimbrough Oller. 2021. "The Origin of Language and Relative Roles of Voice and Gesture in Early Communication Development." *Infant Behavior & Development* 65: 1–13.

Chapter 13

1. Searle, John R. 1995. *The Construction of Social Reality. Ethics*. New York: Free Press.

2. Bornkessel, Ina, Stefan Zysset, Angela D. Friederici, D. Yves Von Cramon, and Matthias Schlesewsky. 2005. "Who Did What to Whom? The Neural Basis of Argument Hierarchies during Language Comprehension." *NeuroImage* 26 (1): 221–233.

3. Tomasello, Michael, and Henrike Moll. 2010. "The Gap Is Social: Human Shared Intentionality and Culture." In *Mind the Gap*, 331–349. Berlin Heidelberg: Springer-Verlag; **Xiao, Yaqiong**, Teresa H. Wen, Lauren Kupis, Lisa T. Eyler, Disha Goel, Keith Vaux, Michael V. Lombardo, Nathan E. Lewis, Karen Pierce, and Eric Courchesne. 2022. "Neural Responses to Affective Speech, Including Motherese, Map onto Clinical and Social Eye Tracking Profiles in Toddlers with ASD." *Nature Human Behaviour* 6 (3): 443–454; **Kano, Fumihiro**. 2023. "Evolution of the Uniformly White Sclera in Humans: Critical Updates." *Trends in Cognitive Sciences* 27 (1): 10–12.

4. Colonnesi, Cristina, Geert Jan J. M. Stams, Irene Koster, and Marc J. Noom. 2010. "The Relation between Pointing and Language Development: A Meta-Analysis." *Developmental Review* 30 (4): 352–366; **Hamilton, Antonia**, Kim Plunkett, and Graham Schafer. 2000. "Infant Vocabulary Development Assessed with a British Communicative Development Inventory." *Journal of Child Language* 27 (3): 689–705; **McGillion, Michelle**, Jane S. Herbert, Julian Pine, Marilyn Vihman, Rory dePaolis, Tamar Keren-Portnoy, and Danielle Matthews. 2017. "What Paves the Way to Conventional Language? The Predictive Value of Babble, Pointing, and Socioeconomic Status." *Child Development* 88 (1): 156–166; **Kendrick, Kobin H.**, Judith Holler, and Stephen C. Levinson. 2023. "Turn-Taking in Human Face-to-Face Interaction Is Multimodal: Gaze Direction and Manual Gestures Aid the Coordination of Turn Transitions." *Philosophical Transactions of the Royal Society of London. Series B, Biological Sciences* 378 (1875): 20210473.

5. Tomasello, Michael. n.d. *A Natural History of Human Thinking*. Cambridge, MA: Harvard University Press. Accessed January 2, 2023; **Buttelmann, David**, Frances Buttelmann, Malinda Carpenter, Josep Call, and Michael Tomasello. 2017. "Great Apes Distinguish True from False Beliefs in an Interactive Helping Task." *PloS One* 12 (4): 1–13; **Krupenye, Christopher**, Fumihiro Kano, Satoshi Hirata, Josep Call, and Michael Tomasello. 2016. "Great Apes Anticipate That Other Individuals Will Act According to False Beliefs." *Science* 354 (6308): 110–114; **Schulkin, Jay**, Alvin Goldman, and Vittorio Gallese. 2000. "Theory of Mind and Mirroring Neurons." *Trends in Cognitive Sciences* 4 (7): 252–254.

6. Price, Tabitha, Philip Wadewitz, Dorothy Cheney, Robert Seyfarth, Kurt Hammerschmidt, and Julia Fischer. 2015. "Vervets Revisited: A Quantitative Analysis of Alarm Call Structure and Context Specificity." *Scientific Reports* 5: 1–11; **Fischer, Julia**, and Tabitha Price. 2017. "Meaning, Intention, and Inference in Primate Vocal

Communication." *Neuroscience and Biobehavioral Reviews* 82 (November): 22–31; **Deshpande, Adwait**, and Klaus Zuberbuhler. 2022. "Socially Learned Arbitrary Call Use in a Wild Primate." *BioRxiv*, April, 2022.04.04.486940; **Suzuki, Toshitaka N.** 2018. "Alarm Calls Evoke a Visual Search Image of a Predator in Birds." *Proceedings of the National Academy of Sciences of the United States of America* 115 (7): 1541–1545; **Morell, Virginia**. 2022. "Dolphins Whistle to Keep in Touch with Distant Friends | Science | AAAS." *Science*, News Plants & Animals; **Scott-Phillip, Thom**, and Christophe Heintz. 2023. "Great Ape Interaction: Ladyginian but Not Gricean." *Proceedings of the National Academy of Sciences of the United States of America* 120 (42):1–10.

7. Ćwiek, Aleksandra, 2022. "The Bouba/Kiki Effect Is Robust across Cultures and Writing Systems." *Philosophical Transactions of the Royal Society of London. Series B, Biological Sciences* 377 (1841): 1–13; **Ćwiek, Aleksandra**, Susanne Fuchs, Christoph Draxler, Eva Liina Asu, Dan Dediu, Katri Hiovain, Shigeto Kawahara, et al. 2021. "Novel Vocalizations Are Understood across Cultures." *Scientific Reports* 11 (1): 1–12; **Winter, Bodo**, Márton Sóskuthy, Marcus Perlman, and Mark Dingemanse. 2022. "Trilled /r/ Is Associated with Roughness, Linking Sound and Touch across Spoken Languages." *Scientific Reports* 12 (1): 1–11; **Fay, Nicolas**, Bradley Walker, T. Mark Ellison, Zachary Blundell, Naomi De Kleine, Murray Garde, Casey J. Lister, and Susan Goldin-Meadow. 2022. "Gesture Is the Primary Modality for Language Creation." *Proceedings of the Royal Society B* 289 (1970): 1–9.

8. Aboitiz, Francisco, 2018. "Voice, Gesture and Working Memory in the Emergence of Speech." *Interaction Studies. Social Behaviour and Communication in Biological and Artificial Systems* 19 (1–2): 70–85; **Froesel, Mathilda**, Maëva Gacoin, Simon Clavagnier, Marc Hauser, Quentin Goudard, and Suliann Ben Hamed. 2022. "Socially Meaningful Visual Context Either Enhances or Inhibits Vocalisation Processing in the Macaque Brain." *Nature Communications* 13 (1): 1–17; **Popham, Sara F.**, Alexander G. Huth, Natalia Y. Bilenko, Fatma Deniz, James S. Gao, Anwar O. Nunez-Elizalde, and Jack L. Gallant. 2021. "Visual and Linguistic Semantic Representations Are Aligned at the Border of Human Visual Cortex." *Nature Neuroscience* 24 (11): 1628–1636; **Damera, Srikanth R.**, Lillian Chang, Plamen P. Nikolov, James A. Mattei, Suneel Banerjee, Laurie S. Glezer, Patrick H. Cox, Xiong Jiang, Josef P. Rauschecker, and Maximilian Riesenhuber. 2022. "Evidence for a Spoken Word Lexicon in the Auditory Ventral Stream." *BioRxiv*, October, 2022.10.09.511436; **Petrides, Michael**. 2023. "On the Evolution of Polysensory Superior Temporal Sulcus and Middle Temporal Gyrus: A Key Component of the Semantic System in the Human Brain." *Journal of Comparative Neurology*, 1–9 (online ahead of print; doi: 10.1002/cne.25521).

9. Gibson, Edward, Richard Futrell, Julian Jara-Ettinger, Kyle Mahowald, Leon Bergen, Sivalogeswaran Ratnasingam, Mitchell Gibson, Steven T. Piantadosi, and Bevil R. Conway. 2017. "Color Naming across Languages Reflects Color Use." *Proceedings of the National Academy of Sciences of the United States of America* 114 (40): 10785–10790; **Forbes, Samuel H.**, and Kim Plunkett. 2023. "Colour Perception

Changes with Basic Colour Word Comprehension." *Developmental Science*, 26 (6): 1–15, e13406.

10. Aboitiz, Francisco. 2017. *A Brain for Speech: A View from Evolutionary Neuroanatomy*. 1st ed. London: Palgrave Macmillan; **Shain, Cory**, Idan A. Blank, Evelina Fedorenko, Edward Gibson, and William Schuler. 2022. "Robust Effects of Working Memory Demand during Naturalistic Language Comprehension in Language-Selective Cortex." *Journal of Neuroscience: The Official Journal of the Society for Neuroscience* 42 (39): 7412–7430.

11. Goldstein, Ariel, Zaid Zada, Eliav Buchnik, Mariano Schain, Amy Price, Bobbi Aubrey, Samuel A. Nastase, et al. 2022. "Shared Computational Principles for Language Processing in Humans and Deep Language Models." *Nature Neuroscience* 25 (3): 369–380; **Klein, Cheslie C.**, Philipp Berger, Tomás Goucha, Angela D. Friederici, and Charlotte Grosse Wiesmann. 2022. "Children's Syntax Is Supported by the Maturation of BA44 at 4 Years, but of the Posterior STS at 3 Years of Age." *Cerebral Cortex*, November: 5426–5435; **Tang, Jerry**, Amanda LeBel, Shailee Jain, and Alexander G. Huth. 2023. "Semantic Reconstruction of Continuous Language from Non-Invasive Brain Recordings." *Nature Neuroscience* 26 (5): 858–866.

12. Watson, Stuart K., Judith M. Burkart, Steven J. Schapiro, Susan P. Lambeth, Jutta L. Mueller, and Simon W. Townsend. 2020. "Nonadjacent Dependency Processing in Monkeys, Apes, and Humans." *Science Advances* 6 (43): 1–9; **Sainburg, Tim**, Anna Mai, and Timothy Q. Gentner. 2022. "Long-Range Sequential Dependencies Precede Complex Syntactic Production in Language Acquisition." *Proceedings of the Royal Society B* 289 (1970): 1–11; **Searcy, William A.**, Jill Soha, Susan Peters, and Stephen Nowicki. 2022. "Long-Distance Dependencies in Birdsong Syntax." *Proceedings of the Royal Society B* 289 (1967): 1–8; **Wilson, Benjamin**, Heather Slater, Yukiko Kikuchi, Alice E. Milne, William D. Marslen-Wilson, Kenny Smith, and Christopher I. Petkov. 2013. "Auditory Artificial Grammar Learning in Macaque and Marmoset Monkeys." *Journal of Neuroscience: The Official Journal of the Society for Neuroscience* 33 (48): 18825–18835; **Huang, Junfeng**, He Ma, Yongkang Sun, Liangtang Chang, and Neng Gong. 2022. "Complex Rules of Vocal Sequencing in Marmoset Monkeys." *BioRxiv*, August, 2022.08.03.502601; **Suzuki, Toshitaka N.**, and Yui K. Matsumoto. 2022. "Experimental Evidence for Core-Merge in the Vocal Communication System of a Wild Passerine." *Nature Communications* 13 (1): 1–7; **Leroux, Maël**, Anne M. Schel, Claudia Wilke, Bosco Chandia, Klaus Zuberbühler, Katie E. Slocombe, and Simon W. Townsend. 2023. "Call Combinations and Compositional Processing in Wild Chimpanzees." *Nature Communications* 14 (1): 1-8.

13. Bortolato, Tatiana, Angela D. Friederici, Cédric Girard-Buttoz, Roman M. Wittig, and Catherine Crockford. 2023. "Chimpanzees Show the Capacity to Communicate about Concomitant Daily Life Events." *IScience* 26 (11): 1–12; Bickerton, Derek. 1990. *Language & Species*. Chicago: University of Chicago Press; **Hopper, Paul J.** 1996. "Some Recent Trends in Grammaticalization." *Annual Review of Anthropology* 25 (1): 217–236;

Murphy, Elliot, Emma Holmes, and Karl Friston. 2021. "Natural Language Syntax Complies with the Free-Energy Principle." *PsyArXiv*, March, 1–76; **Kirby, Simon**, and Monica Tamariz. 2022. "Cumulative Cultural Evolution, Population Structure and the Origin of Combinatoriality in Human Language." *Philosophical Transactions of the Royal Society of London. Series B, Biological Sciences* 377 (1843): 1–10.

14. Wilson, Vanessa A. D., Klaus Zuberbühler, and Balthasar Bickel. 2022. "The Evolutionary Origins of Syntax: Event Cognition in Nonhuman Primates." *Science Advances* 8 (25): 1–12; **Hahn, Michael**, and Yang Xu. 2022. "Crosslinguistic Word Order Variation Reflects Evolutionary Pressures of Dependency and Information Locality." *Proceedings of the National Academy of Sciences of the United States of America* 119 (24): 1–10; **Dehaene, Stanislas**, Fosca Al Roumi, Yair Lakretz, Samuel Planton, and Mathias Sablé-Meyer. 2022. "Symbols and Mental Programs: A Hypothesis about Human Singularity." *Trends in Cognitive Sciences* 26 (9): 751–766.

15. Aboitiz, Francisco, Sergio Osorio, and Rodrigo A Henríquez-Ch. 2020. "A Neural Code for Multimodal Language Processing and Its Origins." *Paradigmi* XXXVIII (2/2020): 187–206; **Lieberman, Matthew D.** 2018. "Birds of a Feather Synchronize Together." *Trends in Cognitive Sciences* 22 (5): 371–372; **Kingsbury, Lyle**, Shan Huang, Jun Wang, Ken Gu, Peyman Golshani, Ye Emily Wu, and Weizhe Hong. 2019. "Correlated Neural Activity and Encoding of Behavior across Brains of Socially Interacting Animals." *Cell* 178 (2): 429–446.e16; **Alderson-Day**, Ben, Susanne Weis, Simon McCarthy-Jones, Peter Moseley, David Smailes, and Charles Fernyhough. 2016. "The Brain's Conversation with Itself: Neural Substrates of Dialogic Inner Speech." *Social Cognitive and Affective Neuroscience* 11 (1): 110–120; **Liu, Lanfang**, Hehui Li, Zhiting Ren, Qi Zhou, Yuxuan Zhang, Chunming Lu, Jiang Qiu, Hong Chen, and Guosheng Ding. 2022. "The 'Two-Brain' Approach Reveals the Active Role of Task-Deactivated Default Mode Network in Speech Comprehension." *Cerebral Cortex* 32 (21): 4869–4884; **Zhang, Wujie**, Maimon C. Rose, and Michael M. Yartsev. 2022. "A Unifying Mechanism Governing Inter-Brain Neural Relationship during Social Interactions." *ELife* 11: 1–36.

16. Cavalli-Sforza, L. L., A. Piazza, P. Menozzi, and J. Mountain. 1988. "Reconstruction of Human Evolution: Bringing Together Genetic, Archaeological, and Linguistic Data." *Proceedings of the National Academy of Sciences of the United States of America* 85 (16): 6002–6006.

17. Schmandt-Besserat, Denise, and Michael Erard. 2008. "Origins and Forms of Writing. Chapter 1." In *Handbook of Research on Writing, Society, School, Individual, Text*, edited by Charles Bazerman, 7–22. New York: Taylor & Francis Group/Lawrence Erlbaum Associates.

18. Dehaene-Lambertz, Ghislaine, Karla Monzalvo, and Stanislas Dehaene. 2018. "The Emergence of the Visual Word Form: Longitudinal Evolution of Category-Specific Ventral Visual Areas during Reading Acquisition." *PLoS Biology* 16 (3): 1–34;

Roux, Franck Emmanuel, Olivier Dufor, Carlo Giussani, Yannick Wamain, Louisa Draper, Marieke Longcamp, and Jean François Démonet. 2009. "The Graphemic/Motor Frontal Area Exner's Area Revisited." *Annals of Neurology* 66 (4): 537–545.

19. Kutter, Esther F., Jan Boström, Christian E. Elger, Andreas Nieder, and Florian Mormann. 2022. "Neuronal Codes for Arithmetic Rule Processing in the Human Brain." *Current Biology* 32 (6): 1275–1284.e4.

20. Dawkins, Richard. 1989. *The Selfish Gene*. New York: Oxford University Press.

Chapter 14

1. Dennett, Daniel C. 1991. *Consciousness Explained.* 1st ed. New York: Back Bay Books/Little, Brown and Company; **O'Regan, J. Kevin**, and Alva Noë. 2001. "A Sensorimotor Account of Vision and Visual Consciousness." *The Behavioral and Brain Sciences* 24 (5): 939–973; **Rudrauf, David**, Antoine Lutz, Diego Cosmelli, Jean Philippe Lachaux, and Michel Le Van Quyen. 2003. "From Autopoiesis to Neurophenomenology: Francisco Varela's Exploration of the Biophysics of Being." *Biological Research* 36 (1): 27–65; **He, Biyu J.** 2023. "Next Frontiers in Consciousness Research." *Neuron* 111 (20): 3150–3153.

2. Chalmers, David. 1995. "Facing up to the Problem of Consciousness." *Journal of Consciousness Studies* 2 (3): 200–219.

3. Sulloway, Frank J. 1979. *Freud, Biologist of the Mind: Beyond the Psychoanalytic Legend.* New York: Basic Books; **Damasio, Antonio R.** 2021. *Feeling & Knowing: Making Minds Conscious.* Pantheon, New York; **LeDoux, Joseph**. 2019. *The Deep History of Ourselves: The Four-Billion-Year Story of How We Got Conscious Brains.* Viking, New York.

4. Seth, Anil K., and Tim Bayne. 2022. "Theories of Consciousness." *Nature Reviews. Neuroscience* 23 (7): 439–452; **Mediano, Pedro A. M.**, Fernando E. Rosas, Daniel Bor, Anil K. Seth, and Adam B. Barrett. 2022. "The Strength of Weak Integrated Information Theory." *Trends in Cognitive Sciences* 26 (8): 646–655; **Engelen, Tahnée**, Marco Solcà, and Catherine Tallon-Baudry. 2023. "Interoceptive Rhythms in the Brain." *Nature Neuroscience* 26 (10): 1670–1684.

5. Hannawi, Yousef, Martin A. Lindquist, Brian S. Caffo, Haris I. Sair, and Robert D. Stevens. 2015. "Resting Brain Activity in Disorders of Consciousness: A Systematic Review and Meta-Analysis." *Neurology* 84 (12): 1272–1280; **Craig, A. D.** 2009. "How Do You Feel—Now? The Anterior Insula and Human Awareness." *Nature Reviews. Neuroscience* 10 (1): 59–70; **Behrendt, Ralf Peter**. 2013. "Conscious Experience and Episodic Memory: Hippocampus at the Crossroads." *Frontiers in Psychology* 4: 1–14; **Seth, Anil K.**, and Manos Tsakiris. 2018. "Being a Beast Machine: The Somatic Basis of Selfhood." *Trends in Cognitive Sciences* 22 (11): 969–981.

6. Gordon, Evan M., Timothy O. Laumann, Scott Marek, Ryan V. Raut, Caterina Gratton, Dillan J. Newbold, Deanna J. Greene, et al. 2020. "Default-Mode Network

Streams for Coupling to Language and Control Systems." *Proceedings of the National Academy of Sciences of the United States of America* 117 (29): 17308–17319; **Chiou, Rocco**, Gina F. Humphreys, and Matthew A. Lambon Ralph. 2020. "Bipartite Functional Fractionation within the Default Network Supports Disparate Forms of Internally Oriented Cognition." *Cerebral Cortex (1991)* 30 (10): 5484–5501.

7. Chalmers, David. 1995. "Facing up to the Problem of Consciousness." *Journal of Consciousness Studies* 2 (3): 200–219; **Dennett, Daniel C.** 1991. *Consciousness Explained*. 1st ed. New York: Back Bay Books/Little, Brown and Company; **Dennett, Daniel C.** 2018. "Facing up to the Hard Question of Consciousness." *Philosophical Transactions of the Royal Society B: Biological Sciences* 373 (1755): 1–7; **Dennett, Daniel. C.** 2023. *I've Been Thinking*. London: Allen Lane.

8. Taylor, Steve. 2018. *Spiritual Science: Why Science Needs Spirituality to Make Sense of the World*. London: World Watkins Publishing; **Fried, Itzhak.** 2022. "Neurons as Will and Representation." *Nature Reviews. Neuroscience* 23 (2): 104–114; **Miyamoto, Kentaro,** Nadescha Trudel, Kevin Kamermans, Michele C. Lim, Alberto Lazari, Lennart Verhagen, Marco K. Wittmann, and Matthew F. S. Rushworth. 2021. "Identification and Disruption of a Neural Mechanism for Accumulating Prospective Metacognitive Information Prior to Decision-Making." *Neuron* 109 (8): 1396–1408.e7; **Patel, Kramay**, Chaim N. Katz, Suneil K. Kalia, Milos R. Popovic, and Taufik A. Valiante. 2021. "Volitional Control of Individual Neurons in the Human Brain." *Brain* 144 (12): 3651–3663.

9. Tononi, Giulio, Melanie Boly, Marcello Massimini, and Christof Koch. 2016. "Integrated Information Theory: From Consciousness to Its Physical Substrate." *Nature Reviews. Neuroscience* 17 (7): 450–461; **Koch, Christof**. 2020. "Hot or Not." *Nature Human Behaviour* 4 (10): 991–992; **Bayne, Tim**, Anil K. Seth, and Marcello Massimini. 2020. "From Complexity to Consciousness." *Trends in Neurosciences* 43 (8): 546–547; **Vigotsky, Andrew D.**, Rami Jabakhanji, Paulo Branco, Gian Domenico Iannetti, Marwan N. Baliki, A. Vania Apkarian, Christof Koch, Todd Parrish, and Lucas Pinto. 2022. "Widespread, Perception-Related Information in the Human Brain Scales with Levels of Consciousness." *BioRxiv*, September, 2022.09.19.508437; **Consortium, Cogitate,** Oscar Ferrante, Urszula Gorska-Klimowska, Simon Henin, Rony Hirschhorn, Aya Khalaf, Alex Lepauvre, et al. 2023. "An Adversarial Collaboration to Critically Evaluate Theories of Consciousness." *BioRxiv*, June, 2023.06.23.546249: 1–68.

10. Baars, Bernard J. 2005. "Global Workspace Theory of Consciousness: Toward a Cognitive Neuroscience of Human Experience." *Progress in Brain Research* 150: 45–53; **Dehaene, Stanislas**, Hakwan Lau, and Sid Kouider. 2017. "What Is Consciousness, and Could Machines Have It?" *Science* 358 (6362): 486–492; **LeDoux, Joseph,** Jonathan Birch, Kristin Andrews, Nicola S. Clayton, Nathaniel D. Daw, Chris Frith, Hakwan Lau, et al. 2023. "Consciousness beyond the Human Case." *Current Biology* 33 (16): 832–840.

11. Hu, Huiqing, Rhodri Cusack, and Lorina Naci. 2022. "Typical and Disrupted Brain Circuitry for Conscious Awareness in Full-Term and Preterm Infants." *Brain Communications* 4 (2): 1–16; **Moser, Julia**, Franziska Schleger, Magdalene Weiss, Katrin Sippel, Lorenzo Semeia, and Hubert Preissl. 2021. "Magnetoencephalographic Signatures of Conscious Processing before Birth." *Developmental Cognitive Neuroscience* 49: 1–10; **Vasung, Lana**, Esra Abaci Turk, Silvina L. Ferradal, Jason Sutin, Jeffrey N. Stout, Banu Ahtam, Pei Yi Lin, and P. Ellen Grant. 2019. "Exploring Early Human Brain Development with Structural and Physiological Neuroimaging." *NeuroImage* 187: 226–254; **Cleeremans, Axel**, Dalila Achoui, Arnaud Beauny, Lars Keuninckx, Jean Remy Martin, Santiago Muñoz-Moldes, Laurène Vuillaume, and Adélaïde de Heering. 2020. "Learning to Be Conscious." *Trends in Cognitive Sciences* 24 (2): 112–123; **Kouider, Sid,** Carsten Stahlhut, Sofie V. Gelskov, Leonardo S. Barbosa, Michel Dutat, Vincent De Gardelle, Anne Christophe, Stanislas Dehaene, and Ghislaine Dehaene-Lambertz. 2013. "A Neural Marker of Perceptual Consciousness in Infants." *Science* 340 (6130): 376–380.

12. Waal, Frans B. M. de, and Kristin Andrews. 2022. "The Question of Animal Emotions." *Science* 375 (6587): 1351–1352; **Andrade, Michaella P.**, Charles Morphy D. Santos. 2021. "Evolutionary Puzzle: Discussing the Evolution of Sentience in Metazoa through a Phylogenetic Perspective." *BioRxiv*, May, 2021.05.21.445182.

13. Nieder, Andreas, Lysann Wagener, and Paul Rinnert. 2020. "A Neural Correlate of Sensory Consciousness in a Corvid Bird." *Science* 369 (6511): 1626–1629; **Ben-Haim, Moshe Shay**, Olga Dal Monte, Nicholas A. Fagan, Yarrow Dunham, Ran R. Hassin, Steve W. C. Chang, and Laurie R. Santos. 2021. "Disentangling Perceptual Awareness from Nonconscious Processing in Rhesus Monkeys (Macaca Mulatta)." *Proceedings of the National Academy of Sciences of the United States of America* 118 (15): 1–9.

14. Birch, Jonathan, Alexandra K. Schnell, and Nicola S. Clayton. 2020. "Dimensions of Animal Consciousness." *Trends in Cognitive Sciences* 24 (10): 789–801; **Irwin, Louis N.** 2020. "Renewed Perspectives on the Deep Roots and Broad Distribution of Animal Consciousness." *Frontiers in Systems Neuroscience* 14: 1–6; **Heyes, Cecilia**, Dan Bang, Nicholas Shea, Christopher D. Frith, and Stephen M. Fleming. 2020. "Knowing Ourselves Together: The Cultural Origins of Metacognition." *Trends in Cognitive Sciences* 24 (5): 349–362.

15. Hameroff, Stuart. 2021. "'Orch OR' Is the Most Complete, and Most Easily Falsifiable Theory of Consciousness." *Cognitive Neuroscience* 12 (2): 74–76; **Chalmers, David J.**, and Kelvin J McQueen. 2021. "Consciousness and the Collapse of the Wave Function—Chapter 1." In *Consciousness and Quantum Mechanics*, edited by Shan Gao, 480. Oxford University Press; **Brüntrup, Godehard**. 2017. *Panpsychism: Contemporary Perspectives*. Oxford University Press—Philosophy; **Goff, Philip**. 2017. *Consciousness and Fundamental Reality. Consciousness and Fundamental Reality*. Oxford University Press.

Chapter 15

1. Lucas, Teghan, Jaliya Kumaratilake, and Maciej Henneberg. 2020. "Recently Increased Prevalence of the Human Median Artery of the Forearm: A Microevolutionary Change." *Journal of Anatomy* 237 (4): 623–631; **Evershed, Richard P.**, George Davey Smith, Mélanie Roffet-Salque, Adrian Timpson, Yoan Diekmann, Matthew S. Lyon, Lucy J. E. Cramp, et al. 2022. "Dairying, Diseases and the Evolution of Lactase Persistence in Europe." *Nature* 608 (7922): 336–345; **Mathieson, Iain**, and Jonathan Terhorst. 2022. "Direct Detection of Natural Selection in Bronze Age Britain." *Genome Research* 32 (11–12): 2057–2067; **Klunk, Jennifer**, Tauras P. Vilgalys, Christian E. Demeure, Xiaoheng Cheng, Mari Shiratori, Julien Madej, Rémi Beau, et al. 2022. "Evolution of Immune Genes Is Associated with the Black Death." *Nature* 611 (7935): 312–319.

2. Zwir, I., C. Del-Val, M. Hintsanen, K. M. Cloninger, R. Romero-Zaliz, A. Mesa, J. Arnedo, et al. 2022. "Evolution of Genetic Networks for Human Creativity." *Molecular Psychiatry* 27 (1): 354–376; **Moriano, Juan**, and Cedric Boeckx. 2020. "Modern Human Changes in Regulatory Regions Implicated in Cortical Development." *BMC Genomics* 21 (1); **Gokhman, David**, Eitan Lavi, Kay Prüfer, Mario F. Fraga, José A. Riancho, Janet Kelso, Svante Pääbo, Eran Meshorer, and Liran Carmel. 2014. "Reconstructing the DNA Methylation Maps of the Neandertal and the Denisovan." *Science* 344 (6183): 523–527.

3. Doan, Ryan N., Byoung Il Bae, Beatriz Cubelos, Cindy Chang, Amer A. Hossain, Samira Al-Saad, Nahit M. Mukaddes, et al. 2016. "Mutations in Human Accelerated Regions Disrupt Cognition and Social Behavior." *Cell* 167 (2): 341–354.e12; **Banerjee, Niladri**, Tatiana Polushina, Francesco Bettella, Vidar M. Steen, Ole A. Andreassen, and Stephanie Le Hellard. 2019. "Analysis of Differentially Methylated Regions in Great Apes and Extinct Hominids Provides Support for the Evolutionary Hypothesis of Schizophrenia." *Schizophrenia Research* 206: 209–216; **Wang, E.**, Y. C. Ding, P. Flodman, J. R. Kidd, K. K. Kidd, D. L. Grady, O. A. Ryder, M. A. Spence, J. M. Swanson, and Robert K. Moyzis. 2004. "The Genetic Architecture of Selection at the Human Dopamine Receptor D4 (DRD4) Gene Locus." *American Journal of Human Genetics* 74 (5): 931–944.

4. Crow, Timothy J. 2008. "The 'Big Bang' Theory of the Origin of Psychosis and the Faculty of Language." *Schizophrenia Research* 102 (1–3): 31–52; **Pryluk, Raviv**, Yoav Kfir, Hagar Gelbard-Sagiv, Itzhak Fried, and Rony Paz. 2019. "A Tradeoff in the Neural Code across Regions and Species." *Cell* 176 (3): 597–609.e18; **Pang, James C.**, James K. Rilling, James A. Roberts, Martijn P. Van Den Heuvel, and Luca Cocchi. 2022. "Evolutionary Shaping of Human Brain Dynamics." *ELife* 11: 1–62.

5. Dawkins, Richard. 1982. *The Extended Phenotype: The Long Reach of the Gene*. Oxford University Press.

6. Bratsberg, Bernt, and Ole Rogeberg. 2018. "Flynn Effect and Its Reversal Are Both Environmentally Caused." *Proceedings of the National Academy of Sciences of the United States of America* 115 (26): 6674–6678.

7. Hutson, Matthew. 2022. "Taught to the Test." *Science* 376 (6593): 570–573; **Gibney, Elizabeth**. 2022. "Open-Source Language AI Challenges Big Tech's Models." *Nature* 606 (7916): 850–851; **Kejriwal, Mayank**. 2021. "An AI Expert Explains Why It's Hard to Give Computers Something You Take for Granted: Common Sense." *The Conversation UK*, August 17, 2021; **Castelvecchi, Davide**. 2022. "DeepMind AI Learns Simple Physics like a Baby." *Nature*, July; **Ruppert, Felix**, and Alexander Badri-Spröwitz. 2022. "Learning Plastic Matching of Robot Dynamics in Closed-Loop Central Pattern Generators." *Nature Machine Intelligence* 4 (7): 652–60; **Stokel-Walker, Chris.** 2022. "Robot That Can Perceive Its Body Has Self-Awareness, Claim Researchers." New Scientist, July 13, 2022. https://www.newscientist.com/article/2328245-robot-that-can-perceive-its-body-has-self-awareness-claim-researchers/; **Sparkes, Matthew**. 2022. "Sentient AI: Has Google's LaMDA Artificial Intelligence Really Come to Life?" New Scientist, Spring 7, 2022. https://www.newscientist.com/article/2323905-has-googles-lamda-artificial-intelligence-really-achieved-sentience/; **Lake, Brenden M.**, and Marco Baroni. 2023. "Human-like Systematic Generalization through a Meta-Learning Neural Network." *Nature* 623 (7985): 115–121; **Lorach, Henri,** Andrea Galvez, Valeria Spagnolo, Felix Martel, Serpil Karakas, Nadine Intering, Molywan Vat, et al. 2023. "Walking Naturally after Spinal Cord Injury Using a Brain-Spine Interface." *Nature* 618 (7963): 126–133; **Chaudhary, Ujwal,** Ioannis Vlachos, Jonas B. Zimmermann, Arnau Espinosa, Alessandro Tonin, Andres Jaramillo-Gonzalez, Majid Khalili-Ardali, et al. 2022. "Spelling Interface Using Intracortical Signals in a Completely Locked-in Patient Enabled via Auditory neurofeedback training." *Nature Communications* 13(1236): 1–15.

8. Capek, Karel. 1920. *R.U.R. (Rossum's Universal Robots)*. Penguin Books; **Asimov, Isaac.** 1950. "Runaround." In *I, Robot*. New York: Doubleday.

9. Koch, Christof. 2019. *The Feeling of Life Itself: Why Consciousness Is Widespread but Can't Be Computed*. Cambridge, MA: MIT Press.

10. Yuste, Rafael, Sara Goering, Blaise Agüeray Arcas, Guoqiang Bi, Jose M. Carmena, Adrian Carter, Joseph J. Fins, et al. 2017. "Four Ethical Priorities for Neurotechnologies and AI." *Nature* 551 (7679): 159–163; **Harari, Yuval N.** 2017. *Homo Deus: A Brief History of Tomorrow*. New York: Harper Collins; **Álamos, María Florencia**, Leonie Kausel, Clara Baselga-Garriga, Paulina Ramos, Francisco Aboitiz, Xabier Uribe-Etxebarria, and Rafael Yuste. 2022. "A Technocratic Oath." In *Protecting the Mind. Ethics of Science and Technology Assessment*, edited by Pablo López-Silva and Luca Valera, 49:163–174. Springer, Cham; **Rabesandratana, Tania**. 2022. "France Introduces Research Integrity Oath." *Science* 377 (6603): 251.

11. O'Connell, Mark. 2017. *To Be a Machine: Adventures among Cyborgs, Utopians, Hackers, and the Futurists Solving the Modest Problem of Death*. New York: Penguin Random House; **Santow, Edward**. 2020. "Emerging from AI Utopia." *Science* 368 (6486): 9; **Kane, Alice E.**, and David A. Sinclair. 2019. "Epigenetic Changes during Aging and Their Reprogramming Potential." *Critical Reviews in Biochemistry*

and Molecular Biology 54 (1): 61–83; **Takagi, Yu,** and Shinji Nishimoto. 2023. "High-Resolution Image Reconstruction with Latent Diffusion Models from Human Brain Activity." *BioRxiv*, March, 2022.11.18.517004:1–11; **Bellier, Ludovic,** Anaïs Llorens, Déborah Marciano, Aysegul Gunduz, Gerwin Schalk, Peter Brunner, and Robert T. Knight. 2023. "Music Can Be Reconstructed from Human Auditory Cortex Activity Using Nonlinear Decoding Models." *PLOS Biology* 21 (8): e3002176: 1–27; **Kneeland, Reese,** Jordyn Ojeda, Ghislain St-Yves, and Thomas Naselaris. 2023. "Reconstructing Seen Images from Human Brain Activity via Guided Stochastic Search," *ArXiv*, May 2, 2305.00556v2: 1–4.

12. Metzl, Jamie. 2019. *Hacking Darwin: Genetic Engineering and the Future of Humanity*. 1st ed. Sourcebooks; **Lovell-Badge, Robin**. 2019. "CRISPR Babies: A View from the Centre of the Storm." *Development* 146 (3): 1–5; **Howard, Heidi C.**, Carla G. Van El, Francesca Forzano, Dragica Radojkovic, Emmanuelle Rial-Sebbag, Guido De Wert, Pascal Borry, and Martina C. Cornel. 2018. "One Small Edit for Humans, One Giant Edit for Humankind? Points and Questions to Consider for a Responsible Way Forward for Gene Editing in Humans." *European Journal of Human Genetics* 26 (1): 1–11.

13. Newman, Sandra. 2022. *The Men: A Novel*. Grove Press; **Alderman, Naomi**. 2017. *The Power: A Novel*. 1st ed. Little, Brown and Company; **Beukes, Lauren**. 2020. *Afterland*. Mulholland Books; **Kleeman, Jenny**. 2020. *Sex Robots and Vegan Meat: Adventures at the Frontier of Birth, Food, Sex, and Death*. Pegasus Books.

14. Blackiston, Douglas, Emma Lederer, Sam Kriegman, Simon Garnier, Joshua Bongard, and Michael Levin. 2021. "A Cellular Platform for the Development of Synthetic Living Machines." *Science Robotics* 6 (52): 1–13; **Reardon, Sara**. 2020. "Can Lab-Grown Brains Become Conscious?" *Nature* 586 (7831): 658–661; **Kagan, Brett J.**, Andy C. Kitchen, Nhi T. Tran, Forough Habibollahi, Moein Khajehnejad, Bradyn J. Parker, Anjali Bhat, Ben Rollo, Adeel Razi, and Karl J. Friston. 2022. "In Vitro Neurons Learn and Exhibit Sentience when Embodied in a Simulated Game-World." *Neuron* 110 (23): 3952–3969.

15. Elhacham, Emily, Liad Ben-Uri, Jonathan Grozovski, Yinon M. Bar-On, and Ron Milo. 2020. "Global Human-Made Mass Exceeds All Living Biomass." *Nature* 588 (7838): 442–444; **Bar-On, Yinon M.**, Rob Phillips, and Ron Milo. 2018. "The Biomass Distribution on Earth." *Proceedings of the National Academy of Sciences of the United States of America* 115 (25): 6506–6511; **Machemer, Theresa**. 2021. "Humans Have Altered 97 Percent of Earth's Land through Habitat and Species Loss." *Smithsonian Magazine—Smart News*, April 20, 2021.

16. Crutzen, Paul J. 2002. "The Effects of Industrial and Agricultural Practices on Atmospheric Chemistry and Climate during the Anthropocene." *Journal of Environmental Science and Health. Part A, Toxic/Hazardous Substances & Environmental Engineering* 37 (4): 423–424; **Leclère, David**, Michael Obersteiner, Mike Barrett, Stuart H. M. Butchart, Abhishek Chaudhary, Adriana De Palma, Fabrice A. J. DeClerck, et al.

2020. "Bending the Curve of Terrestrial Biodiversity Needs an Integrated Strategy." *Nature* 585 (7826): 551–556; **WWF World Wide Fund For Nature.** 2022. "WWF LIVING PLANET REPORT." WWF. 2022; **Waters, Colin N.**, and Simon D. Turner. 2022. "Defining the Onset of the Anthropocene." *Science* 378 (6621): 706–708.

17. Borgatti, L., and M. Soldati. 2013. "7.30 Hillslope Processes and Climate Change." *Treatise on Geomorphology* 7 (January): 306–319.

Final Thoughts

1. Tipler, F. J. 1981. "A Brief History of the Extraterrestrial Intelligence Concept." *Quarterly Journal of the Royal Astronomical Society* 22 (June): 133; **Tarter, Jill.** 2003. "The Search for Extraterrestrial Intelligence (SETI)." *Annual Review of Astronomy and Astrophysics* 39 (1): 511–548; **Champagne, Marc.** 2019. "Diagrams and Alien Ways of Thinking." *Studies in History and Philosophy of Science* 75 (June): 12–22; **Lingam, Manasvi**, and Abraham Loeb. 2019. "Relative Likelihood of Success in the Search for Primitive versus Intelligent Extraterrestrial Life." *Astrobiology* 19 (1): 28–39.

2. Livio, M., D. Hollowell, A. Weiss, and J. W. Truran. 1989. "The Anthropic Significance of the Existence of an Excited State of 12C." *Nature* 340 (6231): 281–284; **McGrath, Alister E.** 2009. *A Fine-Tuned Universe: The Quest for God in Science and Theology*. Westminster John Knox Press; **Davies, P. C. W.** 2007. *Cosmic Jackpot: Why Our Universe Is Just Right for Life*. 1st ed. Mariner Books.

3. Jay Gould, Stephen. 1990. *Wonderful Life: The Burgess Shale and the Nature of History*. W. W. Norton & Company, New York; **Kershenbaum, Arik.** 2021. *The Zoologist's Guide to the Galaxy: What Animals on Earth Reveal about Aliens—and Ourselves*. 1st ed. Penguin, New York.

Index

Note: Page numbers followed by *f* indicate figures.